A Global Enlightenment

SERIES EDITOR
Darrin McMahon, *Dartmouth College*

After a period of some eclipse, the study of intellectual history has enjoyed a broad resurgence in recent years. The Life of Ideas contributes to this revitalization through the study of ideas as they are produced, disseminated, received, and practiced in different historical contexts. The series aims to embed ideas—those that endured, and those once persuasive but now forgotten—in rich and readable cultural histories. Books in this series draw on the latest methods and theories of intellectual history while being written with elegance and élan for a broad audience of readers.

Buddhist cosmology, c. 1789.

A Global Enlightenment

WESTERN PROGRESS AND CHINESE SCIENCE

Alexander Statman

The University of Chicago Press CHICAGO AND LONDON

The University of Chicago Press, Chicago 60637
The University of Chicago Press, Ltd., London

Published 2023
Printed in the United States of America

32 31 30 29 28 27 26 25 24 23 1 2 3 4 5

ISBN-13: 978-0-226-82576-2 (cloth)
ISBN-13: 978-0-226-82574-8 (e-book)
DOI: https://doi.org/10.7208/chicago/9780226825748.001.0001

Library of Congress Cataloging-in-Publication Data

Names: Statman, Alexander, author.
Title: A global enlightenment : western progress and Chinese science / Alexander Statman.
Other titles: Life of ideas.
Description: Chicago : The University of Chicago Press, 2023. | Series: The life of ideas | Includes bibliographical references and index.
Identifiers: LCCN 2022039125 (print) | LCCN 2022039126 (ebook) | ISBN 9780226825762 (cloth) | ISBN 9780226825748 (ebook)
Subjects: LCSH: Enlightenment. | Civilization, Western—Chinese influences. | East and West. | Science—China. | Science and civilization. | Science—History. | Progress—Philosophy—History. | France—Intellectual life—18th century. | China—Intellectual life—18th century.
Classification: LCC B802 .S735 2023 (print) | LCC B802 (ebook) | DDC 190—dc23/eng/20221019
LC record available at https://lccn.loc.gov/2022039125
LC ebook record available at https://lccn.loc.gov/2022039126

♾ This paper meets the requirements of ANSI/NISO Z39.48-1992 (Permanence of Paper).

A bright light illuminated high antiquity; but scarce few rays have reached all the way to us. It seems to us that the ancients were in shadows, because we see them through the thick clouds from which we have just emerged. Man is a child born at midnight; when he sees the sun rise, he believes that *yesterday* never existed.

ATTRIBUTED BY JEAN-PIERRE ABEL-RÉMUSAT
TO A DISCIPLE OF LAOZI, 1825

Contents

Introduction

A Global Enlightenment

The idea of progress frames our modern understanding of understanding itself. It offers a distinctive historical account of the development of knowledge in space and time, according to which the improvement of society inevitably follows upon cumulative advances in the natural sciences. This account has a history all its own. Historians have long maintained that it emerged in Europe over the course of the eighteenth century, a key component of the social, cultural, and intellectual movement known as the Enlightenment. It is not a coincidence that the European Enlightenment has also been seen as the beginning of the modern period. In one very influential formulation, modernity was characterized as the disenchantment of the world, or alternatively the disenchantment of the West. Its watchword was "'progress,' to which science belongs as a link and motive force."[1] The idea of progress with science as its mark and guarantor has often been taken as our inheritance from the very Enlightenment for which it also stood.[2] But more recently, historians have begun to realize that in order to fully understand the Enlightenment, we must also pay attention to the things it often told us to ignore: magic, religion, occult science, and the knowledge of times and places beyond the modern West. That is to say, the Enlightenment was constituted not only by ideas of progress, but also by debates about them.

And, I will show, many of those debates in Europe involved a dedicated, continuous, and transformative engagement with the intellectual traditions of China.

A Global Enlightenment is a book about the idea of Western progress, told through a series of interlinked conversations about Chinese science. Its protagonists are an ex-Jesuit missionary, a French statesman, a Manchu prince, Chinese literati, European savants, and diverse other figures of the late Enlightenment world who exchanged old ideas across cultures and created new ones along the way. They studied ancient astronomical records, contemporary natural histories, gas balloons, electrical machines, stone inscriptions, *yin-yang* cosmology, animal magnetism, and Daoist medicine. They invented esoteric traditions from tarot-card fortune-telling to mesmerist hypnotherapy, contributed to academic disciplines including philosophy and orientalism, and helped establish a new approach to the global history of science. Some of their stories are very well known; others have been long forgotten. In telling them together here, my contention is that Chinese science shaped a signature legacy of the European Enlightenment: the idea of Western progress.

To follow ideas of progress through a crucial period of cultural transformation, I focus on a group of people I call orphans of the Enlightenment: erstwhile members of the Enlightenment family who felt like they were being left behind. Throughout the early modern period, many Europeans had believed in the existence of ancient wisdom. The idea was that perfect knowledge had been given by God to man in remote antiquity but was later forgotten and might never be regained. While the philosophes of Paris were writing ancient wisdom out of their new accounts of progress, orphans of the Enlightenment sought to vindicate it in their own. Joining their ranks from Beijing was the last great scholar of the Jesuit mission to China, Joseph-Marie Amiot. Step by step, with each exchange building on the last, orphans of the Enlightenment drew from China to think through the problems of progress. Seeking to recover knowledge they believed had been lost to modern times, they developed new theories, evidence, and analyses that were the distinctive products of their own. Their contemporaries broadly rejected the global account of knowledge that they created. But those who came after them aimed instead to explain it. When their work on ancient wisdom and Chinese thought was brought back into modern progress theories, the past became a foreign country. Distant times and distant places together were made a window into an essentially different way of knowing.[3]

In telling this story, I make two revisionist interventions in the intellectual history of the early to late modern transition: One concerns the idea of progress, and the other is about Chinese science. Historians have long assumed that the idea of the uniqueness of the West—and the mystery, inscrutability, or otherness of the East—followed inherently from the Enlightenment idea of progress. I argue instead that it had to be invented, and that it was invented through the work of the orphans of the Enlightenment. Challenging core components of modern progress theories, they believed that the knowledge of the past and the East still had value for modern Europe. Their efforts to recover and explain that knowledge reveal, in turn, an unknown story of European engagement with Chinese science. For many years, the historical consensus has generally been that over the course of the Enlightenment, non-Western ideas were banished from European thought. In fact, I find that something much more like the opposite is true. Toward the end of the Enlightenment, Europeans only grew more interested in Chinese science: first as an ancient precursor to modern discoveries; then as an example of what Western science was not; and, finally, as a new kind of distinctively modern alternative knowledge. Each of these conceptions has been lastingly influential, from the eighteenth century through today.

The Enlightenment has long been imagined as an essentially European intellectual movement. It was thought to be both what distinguished the West from the rest, and that by which the West constituted its own distinction.[4] Put another way, the idea that the Enlightenment is Eurocentric is itself an Enlightenment idea.[5] Historians of eighteenth-century Europe have always been aware that the Enlightenment was global in its outlook, at least insofar as new ideas in Europe about history, geography, and anthropology resulted from efforts to comprehend the world beyond it.[6] Yet the global history of Enlightenment ideas remains largely unwritten.[7] The most recent approaches to global Enlightenment—pointedly omitting the definite article—have tended to treat it not as a fixed phenomenon, but rather as a movable discourse. If the Enlightenment had been defined as something that happened in eighteenth-century Europe, then it simply needed to be redefined. There has been no secret about the implications of this approach. Thus we find Islamic and South Asian Enlightenments based on conceptual similarities to the European one, or Haitian and Ottoman Enlightenments genealogically descended from it.[8] In the history of China, one might point to Qing evidential scholarship on the one hand, and the May Fourth movement on the other, as respective examples of each.[9] Historians have embraced and applauded such efforts to move beyond the Enlightenment's supposedly European origins.[10] But what I have found is that at the very root of its European origins, the Enlightenment as traditionally understood was global already.

This book thus presents a global Enlightenment that should look familiar even to historians who still embrace "the" Enlightenment with a definite article. It happened at the right time and place—eighteenth-century France—and it concerned an issue that was central even by its most conservative definition: the idea of progress. What I propose is that Chinese thought continuously informed the thought of the Enlightenment, though to different degrees, at multiple points, and in various ways. First, through contribution. Voltaire and his predecessors took Chinese philosophy as an example of what Europe might progress toward. Then, through tension. Later philosophes like the marquis de Condorcet redefined progress as something that China did not have, while orphans of the Enlightenment pointed to what China had instead so as to contest that theory of progress. Finally, through opposition. Georg Wilhelm Friedrich Hegel studied China to explain what he believed made European progress unique, and those who came after him framed their ideas of progress accordingly. Some of these moments touched upon China only in passing, while others could not have happened without its presence. But all of them reflected genuine engagement with Chinese knowledge.

I use the term *engagement* here to capture the fact that to a much unrealized extent, China was an active agent in European intellectual history. Given the known importance of ideas *about* China in the Enlightenment, it is surprising how little has been said about the role of ideas *from* China in shaping it.[11] Historians have rightly collapsed an old distinction between "influences" and "images" of China, thereby avoiding the positivist trap of being forced to rule on what Europeans got right and what they got wrong.[12] The framework of engagement presupposes neither, while further suggesting that real Chinese texts and people were deeply involved. In an indirect sense, China was Europe's main teacher about itself.[13] Enlightenment thinkers read genuine translations of Chinese texts by ancient writers like Confucius and contemporary ones such as the Kangxi emperor, explicitly citing titles and authors by Romanized Chinese name. They were also in sustained communication with Catholic missionaries, who spent most of their lives in Beijing and sometimes developed a thoroughly cross-cultural sense of personal identity. More directly, non-European figures also played a crucial role in the process of exchange. A few of them, like the Manchu prince Hongwu, can be identified by name. Others, such as a certain Chinese minister at the Board of Rites, are known only by occupation. Many more, including a once-mentioned Tibetan lama and a Khorchin Mongol chief, will remain forever anonymous. Nevertheless, what they did mattered, since without them, Europeans would have engaged with China differently, or, in some cases, not at all.

This study is just one example of a global Enlightenment. It is a model and an illustration. It is neither comprehensive nor exclusive. It is, however, particularly explanatory. For, whether the Enlightenment was a coherent intellectual movement united around a particular narrative of human civilization, or a broad family of thinkers with their characteristic quarrels and resemblances, the importance of eighteenth-century France and the set of ideas articulated there is unquestioned.[14] Indeed, the predicament of orphans of the Enlightenment reveals the tension between these two interpretations: they were a part of the Enlightenment family, but they found themselves at odds with core aspects of the emerging Enlightenment narrative. Furthermore, the particular role that China played in the Enlightenment was uniquely important. Other faraway places, from the Americas and the Islamic world to South Asia and the South Pacific, also had their moments in the European limelight.[15] But only China was ever taken as a model to be emulated. Nowhere else was ever thought to be or have been what Europe could or should become. Bringing these points together, it was in France, the center of the Enlightenment, that discussions of China during this time were the most vibrant and the best informed. Undoubtedly, there are other global Enlightenments for historians to uncover. It is through this particular one that the transformation of the Enlightenment idea of progress is best explained.

A Global Enlightenment begins with a chapter on the cultural and intellectual lead-up to European engagement with Chinese science in the late eighteenth century. For almost two hundred years, Jesuit missionaries had been almost uniquely responsible for all serious intellectual exchange between Europe and China. Their self-fashioning as men of science and religion encouraged them to critique Chinese knowledge of nature and to praise instead the social and political philosophy of Confucius. This imbalance became increasingly important for Europeans when the paramount value of the Enlightenment subtly shifted: from reason to progress. For Voltaire, China was an ancient land of sage philosophy. For Condorcet, it was a static place stuck in time. The later philosophes claimed a monopoly on progress for themselves alone—but not everyone in France was a philosophe. Dissatisfied with the direction that their own society was taking, conservative statesmen and orientalist scholars looked to other ones that were distant in time and space. Once a model for progressives and freethinkers, China was taken up by those who sought to support the ancien régime with a regime more ancient still.

The next chapter shifts focus to the rapidly changing social, cultural, and political conditions in Beijing that laid the groundwork for new engagement with Chinese knowledge. In 1773, the global suppression of the Society of Jesus brought turmoil to the small community of Europeans who were living and working there as scientific and technical experts at the Qing court. Amiot, by then the greatest living missionary scholar of China, became what he called an "ex-Jesuit." Seeking to construct a new identity, he aligned himself with patrons in Paris, especially the minister of state Henri Bertin, and made friends in Beijing, like the Manchu prince Hongwu. He continued to share Europe's newest discoveries with Chinese audiences, from electrical medicine to gas balloons, while interpreting them in the context of *kaozheng*, or evidential-studies scholarship. Surprisingly, the suppression of the Society of Jesus had the ultimate effect of increasing his productivity. It also turned his attention toward natural science. With his identity transformed, Amiot began a new research program that would have been nearly impossible during the days of the Jesuit mission.

The following chapter shows how orphans of the Enlightenment in Paris and Beijing began a conversation about the global history of science. During the 1770s and 1780s, some French savants argued that the scientific knowledge of the ancient world had developed upon the long-lost legacy of an even more advanced primordial people who had preceded all the others. Seeking to recover that legacy with the help of Amiot, they drew from Chinese sources to invent modern esotericism. Ideas had to pass through many interpretive filters in their cross-cultural journey from late imperial China to early modern Europe. Nevertheless, recognizable concepts from identifiable Chinese texts entered into new Enlightenment theories. The Protestant minister Antoine Court de Gébelin cited a rubbing of the Song-period stone Stele of King Yu in the invention of tarot-card fortune-telling. The astronomer Jean-Sylvain Bailly read a Qing travelogue by Translation Bureau Chief Jiang Fan to locate the ruins of Atlantis in far-off Xinjiang. Their ideas were discussed in the academies and salons of Paris, forming a new account against which the philosophes refined their own narratives during a crucial moment in the development of progress theory.

Chapter 4 reveals the previously unknown culmination of early modern Sino-European scientific exchange with a robust, cross-cultural account of natural philosophy. During the 1780s through 1790s, orphans of the Enlightenment deployed an idiosyncratic reading of Neo-Confucian cosmology to develop Franz Mesmer's popular new theory of animal magnetism. The aim of their investigations was to reconstruct ancient wisdom—but the result of their investigations was to link Chinese theories with occult science. In Beijing, Amiot became a devotee of mesmerist cosmology and medicine, leading him to investigate realms of Chinese thought that had been previously unexplored, from esoteric alchemy to religious Daoism. In Paris, an obscure cavalry officer, the comte de Mellet, sought out Chinese sources and used the principles of *yin* and *yang* to develop his own account of electricity and magnetism. Orphans of the Enlightenment argued to the scientific authorities of their day that ancient China had possessed knowledge of nature that modern Europe lacked. They did not win this argument. But they did raise the question of what kind of knowledge China had possessed instead.

The story concludes with the integration of new research on Chinese thought back into modern progress theories. During the 1810s and 1820s, the Enlightenment's children approached their parents' problems in a much-changed world. The preeminent academic philosopher, Hegel, and Europe's first professional Sinologist, Jean-Pierre Abel-Rémusat, both drew extensively from the work of Amiot and the orphans of the Enlightenment to reconceptualize the development of human knowledge over time. Still in conversation with each other, they severed orientalism from philosophy and reframed China as a source of "ancient eastern wisdom." Remoteness in space and time, though already equated, only then became incommensurable. The idea of the modern West, with its own distinctive modes of thought, established new foundations for the academic study of the history of science in China. At the same time, it also created new possibilities for engagement with Chinese knowledge. An alternative way of knowing, lost to the West in its transition to modernity, might still be alive in a non-Western past.

Western Progress

The Enlightenment idea of progress became so powerful in the self-conception of the modern West that we can easily forget it was ever contested. But progress was the story that the Enlightenment told about itself—and that is exactly why it was so hotly debated. What we now call the Enlightenment on the one hand, and the counter-Enlightenment, conservative Enlightenment, and super-Enlightenment on the other, all took shape through shifting engagement with non-modern and non-Western thought. The philosophes wrote for themselves a genealogy of progress that placed its beginning in ancient Greece and its end in modern Europe. But that conception only made sense in contradistinction to the alternative ones of their contemporaries and to those times and places that were not included in the progress paradigm. China was always an integral a part of the conversation, since it appeared as the best or even the only example of a non-European place with a rich and ancient intellectual tradition that seemed broadly comparable to Europe's own. If it was questions of progress, more than any particular view of it, that defined the Enlightenment, then China was crucial for posing them.

The idea of progress is notoriously difficult to define. It is not for lack of effort that the most commonly cited formulation remains one that was proposed more than a hundred years ago: "This idea means that civilization has moved, is moving, and will move in a desirable direction." One might pause to consider, in this definition, what progress is not. What moves is civilization—society, culture, the arts and sciences, even mind or spirit—that is, the human world, as opposed to the natural world that remains at rest. The movement has taken, is taking, and will take place, but typically in a way that is stadial, sporadic, and nonlinear; no sophisticated thinker has likely held that each moment in history is better than the one immediately preceding it. The most influential theories of the late Enlightenment further associated progress with the natural sciences as both its mark and guarantor. It seemed clear that scientific knowledge accumulates over time, and that increase in understanding was supposed to grant power to modify the human condition in beneficial ways.[16] The grounding of progress in abstract and unchanging laws of nature was what made it come to seem universal: not a limited feature of a particular time and place, but the most important feature, perhaps even a metaphysical fact, of history itself.[17] Taken together, this package is something like what historians mean by the Enlightenment idea of progress, though throughout the Enlightenment, most of it was still up for debate.

The point has not been lost to historians. Indeed, many of the most influential accounts of the Enlightenment have been histories of the idea of progress in one sense or another.[18] The first studies to tackle the issue directly date back all the way to the nineteenth century, when faith in progress as both good and true reached a high-water mark in Europe and the United States.[19] Such celebratory views were dealt a shattering blow by the uniquely modern horrors of the First World War. In the historiographical perspective of the early twentieth century, whether progress was a metastasized form of Christian millenarianism or an ahistorical justification for the conditions of the present, the governing attitude was already one of suspicion.[20] The Second World War then prompted a still more influential critique: the quest to dominate nature had found expression in the ambition to master human beings, and totalitarianism, motivated by the illusory dream of progress, was the ultimate Enlightenment project.[21] But the fact that some historians and philosophers critiqued the idea of progress throughout the twentieth century also reveals the extent to which others in the academy continued to embrace it. In the postwar period, a synthesis emerged, holding that the idea of progress in the Enlightenment was neither as universal, nor as universalizing, as it later came to seem. With some modification and elaboration, this intermediate position remains the point of departure for most scholars today.[22]

The history of progress is vexed because it sits at odds with two of the most deeply held commitments of the modern historical profession: contingency and objectivity. The history of the idea of progress has often been narrated as itself progressive—a teleological march from the backward-looking Middle Ages, through the equivocal Renaissance, to the forward-looking Enlightenment. If progress was inevitable, the history of the idea of progress seemed to be, too. Presenting as a kind of just-so story, the topic seems terribly old-fashioned, and it has for quite some time.[23] Furthermore, in an academic climate mistrustful of all historical metanarratives, it is especially difficult for historians to write about the one that sits closest to the heart of the discipline of history itself. The subject has therefore been left to be taken up by sociologists, economists, and psychologists, among others, who pronounce their own beliefs in progress despite the fact that, oddly enough, their personal orientations tend to be conservative, or in any case, not what is now more usually called progressive.[24] Professional historians prefer to leave well enough alone out of fear of being mistaken for true believers or, worse, exposed as such. At this point, I wish to state forthrightly that although my own doubts about the historical reality of progress have deepened over the course of a decade spent thinking about it, I remain deeply ambivalent on the matter. Which is, I think, why I have written about it in the way that I have.

This book is an attempt to narrate a non-teleological history of progress. There is, of course, a sense in which all history is necessarily teleological: the historian's job is to determine what happened and to explain why it happened as it did. But getting past teleology is particularly difficult in intellectual history, where it is hard to deny that ideas are accumulative, at least insofar as we can understand those of the past better than we can predict those of the future. Still, historians have done their best at least to avoid determinism with respect to the present. The usual technique for eschewing teleology has been to argue that historical phenomena were not inevitable, which typically means pointing out how things could have been otherwise. I occasionally employ this mode of analysis, most notably in discussions of how canonical progress theorists such as Voltaire and Hegel, as well as mainstream scientific thinkers in the Académie des sciences and the Paris Academy of Physicians, responded to the nonprogressive proposals of their contemporaries, the orphans of the Enlightenment. But I also develop a more constructive approach, one that is not teleological to the extent that it is instead genealogical, or organic, branching, splitting, and recombining, with traits and typologies that disappeared and reappeared across and between generations. From this non-teleological perspective, I aim to explain in particular a fundamental transformation in progress theory that took place around the turn of the nineteenth century: put simply, from early modern cosmopolitanism to late modern orientalism.

Early modern ideas of progress were broadly cosmopolitan and universalist.[25] They typically held that human nature was constant across time and space, so human societies, too, were fundamentally comparable. All shared a single conceptual origin and destination, differing only in their relative positions on a one-track timeline of unidirectional development. If left to their own devices, they would all arrive at the same point, by the same route. As John Locke famously put it: "In the beginning, all the world was America."[26] Accordingly, Europe was America plus a given amount of time, and America was Europe minus the same. Whether the much-discussed "state of nature" was a harmonious paradise or a savage battlefield, Europeans had once been in it, and others perhaps still were. Much of the world was revealed as a proto-Europe or a West-in-waiting. The idea of progress thus left little for the philosophes to admire in non-European cultures. But it did not require that those cultures be opaque, or even all that different from their own. Non-European and non-modern people were not at all mysterious; quite the contrary, they were simply Europeans as they had once been, or moderns in their infancy. At the end of the Enlightenment, the rest of the world was not yet seen by most European progress theorists as inscrutable—it was simply not very interesting.

And yet, by the early nineteenth century, the full-blown complex of modern orientalism had emerged: both the idea of the East as other and the academic discipline designed for studying it as such. Modern orientalism invented what Edward Said aptly called "the mythology of the mysterious East, notions of Asian inscrutability."[27] Its premise was thus quite at odds with early modern progress theory, and its new importance in European culture was profound. This was the time of the Royal Asiatic Society in Britain and the *Société Asiatique* in France, the era of the decipherment of the hieroglyphs and the translation of the Bhagavad Ghita. Under the aspect of orientalism, progress theory was redeployed as an account of European difference and a justification for European power. What had once linked humanity together now served to single out just one part of it. Only the West had progress—and, vice versa, all progress was Western. Non-European peoples might still experience progress, but only if Europeans brought it to them. What made the modern idea of the primitive different from the early modern idea of the primeval, primordial, or pristine was not that it tied distant times and places together. Progress theory had already done that. Rather, it was that distant times and places now appeared different not only in degree, but also in kind. The past and the East were distinguished from the modern West, for progress was no longer opposed to backwardness, but, rather, to timelessness.[28]

Historiographical convention has obscured what a momentous change in progress thinking this really was. The stark division of the academic profession today across the chasm of the early to late modern divide has made it appear as an epiphenomenon of a phase shift in European culture writ large. Since all its component parts existed for logically necessary reasons, nothing needed to be explained. But in order for progress theory to come together in the way that it did to accommodate the new idea of the modern West, many discrete components had to be already in place. A conception of radical difference between societies across historical time emerged in early eighteenth-century France with the quarrel between the Ancients and the Moderns.[29] Its extension to cultures separated by geographical space was a product of late eighteenth-century German Romanticism.[30] The identification of a Scientific Revolution following Isaac Newton provided one account of what Western thought was.[31] The demarcation of a specific philosophical tradition following the work of Immanuel Kant presented another.[32] There were doubtless other necessary components as well. Still, at the end of the early modern era, there was no comprehensive story of progress that served to bring them all together.[33] No single step was requisite for or determinative of the shape, contours, and configurations that that story eventually took. In sum, the shift from cosmopolitanism to orientalism, and the accompanying transformation of modern progress theories, was by no means foreordained.

The story of the orphans of the Enlightenment explains this development in a new way. Their ultimate ambition was not to discount the knowledge of the present, but to reconcile progress with the knowledge of the past. Some of them were respected members of the intellectual establishment, while others were plainly eccentric. Yet all of them considered themselves to be active participants in what we now think of as Enlightenment debates. Unlike others among their contemporaries, they did not wish to cast their work in opposition to the intellectual currents of their day.[34] To the contrary, they embraced the fundamental assumptions and conceptual vocabulary of progress theory.[35] For them, as for the philosophes, the thought of ancient times and foreign places was different in quality; but it was not, and could not be, different in kind. All knowledge was by definition commensurable—that commensurability was simply what made it knowledge. Where they departed from the philosophes was in their conception of what that knowledge looked like and their theory of how it changed over time. They retained interests and expertise that progress could not easily accommodate, from conservative institutions like the Society of Jesus and the Bourbon monarchy, to obsolete disciplines like alchemy and astrology, and they sought to revisit these things as the Enlightenment was moving on. For this reason, they were cast out of their fields in their own time. But after wandering for a while in the wilderness, the sources, analysis, and ideas that they created were later brought back in. Ideas that had sprouted orthogonally to the main trunk of progress theory rejoined it at another point. In this way, their unorthodox participation in the late Enlightenment eventually formed a part of its legacy.

By contesting new ideas of progress, the orphans of the Enlightenment inadvertently became the true inventors of modern esotericism. They gathered together disparate threads of ancient and modern European culture, from masonry and mesmerism to mysticism and mythography, added new threads from China and elsewhere, and weaved them all together into an outfit that remains recognizable still today. Nineteenth-century theosophy, twentieth-century psychoanalysis, and the contemporary New Age movement each owes something to their work. That it should nevertheless be considered as a part of the Enlightenment might seem contentious. After all, the traditional historiography of the Enlightenment described it as a flowering of reason that entailed the rejection of various kinds of earlier superstition. Romanticism then appeared as a natural reaction with the revival of myth and tradition, fable and folklore, the occult, the ancient, and the obscure. Since the former came first, the presence of the Enlightenment thesis in the Romantic anti-thesis was always taken for granted, though not vice versa.[36] But historians have recently amended this view, shining a light on what might be termed the dark Enlightenment and arguing that *illuminisme* was never foreign to the thought of the *lumières*.[37] Progress locates knowledge in an infinite future, and esotericism locates it rather in an indefinite past. They are flip sides of the same coin. The Enlightenment and its others were minted together.[38]

Much as progress has been thought of as a distinctively modern idea, it has been even more consistently understood as a uniquely Western one. I am aware of no historical study that attributes a conception of progress to any non-European tradition before modern times. Indeed, since the nineteenth century, Chinese scholars and statesmen have themselves embraced it as one of if not the most important Western import for their own intellectual apparatus.[39] The essential Eurocentrism of progress theories, long accepted on its face by their proponents, has become even more interesting of late to their critics.[40] For example, a recent volume on *The Postcolonial Enlightenment* begins on the first page with a citation from Condorcet, the most canonical Enlightenment progress theorist of all. As the argument normally goes, if the rest of the world was construed as backward, primitive, savage, or superstitious, then progress was a self-serving justification for conquest. This point indicates, too, the extent to which postcolonial critiques of the Enlightenment, in focusing on the contested legacy of imperialism, have self-consciously left its European origin story mostly untouched.[41] But the peculiar case of China—largely unaddressed in that body of scholarship—points toward a different kind of postcolonial argument. One influential way of provincializing Europe has been to trace the effects of modern progress myths out and into the rest of the world.[42] Looking the other way around, modern progress myths were originally created in Europe through the embrace of knowledge from the very places they were later deployed to exclude.

This realization calls into question some of what we think we know about the birth of orientalism. It has not been my intention to challenge the conclusions of Edward Said; though his primary interest lay in a different time and place, most of what he said about the European study of the Islamic world became true for the European study of China as well. Nevertheless, there are some surprises here about how the formations he so vividly described first came to be. Recent studies have suggested that academic Sinology, as a branch of orientalism, probably had more to do with its early modern foundations than its later link with modern imperialism would otherwise suggest.[43] In fact, it owed both more and less to its predecessors than one might imagine: less to the Jesuits and philosophes whom the first Sinologists canonized, but more to ex-Jesuits and anti-philosophes whose work they struggled but failed to expel. More broadly, it has often been suggested that the core idea of orientalism was intrinsic to the Enlightenment idea of progress.[44] Yet it appears rather that the otherness of the East was not a necessary corollary of Western progress, but the contingent result of later attempts to distinguish between them. This historical finding does not mean that anyone on either side of orientalist debates approached non-Western cultures from any perspective other than their own. But I am not convinced that any scholar, myself included, has ever been able to do otherwise. In fact, using China to talk about Europe is precisely what I intend to do.

Chinese Science

The history of science, like the rest of the historical profession, is currently undergoing a global turn. It has indeed been especially marked and welcome in this field, since its core subject was long taken—and, outside the field, still often is—to be European by definition. The old story of the Scientific Revolution in early modern Europe starred an inevitable succession of increasingly correct ideas about the world and how it should be studied, like geocentric cosmology and universal gravitation on the one hand, or quantification and experiment on the other. Today, historians instead view science as a contingent feature of human culture and society. They have globalized our understanding of early modern science, mostly by expanding their focus beyond the locus classicus of European intellectual history. Indeed, because many scholars have assumed that European scientific ideas did not incorporate much in the way of non-Western thought, this expansion of purview has perhaps seemed like the only possible way to achieve that globalized understanding. My ambition is to draw from these new insights about the global formation of science to return to early modern European intellectual history from a fresh perspective. In so doing, what I have found is that Europeans engaged with Chinese scientific concepts in order to formulate their own. The result was not only a series of extraordinary cross-cultural scientific theories, but also a lasting conception of what modern Western science actually was.

The term "Chinese science" is even thornier than the idea of progress. To start, it was not an actors' category before the early twentieth century, when the current Chinese word for "science" entered the language via a Japanese neologism. Scholars in late imperial China wrote instead about the "investigation of things," "extending knowledge," and "learning of the way."[45] These categories included some disciplines related to the natural sciences, such as "calendrical calculation," "roots and herbs," and eventually "Western studies" as well; but they also embraced politics, ethics, history, and most other forms of formal intellectual endeavor.[46] Partly for this reason, and partly to sidestep any philosophical claims about the universality of science, historians scrupulously avoid the term "Chinese science," preferring to talk about "science in China" instead. But historically speaking, there was also such a thing as "Chinese science." This was a European term that first referred specifically to the moral and political philosophy of the Confucian literati elite. Only during the Enlightenment did it take on its now more recognizable meaning of the natural sciences of China. It therefore makes sense to distinguish "science in China"—explanations of the natural world that were formulated in China—from "Chinese science," an originally European concept the meaning of which changed radically over time.

When it comes to the history of science in China—as studied in Europe, the United States, and arguably China as well—no one has been more important than the twentieth-century Cambridge scholar Joseph Needham. It is not an exaggeration to say that he founded the field: one would be hard-pressed to find a practicing historian who did not collaborate or study with someone in his lineage, and the series that he oversaw, *Science and Civilization in China,* spanning over fifty years and twenty-seven volumes, remains the go-to reference source on almost every topic within it. Yet both despite and because of Needham's outsize influence, his work is now widely held in suspicion and subjected to relentless critique. The problem is that his name has become inextricably bound up with the so-called "Needham question": Why did China not develop modern science? This remains both the question by which the field enters into broader conversations in the academy and the question that people outside the academy are by far most likely to ask about the field.[47] But according to virtually all specialists of the history of science in China who are working in the United States and Europe today, the question is a canard: like asking why your name did not appear on page 3 of today's newspaper, in the words of one influential critic.[48] The point is that, as a counterfactual, it is really about the history of Europe, not the history of China. In fact, I will argue that this is a point on which Needham himself agreed.[49] But more importantly, I contend that the histories of science in Europe and in China were far more closely connected than they have generally seemed.

I thus aim to reframe the still-important Needham question by revealing the conditions that made it possible to ask. The idea of the uniqueness of European science is not merely a side effect of how we understand progress; it is actually a part of how we came to understand progress as we do. Even if one is not committed to the Scientific Revolution as a discrete period in early modern European history when transhistorical truths about nature were discovered, there is no doubt that the Scientific Revolution has existed in Europe, if only as a narrative, since the end of the Enlightenment.[50] That was the time when both the history of Chinese science and the history of science in China become cognizable as fields of inquiry. The Needham question might be more properly dubbed the Voltaire question, for it was he who first devoted significant attention to it in the 1730s. Needham himself believed that the origins of his project were to be found in the inaugural lecture of European Sinology delivered in 1814. I have incidentally made some discoveries about the history of science in China during this period, concerning electricity and ballooning, for example. Most important, though, when comes to my claims about the history of science in China, is the simple fact that it informed the history of science in Europe.

There is a general imbalance in the history of science: while we know a great deal about what China learned from Europe, we are comparatively ignorant about what Europe learned from China. In a way, the asymmetry is surprising, given that for most of the early modern period, China's overall importance for Europe was undoubtedly greater than Europe's was for China. The extreme disparity in the surviving written sources of each about the other is itself a compelling testament. The point has been amply demonstrated when it comes to commerce, diplomacy, and geopolitics, as well as culture both materially and nonmaterially construed.[51] Yet there are both professional and historiographical reasons for which China's importance for early modern Europe has not yet been demonstrated in the realm of science. Historians with the relevant expertise tend to be trained as specialists of China, not Europe, while European engagement with Chinese science is more properly a matter of European, not Chinese history.[52] And even for historians of China, modern science long seemed to be something that did not, as a historical matter, develop independently there, regardless of whether they believed that the reason why this was so made for an interesting research question.[53] My peculiar professional training has informed my own perspective on the matter. Originally a historian of Europe who only later learned to use Chinese sources, I was taught first to be attentive to new ways of explaining European science—which in turn revealed the unknown role that Chinese science played in its development.

Global approaches to the history of science are literally as old as the profession itself.[54] For a long time, there were two broad paradigms: comparison and diffusion. It is difficult to say which one is more unpopular these days. Comparison can appear to treat non-European science as deficient, lacking, or inferior, while diffusion risks making non-European people into passive receptors of science that was brought to them from elsewhere.[55] Recent studies of early modern science in China suggest that both frameworks might have some life in them yet: the former by emphasizing aspects of science in China that could seem important from the perspective of comparison, the latter by highlighting the agency of Chinese people and practices in processes of diffusion.[56] Yet it is telling that, to my knowledge, no contemporary historian has attempted to resuscitate the terms. Most now prefer a framework that could be called something like "connection," "entanglement," or "circulation," which begins by rejecting the old premise that there ever existed a distinctive thing called Western science, a notional requirement for both comparison and diffusion alike.[57]

My discussion of engagement is essentially an extension of this new approach to the global history of science; yet I also aim to take both diffusion and comparison more seriously than has been fashionable by recasting them in all their historical strangeness. After all, the familiar conception of a unique body of Western science that diffused from Europe to the rest of the world was itself a product of the late Enlightenment. Both comparison and diffusion actually figured prominently in live debates about global science that were taking place at that time—though in ways that are quite surprising. Orphans of the Enlightenment compared the sciences of China favorably to those of Europe, believing that ancient China had preceded modern Europe in discoveries ranging from chemistry and physics to cosmology and psychology. They also told their own story of diffusion with China in a starring role, locating the ultimate origins of all scientific knowledge in North or Central Asia and reversing the direction of its spread from East to West. It was only later, and partly in reaction to their theories, that there emerged the now-rejected consensus that an incomparable knowledge of natural science emerged only in Europe and was bound to diffuse outward from there. In sum, Enlightenment accounts of comparison and diffusion that might seem downright bizarre to historians today informed the very conception of modern science that the more familiar accounts later presupposed.

New frameworks for globalizing the history of early modern European science have had the salutary effect of expanding its horizons. For the sake of convenience, they can be grouped into three categories—open-air sciences, material culture, and sociology of knowledge—keeping in mind that almost every influential study of one has also been about the other two. The first draws attention to disciplines beyond physics and astronomy, including natural history and the earth sciences in particular, which by their nature required global knowledge to construct.[58] The second focuses on objects like instruments, artifacts, images, and print, which write science into one of the most important stories in early modern world history: globalization.[59] The third considers social and institutional contexts, where mediators, go-betweens, informants, and network actors played active roles in creating science both locally and globally.[60] All three of these new approaches are crucial for understanding how the Enlightenment engaged with Chinese science. To mention just a few examples: Europeans were particularly keen on Chinese knowledge in natural history and related aspects of medicine; print matter and laboratory equipment were invaluable for investigations in both Paris and Beijing; and the network connecting them involved just a handful of individuals through whom all exchange was channeled. Without historians having already drawn attention to such issues, this book would not have been possible to write; but, in the end, they are not its focus.

I seek instead to outline a new program for the global intellectual history of science, in two ways: first, by demonstrating the existence of several now-forgotten cross-cultural scientific theories, and second, by showing that they contributed to the evolving definition of modern Western science itself. One might wonder at the outset why a more global history of scientific ideas along these lines has not been attempted before. The answer boils down to the immense legacy of positivism in the history of science. Most historians in Europe and the United States have not believed in the validity of the scientific concepts prevalent in late imperial China; but neither have they put much stock in the European theories in which Chinese ideas were most enthusiastically adopted. By comparison to open-air sciences, material culture, and the sociology of knowledge, the intellectual history of eighteenth-century Sino-Western exchange enters into territory that is just as far outside the bounds of the canonical history of science, if not farther. The orphans of the Enlightenment were roundly condemned by its physicists and physicians, and their theories were categorically rejected by the institutions of establishment science. In this respect, their project must be considered a failure, at least in its own time. Yet it was also a historically significant failure: first, because it represents something quite unknown in the historiography of Enlightenment science, and second, because it informed the science of the Enlightenment all the same.

The empirical finding that some Europeans during the late eighteenth century believed that non-Western people had unlocked the secrets of nature is by itself surprising; still more so is the fact that they successfully incorporated non-Western theories into their own. By tracing original Chinese sources through to their deployment in French debates, I show not only what French thinkers said about Chinese natural philosophy, but also what they did with it. In particular, they focused on two fundamental concepts: *yin-yang*, expressing complementary, dualistic principles such as the sun and the moon, the male and the female, the active and the passive; and *qi*, usually translated as *vital energy, spirit, matter,* or *pneuma,* the stuff of which everything in nature is composed. In China, *yin-yang* and *qi* were used to explain all manner of natural and human phenomena, from astronomy and feng shui to history and politics.[61] In Europe, they found their way into new accounts of gravity, electricity, chemistry, and, most especially, animal magnetism. Through their travels from one context to another, Chinese ideas were transformed, often beyond recognition. But in a way, the distortion is also proof of sincerity, for it reveals the substantial effort and substantive analysis that were involved. Orphans of the Enlightenment treated China not only as a primary source, but also as a secondary source. They drew not only facts and information from China, but also explanations and interpretations, and in doing so they took real Chinese people as scientific authorities—an approach that seemed shocking even at the time.

The result was a reconceptualization of what science was thought to be. Scholars have proposed that historicizing the idea of modern Western science might lead to a more genuinely global history.[62] Such an inquiry should pay special attention to the late Enlightenment, because this was the period when the historiography of science was invented. Through a complex process of definition, debate, exclusion, and redefinition, the modern understanding of science came to resemble our own. But earlier notions of science did not just disappear overnight. Historians long ago discovered that magic and occult science continued to be important well past their heyday in the Renaissance, and, though tracing such modes of thinking into the eighteenth century is in a way the logical extension of that program, their significance in the Enlightenment was very different.[63] The heroes of the Scientific Revolution had practiced alchemy, astrology, and other esoteric disciplines, sometimes quite openly, while also serving as the scientific authorities of their day.[64] This was no longer true of their successors a hundred years later. The orphans of the Enlightenment harbored interests that were no longer mainstream, and that no longer appeared to be; the point is rather that those interests had still been acceptable in some quarters until quite recently. Those who invoked Chinese science during the Enlightenment did so in an iconoclastic effort to revive the discredited knowledge of the European past. Their new accounts were the very ones that the borders of modern Western science were drawn to exclude.

But the Enlightenment's conclusion that knowledge was by definition scientific and progressive created a new problem: what was to be done with the ways of knowing that appeared to be neither? The orphans of the Enlightenment, torn between their conflicting commitments, were among the first to grapple with this problem, but it is still with us now. As of 2016, only 10 percent of North American universities had a specialist of China on the faculty of the Philosophy Department.[65] Instead, one can go to History, Sociology or Anthropology for that. The social sciences were designed in part to deal with the problem of alternative knowledge that was exposed at the end of the eighteenth century.[66] Natural scientists in Europe and China may have little to say these days about *qi* and *yin-yang*. But walk into a bookstore today. You will still find studies of the *Classic of the Way and Virtue* and the *Classic of Changes* right next to books on tarot, Atlantis, and mesmerism—just where the orphans of the Enlightenment left them. Max Weber's famous claim that the Enlightenment concluded with the disenchantment of West takes on a whole new meaning when we consider how it also resulted in the enchantment of the East. The global Enlightenment did not mark the end of European interest in Chinese science; in the modern sense, it was only the beginning.

✷ 1 ✷

The Death of Voltaire's Confucius

China in the Enlightenment

Voltaire once wrote: "I knew a philosopher who had only the portrait of Confucius in his back office." Below the portrait was inscribed a panegyric poem:

> Interpreter of right reason alone,
> Enlightening minds without dazzling the world,
> He spoke only as a sage and never as a prophet:
> Yet he was believed even in his own country.[1]

If this "philosopher" ever really existed, he was no one other than Voltaire himself. For the most famous man in modern Europe, the sage of ancient China was the prototype of an Enlightenment philosophe.[2] By embracing a philosophy of reason, Voltaire held, China had become "the wisest and most civilized nation in the universe."[3] Remarkably, he believed that a philosophy of the sixth century BCE should be actively studied by the philosophers of eighteenth-century France because it provided a model that Europe might follow. His final interest was always in his own time and place. Yet if China was for him only a mirror or a cipher for Europe, then this was exactly why he studied it with such care.[4] He returned to the subject again, from his debut in the Republic of Letters in the 1720s almost to his death in 1778.[5] During this period, Chinese philosophy drew more enthusiasm in Europe than it ever had before, or arguably has yet since.[6]

There can be little doubt that China was the favorite foreign place of the high Enlightenment.[7] Admiration for China was the rule, not the exception, cutting across national and confessional divides and finding expression across the spectrum of European culture. Churchmen studied its rituals to learn about the primitive church. Freethinkers interpreted them to challenge the modern one.[8] Merchants traded fortunes for its silks, satins, painting, and porcelains. Artists imitated them, culminating in the

FIGURE 1.1 *Confucius*, in Couplet et al., *Confucius Sinarum Philosophus*, 1687. Courtesy of the Getty Research Institute.

craze known as *Chinoiserie*.[9] Chinese tea changed Europe's tastes, while Chinese gardens changed its landscape.[10] Most admired of all was China's ancient written tradition: its history, philosophy, and religion, all thought to be broadly comparable to Europe's own. Enlightenment thinkers wrote admiringly about some other places too—from Polynesia and India to Mexico and Peru—but their interest in China was more widespread and better-informed, rooted not just in the observations of merchants and missionaries, but in the words and thoughts of real ancient and contemporary Chinese figures.[11] In one crucial respect, their interest was unique: Europeans not only admired China, but regarded it as a place that Europe could and should actually become more like.[12] In sum, "sinophilia" was the best example of what has been called the cosmopolitanism, or universalism, of the high Enlightenment.[13]

Few transformations in Enlightenment thought, then, were as marked as the philosophes' reversal of opinion on China. One might take 1773 as the watershed year, marking both the global suppression of the Society of Jesus and the publication of a monumental anti-Chinese diatribe by a well-known philosophe.[14] Some historians place the shift a little earlier; others a little later.[15] But in any case, by the early nineteenth century, European critics were already aware of the "generally little favorable opinion that one conceives of the Chinese in these recent times."[16] Just as praise for China had once centered on its ancient written tradition, condemnation was based on its wholesale rejection. Confucius, still the paragon of that tradition, bore the brunt of the attack. Once a prophet of reason, he was recast as an insurmountable barrier to progress. For the philosophes of the late Enlightenment, there was little more to be said on the matter. And largely for this reason, no broad study of China in Europe has much considered what happened after the death of Voltaire, except to address what has seemed like a new consensus.[17]

There has been much debate over how to explain the reversal.[18] Many proposals have involved a methodological assumption that the intellectual transformation must have been caused by a material one: namely, the "Great Divergence." A growing economic disparity undermined admiration for China's pre-industrial agrarian economy.[19] New cultural tastes in arts and manufactures reduced the appeal of exotic Chinese luxuries.[20] Expanding geopolitical strength highlighted the weakness of the once-mighty Qing state.[21] Evidence for all these claims can indeed be found, and there is no question that a shift in the broader socioeconomic relationship between Europe and China was taking place around this time. Nevertheless, it is a little too easy to read later developments back into earlier times—and comparatively difficult to show contemporary awareness of them. As late as 1773, it was not at all clear to observers that a macroscopic realignment between the Europe and China had yet taken place, if indeed it really had.[22] Chinese goods remained a driving force in global trade.[23] None could doubt that the Qing empire was one of the largest and most powerful in the world.[24] Moreover, these domains—economic and political might—were never at the center of European interest.[25] In sum, even if the Great Divergence was already underway, it cannot fully account for the shift in Enlightenment attitudes toward China, because most people had not yet noticed it.

Another set of explanations has pointed to a different story of European exceptionalism: the idea of historical progress. Toward the end of the Enlightenment, progress was increasingly thought to be what divided moderns from ancients, and Europeans from everyone else.[26] On the whole, the later philosophes still described China in the same way that the earlier ones had: as ancient and enduring. But their views of change itself were changing. Accordingly, China was reframed, from stable to static.[27] China became an "immobile empire," stuck in the past as Europe progressed toward the future.[28] Yet here, too, it is easy to be fooled by later developments. Many historians, themselves tacitly self-identifying as heirs to the philosophes, have not appreciated the extent to which ideas of progress themselves were in flux. Throughout the eighteenth century, the elements of Chinese knowledge that were most admired—history, religion, and ethics—were valued not despite but because of their antiquity. In fact, China's long history, pristine religion, and enduring ethics played a significant role in the thought of canonical progress theorists from Voltaire through Turgot. Missing too from this account is the centrality of the one domain in which new narratives of progress did become determinative: the natural sciences. In sum, a long-standing focus on the likes of Diderot and Condorcet conceals a more interesting, and ultimately more influential, story.

What actually took place toward the end of the Enlightenment was not the expulsion of Chinese models, but their realignment. The philosophes' turn toward progress thinking spelled the death of Voltaire's Confucius.[29] A new critique of China as lacking progress focused on its supposed failure to develop the natural sciences. But it was not always clear that Europeans should find the natural sciences of China to be lacking. That conclusion was the result of selective Jesuit reporting on the natural knowledge of China, made over the course of nearly two centuries that coincided quite closely with the Scientific Revolution. As recently as the early eighteenth century, aspects of Chinese natural science, including natural philosophy, natural history, and some empirical sciences, had been widely embraced.

It was only late in the Enlightenment that a unified story about the natural sciences and their cumulative development over time emerged. This was a story in which China and Europe were thought to have played roles of unequal significance; but both were still thought to have played a part.

While the later philosophes cast Confucius aside as a hidebound conservative beholden to ancient tradition, others embraced him for the very same reason. The model of China persisted, co-opted by conservatives who were skeptical of the philosophes and their views of progress. The sinologue scholars and physiocrat statesmen who took it up were not troubled by the question of China's natural science, because they did not believe it was the only metric of progress, nor that change was desirable at all. The philosophes helped to both build and then destroy early modern interest in Chinese knowledge. But they were not responsible for what became the modern approach to engagement. It was indeed just because China ceased to be a viable model for them that China began to be put to new uses around this time. From the 1760s to the 1790s, intellectual exchange between France and China was effectively co-opted by the minister of state, Henri-Léonard Bertin. Seeking to prop up the ancien régime with a regime more ancient still, he put the surviving French missionaries in Beijing on life support and acted as the sole gatekeeper between them and the rest of the Republic of Letters. The result of their efforts, the *Mémoires concernant les Chinois*, was a mountain of new scholarship that recast China in Europe while placing new demands on those Europeans who were still in China.

The Jesuit Mission and Chinese Science

In the history of early modern intellectual exchange between Europe and China, the importance of the Jesuit mission is difficult to overstate. This is for a very simple reason: the missionaries were, with no known exceptions, the only Europeans who ever became fully literate in Chinese until the nineteenth century. Although there were other missionary orders that were also active in China, their clerics produced little significant scholarship after the end of the sixteenth century.[30] And although there were certainly some Chinese people who learned European languages, few of them left any literary legacy, while those who did worked under the auspices of the Jesuit mission.[31] For a period of more than a century, the Jesuit missionaries taught each side virtually everything that it knew about the written traditions of the other. They were ultimately responsible for drawing the picture that the philosophes would first celebrate and later condemn: one that praised Chinese moral philosophy, questioned its natural philosophy, and associated both with the ancient past.

The Jesuit mission began near the very start of the Scientific Revolution; at this time, there was no consensus about what counted as science in Europe, much less in China. Matteo Ricci was born in 1552, twelve years before Galileo. His worldview was grounded in the Catholic Aristotelianism of the late Renaissance.[32] The disciplines then relevant to natural science can be divided a thousand ways; suffice to say only that they were disaggregated. Natural philosophy referred to causal explanations of nature, including aspects of physics and metaphysics. Mixed mathematics involved the quantification of the physical world, including exact sciences such as optics and mechanics. Natural history meant the description and classification of natural objects, relevant to both medicine and industry.[33] These categories were far from fixed in Europe, and in China, none of them applied at all. The upshot was that at the beginning of sustained exchange, the missionaries could not have had a unified conception of Chinese science. They engaged with local natural knowledge separately and across domains, with frequent disputes emerging between each other and over time. It was not until later in the eighteenth century that a coherent narrative about Chinese science began to emerge.

From the start, the Jesuits in China realized that their expertise in certain mathematical disciplines was an invaluable asset. Ricci first entered the Forbidden City in 1601. By this point, he had already issued a Chinese version of Abraham Ortelius's world map in collaboration with the Ming mathematician and astronomer Li Zhizao, and he soon began working with another, Xu Guangqi, to translate Euclid's *Elements* into Chinese.[34] These efforts won him major literati converts and eventually the patronage of the Wanli emperor. From then on, the Jesuits saw science as crucial for the success of their mission.[35] A string of famous missionary-scholars followed: Johann Adam Schall von Bell, Ferdinand Verbiest, and Antoine Gaubil especially eminent among them. Some of them earned formal positions in the Ming and Qing bureaucracies for their expertise in mathematics, astronomy, and geography, and later also in adjacent fields such as engineering, music, and painting.[36] So when it came to these disciplines, the missionaries never had much reason in China to tout their indigenous competition. Indeed, they openly claimed that their mathematical and astronomical techniques were superior to those previously employed by officials serving essential functions of the Ming and Qing government.[37] Some eighteenth-century commentators even "suspected the Jesuits of having kept the Chinese ignorant in order to further their credit in the

FIGURE 1.2 *Johann Adam Schall von Bell, Ferdinand Verbiest, Candida Xu, and Paul Siu (Xu Guangqi)*, in du Halde, *A Description of the Empire of China*, volume 2 (1741). Courtesy of Boston College University Libraries.

Court of Beijing."[38] Whether or not that suspicion had any merit, it demonstrates where their interests at the time were perceived to lie.

Facing a European audience, however, the missionaries' interests were rather different; it was generally preferable there that Chinese knowledge seem to be valuable, since the missionaries held a near-total monopoly on it. The missionary-man-of-science quickly soon became a stock figure of Jesuit propaganda in Europe, as he had in China. The Jesuits promoted their expertise in mathematical sciences and Chinese learning as dual components of their learned identity, in the salons, museums, academies, and colleges, both in person and in print.[39] In particular, they focused on astronomy.[40] The history of celestial observation in China was almost as old as the history of China itself, with records of eclipses and comets going back at least as far as the early first millennium BCE.[41] The Jesuits discovered these records almost immediately and deployed them as evidence of the accuracy and ancientness of Chinese knowledge. Their argument was broadly successful. By the end of the seventeenth century, luminaries of European science including Robert Hooke and Jean-Dominique Cassini had taken notice of Chinese astronomy in the pages of publications like the *Philosophical Transactions* and the reports of the *Académie des sciences*.[42] But the Jesuits promoted it only to a certain point. For example, they took little interest in theories about the structure of the heavens and the earth, which were then a matter of some dispute in China as they were in Europe.[43] But in both traditions, observational astronomy was somewhat independent from natural philosophy.[44] And for the missionaries, the latter was more important by far, because it struck at the heart of their ultimate commitments.

The orthodox doctrine of the late imperial literati elite was a philosophy known in Chinese as "Song learning," "principles learning," or "Song-Ming principles learning," more commonly called "Neo-Confucianism" by later historians.[45] From the Six Dynasties to the Tang, Buddhist and Daoist teachings spread widely, presenting appealing transcendental religions as well as sophisticated metaphysical visions. In response, a series of Confucian philosophers, writing mostly during the Song (960–1279 CE), fashioned a new cosmology to buttress what had originally been an essentially moral and political teaching. The process culminated in the twelfth century, when the influential philosopher Zhu Xi canonized the Four Books of Confucius and wrote authoritative commentaries expanding upon their natural-philosophical content based upon the earlier writings of Zhou Dunyi and the brothers Cheng Yi and Cheng Hao.[46] Zhu Xi's theory begins with "non-polarity" and "supreme polarity," or *taiji*, which he identified with his preferred term *li*, literally "form" or "principle."[47] *Taiji* or *li* creates the dual principles of *yin* and *yang*: dark and light, female and male, passive and active.[48] *Yin* and *yang* in turn interact to produce *qi*, which can be translated variously as "air," "spirit," "ether," "energy," or "pneuma."[49] *Qi* is the stuff of which the natural world is composed. Working back up the metaphysical ladder, *qi* operates according to the principles of *yin-yang*, which are in turn governed and set in motion by *li* or *taiji*.[50] This synthesis proved powerful and persuasive. It was soon endorsed by the imperial state and became the backbone of all formal education—which it remained until the abolition of the civil-service examinations in the early twentieth century.[51]

This form of Neo-Confucian, sometimes called the "Cheng-Zhu" school after Zhu Xi and another of its principal architects, presented an immediate problem for the missionaries because it was undeniably naturalistic. Mostly, they engaged with it through secondary explications written long after the end of the Song, such as the *Great Compendium on Nature and Principle*, compiled in 1515.[52] Centrally at issue for the missionaries were the concepts of *li*, which they usually translated as "reason," and *qi*, which they often glossed as "matter." Although Zhu Xi did ascribe *li* with a kind of formal priority over *qi*, the relationship between them was explicitly ambiguous: in his words, "We cannot say between *li* and *qi* which fundamentally comes before the other. If we want to deduce their origins and development, then we must say that *li* comes first. But *li* is also nothing other than an entity, so it exists among *qi*." Moreover, Zhu Xi's conception of *li* itself was anything but spiritual: "*li* has no feelings, no plans, no intentions."[53] The upshot for the missionaries was that it was plainly impossible to identify the first principles of contemporary Chinese natural philosophy with the Christian God. Over the course of the seventeenth and eighteenth centuries, they wrote a great deal about Neo-Confucianism, warping it beyond recognition by putting to it a host of questions with which it was never designed to deal. Yet they also managed to extract from it a set of answers on which they all more or less agreed: the natural philosophy of late imperial China was materialist at best or atheist at worst, and thus incompatible with Christianity.[54]

And yet, very early on, a conflict emerged among the missionaries that made it crucial to study all the same. Virtually all of them agreed that the dominant doctrine of modern Chinese philosophers was incompatible with Christianity, but they disagreed as to whether that of the ancients had also been. The canonical Confucian texts—the Four Books, written down by Confucius's early disciples in the fifth through third centuries BCE, and the Five Classics, now dated to the middle of the first millennium BCE but then thought to be much older—had little to say on such matters.[55]

The missionaries exploited their silence. Ricci himself first argued that the original doctrine of ancient China had been a natural religion, lacking only for divine revelation. His successor as superior general of the Jesuit mission, Niccolò Longobardo, rejected this view, holding instead that the beliefs of ancient China were obscure and indistinguishable from those of his own day. Crucially, the disagreement was largely historical, in that it pertained not to the beliefs of living Chinese people, but rather to the ones ascribed to those of thousands of years earlier.[56] Each position found its European supporters. Ricci's won out among the Jesuits, who sought to proselytize among the literati elite, while Longobardo's became the dominant view of the Dominicans, Franciscans, and others who worked mostly with the poorer and less educated. By the turn of the eighteenth century, both positions were widely known in Europe.

On the one side, the Jesuits promoted an approach that historians have come to call "accommodationism"—the idea that Christianity could and should be accommodated to local Chinese practices and beliefs.[57] Accommodationism reached a high point in 1687 with the publication of *Confucius, Philosopher of the Chinese,* the springboard for the first wave of Enlightenment sinophilia.[58] Along with the first substantial translations of the Four Books, the Latin publication included a long introduction to the history of Chinese philosophy.[59] The most ancient Chinese texts, the Five Classics, had invoked the worship of a deity named *Tian,* "heaven," or *Shangdi,* "Supreme Emperor."[60] According to the Jesuit commentators, this was a stunning testament of a "Pristine Age" of universal piety and perfect knowledge.[61] They sincerely believed that "the name by which the first Chinese called the true God was *Shangdi.*"[62] Unfortunately, they maintained, the pristine religion alluded to in the Classics had been abandoned long ago, perverted most of all by the Neo-Confucians, or "Neoteric interpreters," whose doctrine predominated into present times.[63] Jesuit treatments of concepts like *taiji, li, qi,* and *yin-yang* therefore had to tread a very fine line: rejecting them as modern corruptions, but without impugning the ancient foundations upon which they had supposedly been built.

On the other side from Jesuit accommodationism, the position that the natural philosophy of ancient and modern China was all incompatible with Christianity was rather more straightforward. Longobardo explained it in an essay that became influential later through a 1701 French translation as the *Treatise on Several Points of the Chinese Religion.*[64] Himself a Jesuit, he began with a personal story of how he had come to doubt his superiors' position that "Shangdi was our God."[65] Since the ancient texts were obscure and corrupt, he claimed, in determining their real meaning, "one should opt for the commentaries"; that is, the works of the Neo-Confucians. Even if the Classics were ambiguous on the nature and power of Shangdi, "the Interpreters attribute all this to Heaven, or the universal substance and nature, which they call *Li*."[66] For him, there was no true distinction between ancient and modern doctrine; only the transhistorical views of "the sect of the literati." The explication of Neo-Confucian natural philosophy that Longobardo gave in service of this argument was far more detailed than any that the Jesuits had put forward and, one might say to his credit, had the advantage of taking contemporary Chinese scholars more directly on their word. For these reasons, it became the preferred source on what Europeans took to be Chinese natural philosophy—including cosmology, physics, metaphysics, and ontology—even for those later writers who found much to admire in the views that it had been written to critique.

As the battle lines around the two camps took shape in Europe toward the turn of the eighteenth century, a third position arose among a small group of French Jesuits living in Beijing; variously called the "Yijingists," "Mythologists," and "Fuxi-Enochists," the name that eventually stuck was the "Figurists."[67] Their central idea was that the Classics concealed hidden elements, or "prefigurations," of divine revelation—"prefigurations" because the events they referred to had not yet occurred. In one sense, this was a reconciliationist position between the other two: Christianity did not have to be accommodated to Chinese beliefs.[68] But in another, it was an extreme form of the Jesuits' main theses: these two things were, in the end, identical.[69] The Figurists further believed that the Classics and the Bible alike contained elements of a mysterious *prisca sapientia*, ancient wisdom, or *prisca theologia*, ancient theology, granted by God to humanity before the Flood. In their ambition to recover it, they became more interested in Chinese natural philosophy than any Europeans before them had been. They scoured the *Yijing*, or *Classic of Changes*, and even some Daoist texts in search of the antediluvian knowledge they believed had been preserved by the figure known as Fuxi in in the Classics, or Enoch in the Bible.[70] Their extraordinary erudition was much admired by the later missionaries and attracted attention in their own day from Gottfried Wilhelm Leibniz and the Kangxi emperor alike.[71] But Figurism was never widely accepted, either in Europe or in China. No major exposition of it was even published until the nineteenth century, and if it managed to capture the imaginations of generations of historians, this is largely because it was always seen as controversial.[72]

The internecine Catholic dispute over Chinese beliefs finally came to a head in the early eighteenth century with the conclusion of what was already known by then as the "Chinese Rites controversy." Fundamentally at issue was whether or not Chinese Christians should be permitted to perform ritual sacrifices to Confucius and their ancestors. The Jesuits said yes: the rites had their origins in the practices of pious antiquity, and in modern times they could be seen as civil rather than religious observances. Almost everyone else said no: the ancient practices were no different from the modern ones, and all of them were idolatrous. Historians have been mostly sympathetic to the Jesuit position, on account of its apparent spirit of toleration, but it is hard to deny that it also involved a certain degree of casuistry, and contemporaries had no compunctions pointing this out. In 1704, Pope Clement XI issued a decree condemning the Chinese rites and sent a legate to communicate his decision to the Kangxi emperor. The results were disastrous. The emperor, outraged at what he took to be outside intervention in domestic affairs, banned Christianity outright in 1724. The pope, furious at what saw as Chinese-Jesuit intransigence, confirmed his earlier decree with a papal bull in 1715. From then on, accommodationism in China was technically proscribed by the Church. Christians were permitted to refer to God only by the neologism *Tianzhu*, "Lord of Heaven," and the ritual sacrifices to Confucius and the ancestors were forbidden. The Society of Jesus never really recovered from the blow.[73]

The Jesuits did, however, win a victory of their own in the Republic of Letters. The conflict had inspired the missionaries to conduct a great deal of new research for polemical purposes. Much of it eventually worked its way back to Paris, where it was collected and edited by the Jesuit Jean-Baptiste du Halde. Though he had never been to China and could not read Chinese, he is known to history for almost nothing else. In 1735, du Halde published the four-volume folio *Geographical, Historical, Chronological, Political, and Physical Description of China and Chinese-Tartary*, more commonly known as the *Description de la Chine*. As the title suggests, the contents were wide-ranging: religion, economics, manufacturing, and the arts all could have been mentioned as well, and much of the material was presented in direct translations from late imperial Chinese sources.[74] But the central focus of the work was "*la Science Chinoise*," or "the science most fitting to man, since it directly regards his conduct, and the ways to perfect it fitting his state and condition"—that is, not the natural philosophy of earliest China, but rather the moral, ethical, and political philosophy of Confucius.[75] The sumptuous Paris edition was followed by a pirated version at The Hague and a complete English translation, both within the year. The *Description* immediately became the single most important work on China for the high Enlightenment.[76]

Though far from its main focus, the discussion of Chinese science in the *Description* set the tone for the next half century of European conversations. The general argument was that it was valuable, but only insofar as it was empirical rather than theoretical. With the Chinese Rites debacle still fresh, the Jesuits mostly ignored natural philosophy. The only explicit discussion of the topic, attributed to a certain "Chinese philosopher," was in fact a translation of a satirical scene from an early Qing short story.[77] Instead, attention shifted to other fields of natural knowledge: in particular, open-air sciences like natural history and geography, where observations were useful independently of the causal explanations given for them. Indicative too was the very short section entitled "Knowledge of the Chinese in other sciences"—that is, sciences "other than" moral philosophy, including the quadrivium of music, astronomy, arithmetic, and related disciplines such as optics, mechanics, and pneumatics—all of which was framed almost exclusively as the accomplishment of the missionaries.[78] For du Halde, the most important thing about the history of the exact sciences in China was the simple fact that Europeans had brought them there: "It was only a little more than a century ago, since the entry of the first Jesuit missionaries in their empire, that they became aware of their ignorance."[79] This story of diffusion has been be retold countless times ever since.

Also, in presenting this early version of it, the *Description* articulated a puzzle that would animate European historians of Chinese science for the next three centuries. As du Halde put it: "They made discoveries in all the Sciences; and they did not perfect any of those which we call speculative."[80] The assertion was in fact a paraphrase of a 1730 letter written by the Jesuit Dominique Parrenin to the astronomer Jean-Jacques d'Ortous de Mairan, in which the missionary had added: "This seems to me, as it does you, almost incredible."[81] Du Halde, by contrast, claimed that "it is not surprising that these kinds of more abstract Sciences are neglected by the Chinese." He proposed that "two main obstacles are opposed to the progress that they could have made." First, China had no neighbors to compete with or to emulate. Second, achievements in this domain were not rewarded by society or tested on the civil-service examinations. One might thus consider the *Description* to include the first published treatment of the Needham question—and to have answered it in the space of a single paragraph.[82] For the missionaries up until this point, the supposed underdevelopment of Chinese science was simply not an important problem, because knowledge of nature was not uniquely privileged, and change per se was by no means desirable. But for some among the philosophes, it was starting to become one.

From Land of Reason to Place without Progress

Like nearly everyone else in early modern Europe, Voltaire learned about China from the missionaries. The young François-Marie Arouet, born in Paris in 1694, was educated at the Jesuit Collège Louis-le-Grand from 1704 to 1711. During this time, he would have heard much about events in China, from the achievements of the French Jesuits bringing glory to the mission and the Kangxi emperor, to the Chinese Rites controversy then reaching fever pitch. He might have enjoyed the rarer opportunity to discuss them with people who had actually been there, including the Figurist Jean-François Foucquet and perhaps a Chinese Christian novice. He went on to read nearly all the major publications on the topic, still fairly small in number. He never forgot the lessons of his Jesuit schoolmasters, even as his work later fed the reactions against them.[83] Voltaire was by far the greatest philosophe to draw such inspiration from China; he was also, in a sense, the last.[84] His followers abandoned the model of China in favor of another teaching in which he had played no small part: the idea of historical progress as rooted in the cumulative development of the natural sciences.

At the height of what was once known as the "Age of Reason," Voltaire identified the teachings of China with this paramount value. Historians have argued about the foundations of his interest in China: some emphasize its religion, others its history, still others its government.[85] The thread that linked them all together for Voltaire was their shared basis in what he took to be a Confucian doctrine of reason identical with his own. Its religion, cast as a rationalist Deism, was "simple, wise, and noble, free of all superstition and all barbarism."[86] Its history, composed, as he put it, "with pen and astrolabe in hand," showed that "as soon as these people began to write, they wrote reasonably."[87] Its government, reportedly a meritocracy of bureaucrat-scholars, began from the philosophical principle that "to learn to govern, you must constantly correct yourself."[88] Missing from Chinese religion, history, and government were reason's counterparts in each domain: scriptural doctrine, allegorical fables, and arbitrary aristocracy. Perhaps the most general opposite of reason was superstition, and as it appeared to him, "superstitions seem to be established among all the nations, except the literati of China."[89] Chinese and French philosophers were cut from the same cloth. For superstition took many forms; reason came in only one.

There was, however, a difficulty with this account, and it troubled Voltaire greatly: "What is particularly surprising," he wrote, "is that, having for so long cultivated all the sciences, they remained at the point where we were in Europe during the tenth, eleventh, and twelfth centuries."[90] Here, too, the Chinese had been precocious, beating modern Europeans even to the three inventions by which Francis Bacon had distinguished them—printing, gunpowder, and the compass.[91] But despite their head start in mechanics, mathematics, medicine, and astronomy, they had somehow fallen behind: "Their great ancient progress and their present ignorance present a contrast for which it is difficult to account."[92] In the 1750s, Voltaire proposed several explanations, ranging from geography to language.[93] By the 1770s, he had settled on just one: "I have always thought that their respect for their ancestors, which is for them a kind of religion, was a paralysis that hindered them from advancing in the career of the sciences."[94] In other words, conservatism was fundamentally opposed to the improvement of the sciences; or, conversely, the history of science was essentially progressive. This position was to be broadly adopted among the philosophes, but they were still working to articulate it. Not long before, Chinese science had seemed not only interesting to their predecessors, but potentially even correct.

No one in Europe had taken a greater interest in the natural philosophy of China than Leibniz.[95] Between 1697 and 1707, he exchanged some fifteen long letters with the leading Figurist missionary in Beijing, Joachim Bouvet. Leibniz identified Chinese doctrine as nearly the same as his own. In the ancient *Classic of Changes*, with its sixty-four hexagrams composed of broken and unbroken lines, he recognized his system of binary mathematics. In the Neo-Confucian conception of *li*, a formal principle existing in unity with matter, he identified his theory of entelechies or monads.[96] While engaged in an epistolary debate with a surrogate of Isaac Newton, he completed a detailed letter interpreting Chinese natural philosophy as consistent with his somewhat idiosyncratic positions in cosmology and metaphysics: space and time as relative, causation as a pre-established harmony, and God as a supramundane intelligence, among others.[97] Like the missionaries, he was less interested in the exact sciences of modern China than their ancient natural-philosophical foundations, but he maintained that there was much to be learned still in some fields, including medicine and even experimental physics.[98] He imagined a future exchange "which could give to us at once their work of thousands of years and render ours to them, and to double so to speak our true wealth for one and the other." At the beginning of the Enlightenment, Leibniz dreamed of what he called "a commerce of light."[99]

Many of the more radical thinkers of the time also took an interest in Chinese natural philosophy, though for a very different reason: if the

Confucian literati were Epicureans, Stoics, or Spinozists, as most of the missionaries had maintained, then all the better, for so as indeed were they.[100] Just as the missionaries' presentation of Chinese history unintentionally called into question Christian chronology, so too did their accounts of Neo-Confucian philosophy provide ammunition for an attack on Christian metaphysics.[101] For Pierre Bayle in the Netherlands, Christian Wolff in Prussia, and Nicolas Fréret in France, Confucianism might have been deistic or even atheistic; it was in any case preferable to Church doctrine. Bayle, a forerunner to the Encyclopedists, interpreted Confucius as a Chinese Spinoza, using each to support the other.[102] Wolff, a preeminent disciple of Leibniz, in a famous lecture "On the Practical Philosophy of the Chinese," upheld Confucius as proof that an ethical society did not require a transcendental religion.[103] Fréret, who was secretary of the Académie des inscriptions, maintained that Confucius had been a kind of Deist, as French philosophers should be as well.[104] All three were widely seen in their own time as freethinkers or libertines, or at least sympathetic to them; that is to say, they were not conservatives. Yet they also believed that modern Europe could learn from ancient China about even more than its famous moral philosophy.

Thus, when Voltaire began to write about it in the 1720s, the status of China's knowledge of nature was much discussed and far from settled. There were pressing questions as to its natural philosophy: whether the ancient and modern doctrines were the same, and whether either or both were consistent with Christianity. There was more consensus in other fields: its exact sciences were underdeveloped, but its observational sciences had value. Yet across their differing opinions, Jesuit missionaries, Church authorities, and early philosophes shared something in common. They attempted to explain ancient Chinese science with classical and contemporary references in the same breath, and all accepted in principle the possibility it could have been correct. This was true for the Jesuits and the Dominicans, who believed in a pristine age of divine knowledge; but it was also true for Leibniz and Fréret, who themselves held the philosophy of the ancients in high regard. Europeans were not yet troubled by the question of why China had not developed natural science, because they had no reason to accept that premise according to their various conceptions of progress.

Voltaire's attempts to answer the question over the course of half a century reflect his thinking through his own. As an accomplished and iconoclastic historian, he overtly aimed to shift the focus from Europe to Asia. China was quite literally at the beginning of his universal history.[105] Its story could be celebratory because natural philosophy was not determinative. While the improvement of natural science played an important role in the story of progress, it was not the starring one.[106] Progress was guided rather by the more general application of universal human reason.[107] Confucius, as the "interpreter of right reason alone," was proof that that "one can be a very bad natural philosopher and an excellent moralist." The two capacities were independent and causally distinct. One kind of knowledge necessarily improved over time, but the other did not: this difference explained why it was possible that "the Chinese, remaining for more than two thousand years within limits they had attained, remained mediocre in the sciences, and the first people in the world in morality and in government, as the most ancient."[108] Moreover, it was clear which kind of knowledge mattered more: "The Chinese have not perfected any of the arts of the spirit except morality, but they enjoyed profusely from what they did know, and ultimately they were happy as far as human nature permits."[109] The improvement of society was not dependent on the development of natural science, and there was no metahistorical law dictating that the future be preferable to the past.[110]

But for some of Voltaire's later contemporaries, the cumulative advances of natural science over time was increasingly seen as the sole engine for improvement for society as a whole, and during the 1750s, they revisited the subject of Chinese natural philosophy, condemning it by association with the very past that they wished to leave behind.[111] In the *Encyclopédie,* the short article on "China," dealing with its art, industry, and commerce, was quite complimentary; the long article on the "Philosophy of the Chinese" was not.[112] Both articles were probably written by Denis Diderot. His discussion of natural philosophy was taken almost word for word, though via a third-party intermediary, from Longobardo's hundred-year-old treatise, which had been written precisely in order to make it look bad.[113] The section was introduced as an overview of the "principles of the Chinese philosophers of the Middle Ages and the Literati of today," casting *taiji, li, qi,* and *yin-yang* as conceptually medieval. The problems of the European past could be found in the Chinese present: "All this doctrine must be expressed by symbols, enigmas, numbers, figures, and hieroglyphs." The conclusion was that "the morals of Confucius are, as one sees, very superior to his metaphysics and his physics."[114] For the earlier philosophes, suspicious of metaphysics of all kinds, that had been a point in his favor.[115] For the later ones, seeking to rebuild metaphysics on the foundations of the natural sciences, it was no longer.[116]

Diderot did not go looking for natural science in China because his view of progress gave him no reason to expect to find it there. In marked contrast to Voltaire, he held that "it is not surprising that although the

Chinese may be more ancient, we have passed them by so far." Chinese natural philosophy was retrograde not despite its antiquity, but because of it: "If this system is as old as one claims," he remarked, "one cannot be too surprised at the multitude of abstract and general expressions in which it is conceived." A preoccupation with the past carried certain advantages in moral philosophy—"it should make particularly in China the usages more constant, the government more uniform, the laws more durable"—but it was harmful across the arts and sciences, which demanded "a more unquiet activity, a curiosity that hardly rests from searching, a sort of incapacity to satisfy itself." Where early philosophes had praised constant Confucian reason over restless metaphysical speculation, Diderot repeated the observation with an evaluative twist: "The spirit of the east is more tranquil, more lazy, more enclosed by essential needs, more born to that which it finds established, less avid for the new than the spirit of the west." Although his observation was not that different from Voltaire's in its substance, Diderot developed his analysis in the direction of a new kind of Eurocentrism that had been quite uncharacteristic of the early Enlightenment: "In a word," he wrote, "they do not have the genius of invention and of discoveries that shines today in Europe." Accordingly, the beginning of sustained communications between Europe and China was "the event that marks the epoch of the modern philosophy of the Chinese." What happened then was quite straightforward: as the missionaries had suggested, "we brought our knowledge to them."[117]

European sinophobia came into full bloom in 1773 with the Dutch philosophe Cornelius de Pauw's *Philosophical Research on the Egyptians and the Chinese,* which offered a fresh reading of most of the very same missionary sources that had sparked the interest of Voltaire. This two-volume treatise displayed in stark relief the ambiguities of late Enlightenment thought. De Pauw was strikingly rationalist yet plainly bigoted, an advocate of secularism and democracy but also racist, misogynistic, and openly anti-Semitic. His condemnation of China was extreme to the point of ridiculousness. Its history was a "miserable little chronicle"; its religion an "absurd doctrine of spirits or Manitous"; its government "despotic" all the way down.[118] Confucianism appeared no longer as a philosophy of reason, but rather as a "gross superstition," replete with magic spells and dowsing rods."[119] Everything previously said in China's favor was turned on its head; but the harshest criticism was still reserved for natural science. De Pauw believed that China had never had any knowledge of "physics and real science" at all.[120] There was no head start even in the distant past: "Physics and natural history are the two subjects against which the canonical books of the ancient peoples of Asia most grievously sinned."[121] The

Jesuits had lied about the history of Chinese astronomy, as about everything else; judicial astrology was all they had ever known.[122] In modern times, the literati had no idea of latitude or longitude and still believed that the earth was square: "It really is a joke to say that such a people were in such a state as to write their annals with astrolabe in hand, as claim the enthusiasts, and verify the history of the earth by the history of the heavens."[123] De Pauw's anti-Jesuitism was de rigueur; but here he took a swing directly at fellow-philosophe Voltaire, who had made this claim just a few years earlier.

Unsurprisingly, the book met with impassioned resistance from almost everyone with any knowledge of the subject.[124] It is telling that Voltaire himself felt compelled to respond. By this time, his opinion might have shifted somewhat since he first wrote on the topic half a century earlier: "We are not enthusiasts of distant places and antique times," he conceded to the spirit of his own; "we know that the entire Orient, far from being today our rival in mathematics and the beaux arts, is not worthy of being our student."[125] Nowhere was more distant nor more ancient than China. Yet, for his part, he continued to defend it: "The body and soul of this country, the true and the false, inspire me with such curiosity, such interest, that I will write to M. de Pauw here and now; I hope he will remove all my doubts."[126] It does not seem that his doubts were ever dispelled.[127] Still, there was one point on which Voltaire and de Pauw did agree, a fundamental commonality connecting the sinophilia of the early Enlightenment with the sinophobia of its end. In 1756, Voltaire had called Asia "the cradle of all the arts"; in 1773, de Pauw placed China in a state of "eternal infancy."[128] The metaphor was exactly the same; only the evaluation had been reversed. The shift from sinophilia to sinophobia reflected little new knowledge of China but changing notions of progress in Europe.

Voltaire's question as to the possibility of uneven moral and natural philosophical development finally found a stable answer not long after: China made sense within the new progress paradigm as a place that had gotten stuck in time. The short article in the *Supplement to the Encyclopédie,* nominally on "Chinese literature," was in fact almost entirely about "the causes that have retarded the progress of the sciences in China," as stated in the very first sentence. The only example given was astronomy. Taking as fully demonstrated "the state of languor in which the sciences in China are in," the matter at hand was merely to explain it. The author gave some credence to Voltaire's theory of Chinese conservatism: "too much attachment to antiquity" had been "the poison of the sciences." But he also proposed a new reason, one that Voltaire had explicitly denied: "We will not be afraid to say it, it is mainly the lack of that inventive genius, which particularly distinguished the Greeks in antiquity, and which seems to belong now for some time to Europeans."[129] Progress, characterized by the rejection of the past and the development of "the sciences," was the provenance of Europe alone. The most important thing task in thinking about elsewhere was only accounting for its absence.

Progress theory thus did bring about an end to the philosophes' enthusiasm for the East; but it did not imply the full package of modern orientalism. For those late Enlightenment thinkers who developed the most sophisticated ideas of historical progress, from stadial theorists like Adam Smith to evolutionists like Georges-Louis Leclerc, comte de Buffon, China was not at all beyond comprehension—it was simply not very interesting. In the most canonical statement of Enlightenment progress theory, Condorcet would only briefly mention "this people, who seem to have preceded all others in the arts and sciences, only to see themselves successively erased by them all"; the context was to provide an example of a place that had failed to progress "even without the aid of superstitious terrors."[130] The absence of superstition, formerly the negative counterpart to philosophical reason, was no longer as important as the development of arts and sciences, now the positive hallmark of historical progress. China's apparent lack of natural science had once been a hard problem for Enlightenment thinkers; progress theory made it into an easy one, and, by extension, a boring one.

Academicians and Physiocrats

And yet, the European renunciation of the model of China was not as universal as it later came to seem. Historians have generally taken the criticism or disinterest of philosophes like Diderot and Condorcet as representative of a new consensus. But while they were giving up on it, others were beginning to take it up. The French academicians and social theorists who carried on the study of China into the last decades of the eighteenth century were not the clear inheritors of the missionaries and philosophes who had preceded them, nor were they the obvious testators to the orientalists and philosophers who would come after. With only a few exceptions, their work on the subject has not been seen as contributing much either to the positivist accumulation of true Sinological knowledge or to the development of late modern thought. Largely for these reasons, historians have not paid them much notice. Doing so reveals that Enlightenment engagement with Chinese thought did not end; it was only transformed.

Despite its weaponization by philosophical radicals and political reformers, the study of China in France had been tied since its very beginning more or less directly to the Bourbon monarchy. In 1685, Louis XIV finalized a plan to extend his control over the Catholic church at home while at the same time asserting French commercial and diplomatic interests in Asia. He commissioned the selection of six Jesuits for the China mission and designated them as "Mathematicians of the King": Jean de Fontaney, Joachim Bouvet, Guy Tachard, Claude de Visdelou, Louis le Comte, and Jean-François Gerbillon. This was an elite and talented group. Several of them were chosen specifically for their skill in such fields as astronomy and geography. Some would go on to establish the Figurist school and others to write works of lasting Sinological influence. After completing specialized training at the Académie des sciences, they were given instructions by the King "to draw as much knowledge as possible" from China. They set out for the mission field and arrived at the Qing court in 1688.[131]

Their timing was favorable, for Beijing was then experiencing an efflorescence of Sino-European exchange.[132] The Kangxi emperor, who had assumed the throne in 1666, spent the first several decades of his reign putting Western learning in the service of the Qing state. With the help of missionary men of science such as Ferdinand Verbiest and Antoine Thomas, he promulgated a major calendar reform and sponsored a comprehensive cartographical survey of his expanding domains.[133] The fresh French missionaries were well positioned to take advantage of the emperor's good favor. Bouvet became his personal tutor in Western studies. Gerbillon helped him negotiate a major diplomatic treaty with the Russians. Fontaney reportedly cured his malaria with South American cinchona bark. In 1699, the emperor granted them a plot of land inside the walls of the Imperial City as a token of his appreciation. There they established a new home, the *Beitang*, or North Church, in 1703. Now with their own residence, the French Jesuits became semi-independent from both the Portuguese mission and the diocesan authorities. From then, on, they reported directly to the Jesuit Visitor, putting them on equal footing with the Vice Province of China, while their scholarship was sent directly back to France.[134]

Their work thus contributed to the glory of the Sun King, with whose support Paris, too, became a center of Sino-European intellectual exchange. The origins of the Chinese book collection at the Bibliothèque du roi went back to a gift of four volumes from the private library of the Cardinal Jules Mazarin in 1661.[135] After the establishment of the French mission, it grew exponentially. In 1697, Bouvet returned to France on a recruiting tour with twenty-two Chinese titles in forty-two cases as a gift for the King, including complete copies of the Four Books and Five Classics.

In 1722, his confrere Foucquet also returned, bearing 1800 more volumes for the library.[136] It was around this time that the Bibliothèque du roi was relocated to its present site at the Rue de Richelieu. With over a thousand discrete titles, many of them printed in the finest imperial editions, it was without question one of the best collections of Chinese books in Europe.[137] There was only one problem—nobody there could read them.

Progress toward overcoming that difficulty began in 1711, when a Chinese Christian, Arcadio Huang, arrived from Beijing to continue his training at the Jesuit seminary in Paris.[138] With his help, two renowned scholars, the classicist Nicolas Fréret and the orientalist Étienne Fourmont, learned some rudimentary Chinese and made early efforts toward compiling a dictionary. Huang died prematurely in 1716, after which they were aided in their research by the French missionaries directly from Beijing. In the following years, Fréret published prolifically on Chinese history and chronology, bringing both subjects to the attention of the philosophes.[139] Fourmont completed a catalogue of the library's Chinese collection, which remains a useful resource for researchers even today.[140] Both scholars earned some renown for their knowledge of China. Yet neither was primarily interested in the subject, nor did either really master the language. For these reasons, nineteenth-century orientalists judged their work harshly.[141] But they did establish a rudimentary framework for Chinese studies in France, and they also trained two students who achieved many of their early ambitions.

Joseph de Guignes and Michel-Ange le Roux Deshauterayes were among the first people to attain a passable reading knowledge of literary Chinese in Europe without ever having left it. Their contemporaries, at least, considered them unique in this regard, which lent considerable authority to their work. They guarded it jealously, mostly by controlling access to what was still the only significant library collection on the topic in France.[142] Their engagement with China was therefore both atypical and paradigmatic: atypical, because there were only two of them, but paradigmatic, because there was no one around who could contradict them. Even so, as had also been true of their mentors, neither was a China specialist. Deshauterayes' primary appointment was in Arabic, while de Guignes's was in Syriac. Both scholars therefore strove to put China in conversation with other ancient Eastern places, with which they were more generally familiar. The result was a body of scholarship that has been interesting to historians less as a contribution to Sinology than as an example of the orientalism of the high Enlightenment.[143]

Both sinologue scholars spent their entire professional lives in the pay of the French state. Their careers progressed in tandem, and almost inevitably they became rivals. De Guignes was born in 1721, Deshauterayes in 1724. In the 1730s they studied together at the Bibliothèque du roi; in the 1740s, both were promoted to interpreter there; and in the 1750s, each was appointed to a chair at the Collège royal.[144] Around this time, their rivalry became public, when de Guignes revived an old theory that China was originally a settler colony of ancient Egypt, and Deshauterayes contested it.[145] De Guignes eventually lost this argument, but not before effectively marginalizing Deshauterayes—who did not really win it either, despite receiving a couple letters on the topic from Voltaire in consolation.[146] Thus for the following half century, the foremost authority on China who was physically present in France was de Guignes. In 1753, he was inducted into the Académie des inscriptions, and in 1772 promoted to pensionary member. He continued to publish new studies on the history, geography, philosophy, language, religion, and science of China almost to the year of his death in 1800.[147]

Yet for the duration of his half-century-long career, de Guignes remained always an amateur in comparison with the missionaries. Their studies, which often found their way into his hands while still in manuscript, presented him with a double-edged sword, slicing open the thickets of Chinese texts, but also undercutting his fragile authority. He decried the missionaries as unobjective; they, in turn, questioned his competence.[148] Yet though the relationship was at times contentious, it was also mostly respectful. Mutual praise was common, especially in the works that were published.[149] In 1763, the Académie des inscriptions designated de Guignes as its official correspondent with the missionaries in Beijing.[150] Sometimes he asked them for their help with his research; more often, he just took it. Not a few of his publications would now be considered plagiarism, and they were judged as such even by the laxer standards of the day.[151] But his liberal use of their labor does not seem to have bothered the missionaries as much as one might expect. Just as he relied on them, so too did they rely on him. During the early 1770s, de Guignes brought several of their works to press, acting as both editor and agent. The occasionally contentious relationship was thus ultimately to the benefit of both sides.

The sinologue scholars were not missionaries, but nor were they philosophes.[152] De Guignes was certainly not a Jesuit apologist. Although he claimed neutrality on the Chinese Rites controversy, he tacitly accepted the anti-accommodationist position on ancient Chinese philosophy.[153] His Sino-Egyptian hypothesis and his magnum opus on the history of the Huns were both framed against the long-standing views of the missionaries. But he was no great friend to the philosophes, either, regardless of their position with respect to China. Voltaire casually dismissed the

Huns as a ridiculous choice of topic, while de Pauw began his own study of China explicitly to disprove de Guignes's Sino-Egyptian thesis.[154] All evidence suggests that de Guignes was a faithful and practicing Catholic. His Egyptian-origins theory, for example, has been seen as an attempt to shield the Bible from the challenge of Chinese chronology, and his study of the Huns fully accepted the story of Genesis as its starting point.[155] What he was, then, was a professional scholar, fully a creature of the intellectual institutions of the ancien régime.

Like many academics, de Guignes began his projects by looking for gaps in the literature, which led him to take an unprecedented interest in Chinese philosophy beyond Confucianism. The missionaries had universally discounted Buddhism as rank idolatry.[156] Daoism had fared only slightly better, with its practitioners often condemned as alchemists or charlatans and its preeminent figure, Laozi, likened to Hermes Trismegistus. The exception that proved the rule were the Figurists, who were interested in Daoism precisely because their own beliefs were also a bit Hermetic.[157] Nor did the philosophes extend their love for Confucius to Laozi and the Buddha; they thought that the popular religions of China were no less superstitious than those of Europe.[158] Surveying this terrain, de Guignes spotted his opportunity. When it came to Chinese philosophy, he observed, "the missionaries have given us only very little knowledge on this subject, and they only speak, as it were, about Confucius and his doctrine."[159] As a corrective, he put out one of the longest studies of Daoism published in Europe to date, as well what some have considered the first significant European-language translation of a Buddhist text, the *Forty-Two Sections Sutra*.[160] De Guignes was by no means sympathetic to either tradition; but, then again, he was not particularly sympathetic to Confucianism, either.[161] In any case, his work was supposed to be scholarly, rather than partisan, and to some extent it did manage to introduce new aspects of Chinese thought into the Enlightenment.

In order to make it appealing in a way that his contemporaries might still accept, de Guignes developed a position on China's natural science that had been suggested in passing by the missionaries: while its natural philosophy could be comfortably ignored, its empirical traditions might still provide new knowledge of scientific value. Along with the philosophes, he waved off Neo-Confucianism as a "scholastic philosophy." Adding his own historical flourish to the proposal, he suggested that Aristotelianism had perhaps entered China via the Silk Road, thus giving the Neo-Confucians and the Schoolmen a shared genealogy. Since Chinese philosophy was built on the very same foundations as that of medieval Europe, it was not surprising to de Guignes that "the Chinese never made great progress in physics."[162] Nevertheless, he maintained, there was "knowledge that one can learn from the Chinese," particularly in history and geography, where their wrongheaded theorizing was less likely to get in the way.[163] From the 1750s through the 1770s, de Guignes's limited argument that correct empirical information could be drawn from China was surprisingly successful. His studies of Chinese geography led leading cartographers to amend their maps of the North Pacific, while his analysis of Chinese history convinced historians that events in Central Asia could help explain the fall of Rome.[164] De Guignes thus achieved his essential ambition to contribute to Enlightenment debates through the position that Europeans could still learn from China; and in fact, some did.

By far the most illustrious among them were the small group of social theorists known first as the "economists" and later as the "physiocrats." It has gone surprisingly little-noted by historians that the physiocrats were the purest sinophiles of early modern Europe, or at least the most self-conscious.[165] As Alexis de Tocqueville remarked: "I do not exaggerate when I affirm that every one of them wrote in some place or another an emphatic encomium on China."[166] As one of them himself acknowledged: "Our bias in favor of the literati of China is one of the excesses for which we have been most reproached: May God grant that one never finds in us anything to blame but this extreme admiration for all that holds the character of honesty, of justice, and of utility!"[167] Historians have generally treated the sinophilia of the physiocrats as a minor corollary to that of the philosophes, but it was in fact quite distinct: a little later, a lot more extreme, and propelled by a very different motivation.[168] To borrow again from Tocqueville, the physiocrats were "reformers, not revolutionaries."[169] The Chinese model was reconstructed according to their own doctrine as an effective political monarchy and a flourishing agrarian economy. And their royal patrons paid attention. Bringing these two main themes together during the 1760s, both the Holy Roman Emperor Joseph II and the Dauphin Louis (later XVI) performed plowing ceremonies directly inspired by the Qing imperial grain sacrifices.[170] In their ambition to make Europe more like China, the physiocrats briefly succeeded.

In 1767, François Quesnay, doyen of the physiocrats and physician to the royal family—"the Confucius of Europe," as his disciples affectionately called him—published *The Despotism of China*.[171] Europeans mostly agreed already that China was a despotism: the main question was whether it was an "oriental" one or an "enlightened" one.[172] As the philosophes were rejecting despotism of all kinds, the physiocrats moved to defend it. Drawing from the missionaries as well as scholars like Fréret and de Guignes, Quesnay argued that government by "legitimate despots" had always made China powerful and wealthy, even as he noted some of the emerging social problems caused by recent overpopulation.[173] In particular, he lauded "the political and moral constitution of the vast empire of China, founded on science and on natural law."[174] What he had in mind here was Confucian philosophy; or, "the Chinese doctrine that merits serving as a model for all States."[175] Only in this context was China ever explicitly referred to as a "model" during the high Enlightenment. But as for the natural philosophy that had fascinated Leibniz, confounded Voltaire, and offended Diderot, Quesnay simply ignored it. According to him, China had neglected natural sciences in favor of moral ones, while the situation in Europe was the reverse: and "it is this that makes one prefer China." So far, Voltaire would not have disagreed. But it is telling that Quesnay conceded the attendant flaw: "In China, where the speculative sciences are neglected, men are too given to superstition."[176] The conclusion that Chinese doctrine was superstitious was exactly what Voltaire, in elevating its philosophy of reason, had denied.

Still, for the physiocrats as for Voltaire, a growing interest in progress was not reason to discount Chinese knowledge, even in more technical fields. In 1750, Anne Robert Jacques Turgot, Quesnay's disciple and Condorcet's patron, had delivered a lecture at the Sorbonne, "Discourse on the Successive Progress of the Human Mind." In this short work, called by one historian "probably first full and complete statement of progress," Turgot sought to explain the improvement of society in stages over time based primarily on developments in the natural sciences. Yet he also maintained that progress was neither universal nor inevitable, and that it remained always tied to the hand of providence.[177] He noted in passing that in China, the sciences had been "retained forever in mediocrity." Yet he did not discount China's knowledge tout court.[178] In 1765, he was avid to learn more not only about China's economy, society, and agriculture, but also its paper, printing, and porcelain manufacture.[179] Indeed, it was a personal encounter with two Chinese students that led him to compose a foundational statement of physiocratic theory, *Reflections on the Formation and Distribution of Wealth*, in 1766. By this time, too, he had broken with some of his former friends among the philosophes: "I am not an Encyclopedist," he is reported to have said, "because I believe in God."[180] The seminal progress theorist proved to be something of a conservative, under which sign his interest in China continued.

At the pivotal year of 1773, then, China had not disappeared from the French Enlightenment, but it had taken refuge in new quarters. Though it was no longer a model for most of the philosophes, who were increasingly interested in science and progress, it remained one for those to whom science and progress were not the primary concern. The sinologue scholars and physiocrat economists of the late Enlightenment continued to regard China in much the same way that their predecessors had: as an ancient and unchanging society that was in most respects quite comparable to their own. The former continued to conduct new academic research in the vein of the missionary specialists, while the latter further developed the comparative criticism of the philosophe generalists. What both groups had in common was the full support of the French crown. And increasingly, that support was channeled through a single individual who sat very near the throne.

The Last Enlightenment Sinophile

The last sinophile of the French Enlightenment was not a Jesuit, a philosophe, an academic, or a physiocrat. Henri-Léonard Bertin was for a time one of the most powerful men in France; he was also the person almost solely responsible for the support of Chinese studies there.[181] Between the Suppression of the Society of Jesus and the outbreak of the French Revolution, he picked up many of the threads that already ran between France to China and connected them all together. He put physiocrats, philosophers, academics, and Jesuits alike to work in support of his own political vision. Under his patronage, the intellectual realignment of China in France was completed.[182]

Bertin was both a professional politician and a participant in the French Enlightenment. Born in 1720 to a prominent family of the Périgord, he attended Jesuit schools and studied law. During the 1750s, his career advanced rapidly, from intendant of Roussillon and Lyon to lieutenant general of police, then controller-general of finances. By 1763, he had won the personal favor of King Louis XV, who put him in charge of industry and infrastructure as well as of the Compagnie des Indes, and who created a new title, "Secretary of State," just for him.[183] Bertin sought to become a better minister by cultivating the skills of a savant.[184] In 1761, he was named an honorary member of the Académie des sciences, where he followed the latest developments in natural science, such as Benjamin Franklin's studies of music, William Herschel's observations of the moon, and Joseph Priestley's chemical experiments.[185] In 1772, he joined the Académie des inscriptions, where he took particular interest in history and religion and wrote extensively on the topics himself.[186] After the death of Louis XV in 1774, his political star waned and his scholarly one waxed. From then on, he spent much of his time pursuing what had clearly become a genuine personal passion: the arts and sciences of China.

Bertin probably became interested in China originally through his association with the physiocrats. He had risen alongside them as a follower of Quesnay, a friend of Turgot, and a protégé of their patron, Madame de Pompadour. Their vision of centralized rule and agrarian society guided his early and somewhat successful program of political and economic reform.[187] During the 1760s, he, like them, saw China as a model for the improvement of French agriculture and industry, both of which fell within his official purview. In particular, he thought Europe might learn from China about road and bridge construction, silk and porcelain manufacture, and wet-rice irrigation techniques.[188] The Qing state presented geopolitical opportunities as well. After the Seven Years' War concluded in 1763 with the cession of French possessions in the Americas to Britain—not to mention recent defeats in India and the expanding Romanov empire further to the north—closer relations with the Qianlong emperor may have seemed like an opportunity for a pivot to East Asia.[189] In sum, China offered practical solutions to the social, economic, and political problems of the ancien régime.

His work on behalf of its signature institutions, the Crown and the Church, led Bertin to develop a deeper interest in Chinese history and religion as a scholar as well. During the 1760s, he had learned the value of history for government when he commissioned researchers to travel around France on a mission to collect documents for a central state archive.[190] With the Encyclopédistes then at the height of their influence, he became increasingly concerned with defending French Catholicism against the perceived threat of "atheists" and "philosophes."[191] China's state-sponsored histories and ancient doctrines were an apparent source of inspiration. To learn more about them, he began corresponding with the local expert, de Guignes, whom he called "one of the most enlightened modern savants." Already, his interest was linked to the rejection of key elements of the idea of progress. He believed in a simpler and more pious past, when humanity had lived in a state of complete innocence and perfect knowledge; "how interesting it would be for the history of the world, and for religion," he dreamed, "to find in China certain monuments of this deep antiquity."[192] Bertin developed an interest in China not despite its religion and antiquity, but because he himself was a Christian and a conservative.

The opportunity to pursue that interest in earnest arrived in the early 1760s, in the form of two Chinese students. Aloys Ko and Étienne Yang were born to Christian parents, brought up by the French missionaries in Beijing.[193] In 1751, still in their late teens, they set out to see the splendors of Christendom and complete their training to join the clergy. After significant delays in Canton and Macau, they arrived in France in 1754. They spent the next six years at the Jesuit college at La Flèche, where they studied French and Latin as well as logic and theology. In 1760, they moved to Paris to complete their education at the Jesuit novitiate. Unfortunately for them, this was not a good time to join the Society of Jesus. In 1762, the Jesuit colleges of France were closed down, leaving the two young novices somewhat stranded and very much alone.[194]

Bertin took Ko and Yang under his wing, assuming personal responsibility for the completion of their education. He commissioned members of the Académie des sciences to teach them physics, chemistry, and natural history. He engaged his own mentor, Turgot, to instruct them in politics and economics.[195] When it became clear that they intended to return to China, he made plans to ensure that their relationship with him would survive the separation. He sent them on a tour of southern France to observe local printing, dyeing, and sericulture, so that they might better comment from China on industry and agriculture in the future. He encouraged Turgot to draw up a list of questions for them to investigate on related topics; the task was reportedly what inspired Turgot to write his magnum opus on economic theory.[196] Bertin wrote out his own instructions for them as well. The most important was that they report back to him regularly.[197] Ko and Yang returned to China in 1766. They never saw Bertin again. Yet it is clear from the many letters they exchanged over the following decades that a personal, almost paternal bond had been formed.[198]

It was at this point, with return of the Chinese students, that a sustained correspondence between Bertin and the French missionaries in Beijing began. With the Society of Jesus expelled from France and more bad news visible on the horizon, the missionaries realized that the future of their mission, and indeed their very livelihood, was at stake. In 1766, Joseph-Marie Amiot, one of the most senior missionaries at the time, wrote to Bertin. Introducing himself as "a Frenchman, transplanted for fifteen years in the capital of the Chinese Empire," he invoked the long history of royal patronage for the mission and solicited its continuation with what he called a "tribute" for the minister: the first Western-language translation of Sunzi, or Sun Tzu's *Art of War*.[199] From that point on, Bertin received the vast majority of the French missionaries' works directly. He was thus already well positioned to move on their behalf when the circumstances of the Jesuits went from bad to worse to nonexistent.

By the middle of the eighteenth century, the Society of Jesus was in a state of crisis. In France, there was agreement across the political spectrum that the Jesuits had no business occupying such high positions in society as they did. Philosophes decried them as meddlesome clerics interfering in secular affairs, while Jansenists denounced them as clandestine secularizers dressed in priests' robes. French authorities were broadly sympathetic to any argument that would curb ultramontane influence. A scandal involving a Jesuit general in the colony of Martinique finally marked the beginning of the end. In 1761, the Parlement of Paris declared the Society of Jesus corporately responsible for all debts that had been incurred there overseas, opening it up to attack on multiple fronts. In 1764, Louis XV expelled the organization from all French domains.[200] The story played out a little differently in each Jesuit province, but within a few years, all the major Catholic powers had done likewise. It was only a matter of time before Rome followed suit. In 1773, Clement XIV issued the papal brief *Dominus ac Redemptor*, officially declaring the universal suppression of the Society of Jesus.[201]

Bertin immediately identified it as an opportunity, and he began working with the king and the minister of the navy to put the French mission directly under royal control. In 1774, he secured 12,000 livres of emergency funding for its operating expenses.[202] This was a fairly small price to pay; by comparison, the education of Ko and Yang in France had cost 22,000 livres in one year alone.[203] When Louis XV died later that year, Bertin had little trouble convincing the new king to continue paying it. From then on, the missionaries at the North Church were effectively royal pensioners. Bertin also established a separate fund specifically designated to "maintain a literary and artistic correspondence with the principal missionary savants and artists who are in Beijing." Its stated purpose was twofold: for the missionaries to contribute to "the progress of Christianity in China" and to be "useful to their country." Tellingly, the first draft had read "useful to Europe"; Bertin corrected it in his own hand.[204] In principle, this was a French royal project. And it seems that Louis XVI did take some occasional interest in it, reviewing letters on the Canton trade, Russian envoys, and imperial administration.[205] "For my part," Bertin explained, "I am charged by the king with the pursuant Correspondence concerning the progress of the sciences and the arts in Europe."[206] Everyone always knew that the project was truly Bertin's. So began an arrangement that came to be called the "literary correspondence" between Paris and Beijing.

Getting an object halfway across the world was not a trivial affair. Europeans were broadly forbidden from entering the Chinese interior, and there was no formal postal service. In the past, the Jesuits had maintained a regular but informal one, and Bertin effectively took it over. Parcels were sent overland to the port of Lorient in Brittany, whence they sailed out with the yearly ships of the Compagnie des Indes. About nine months later, they arrived by way of the Cape of Good Hope at Guangzhou, or Canton, the only city in China that was legally open to European traders. After passing a customs inspection, they were handed over to the "procurer" of the Beijing mission—a position filled in 1776 by Bertin's protégé, now Father Étienne Yang.[207] He then set out from Canton, traveling overland to Nanjing and from there by barge up the Grand Canal all the way back to Beijing, where the packages were finally distributed to their addressees at the North Church. The whole journey required coordination between many different parties, including French and Chinese merchants as well as Bourbon and Qing bureaucrats, both in the provinces and the metropoles.[208]

The logistic situation presented numerous difficulties for would-be correspondents, but it was a blessing for Bertin, because the upshot was that it became virtually impossible to move anything between Paris and Beijing without his knowledge and consent. Every year, he exchanged many long letters with several of the missionaries, sometimes more than a dozen, setting initial research agendas and coordinating subsequent efforts. When someone in China had a question about France, he would find the appropriate person to ask; when someone in France had a question about China, he would often instruct a missionary to answer it. He vetted all the letters on both sides, sometimes even sending one back to its author for revisions. He further solidified his control with textbook patronage moves. He secured a post for de Guignes's son with the Compagnie des Indes in Canton, and another for Amiot's nephew in the office of the intendant of Aix-en-Provence.[209] In 1780, Bertin gave up his position as secretary of state, relinquishing his formal responsibilities for the French mission.[210] But in one of his last acts in government, he obtained specific permission from the king to direct the literary correspondence upon his "retirement," which he continued to do for the following ten years.[211]

Besides hundreds of books and letters, the literary correspondence also included plant seeds and animal specimens, rare gems and minerals, objects of human art and artifice—everything that was "new in Europe, and rare in China."[212] Bertin offered to pay the missionaries' costs out of his own pocket, though he asked that purchases be cleared with him in advance "if the price is considerable."[213] The only explicit condition he set was that all items he received should have some scholarly value. The demand was apparently a response to the habit the missionaries had developed of sending him little "presents": a lacquer box here, a bamboo pagoda there.[214] When Bertin ordered that they should "absolutely cease," the missionaries ignored him.[215] They pointed out that even if he possessed similar luxuries already, the ones they had access to in Beijing were of higher quality than the cheap export goods flowing out of Canton. Furthermore, much of what they sent was consumable, and therefore needed to be replaced: fireworks, ginseng, and honey, for example, all of which were obviously intended for personal use.[216] Bertin eventually gave up the pretext. After hearing about a lavish party thrown for the emperor in 1786, he requested "just two bottles" of Chinese wine to sample as a mere "curiosity."[217] The missionaries seem to have understood all along that what Bertin really meant was not that he did not want presents, but that he wanted only nice ones.[218]

Through his unique position, Bertin soon acquired one of the finest collections of Chinoiserie in France if not all Europe.[219] His "Chinese cabinet" was famous enough to be included in a popular guidebook to Paris, which singled out in particular its holdings of costumes, jade stones, porcelains and bronzes, paintings, and musical instruments.[220] He was quite proud of his collection and showed it off regularly. By 1785, it had outgrown its display. Bertin began construction on a "Chinese garden and manor" at his château in Chatou, complete with a lily pond.[221] Seeking to make it the setting as authentic as possible, Bertin consulted the missionaries for their advice.[222] One replied with a hand-drawn illustration of instructions for a Chinese study. In the most "honorable" location, above a table set with a bronze vessel, ink, and incense, was to be hung a banner in a golden frame inscribed with a phrase from the *Analects* of Confucius: "The gentleman is not an instrument." Rather, he should be adept at many things—like Bertin, both minister and savant.[223]

In return for the finest of China, Bertin sent the missionaries the best of France. Life in Beijing did not afford them many shopping opportunities, and with the nearest international trade entrepôt over a thousand miles away in Canton, European goods were particularly hard to come by. Bertin invited his correspondents to ask for whatever that they might need in the course of their research and seems to have fulfilled their every request for scholarly paraphernalia.[224] For his part, Bertin too sent presents: a tobacco grater here, a pair of spectacles there, and, especially, alcohol.[225] In 1784, he shipped fifty bottles of wine, plus a few bottles of eau-de-vie, for Amiot alone.[226] In the next shipment, he upped the count to 150 bottles of champagne wine, plus a few bottles of twenty- and thirty-year-old cognac, no questions asked—except whether the missionary preferred his Bordeaux red or white.[227] Items like these were a "little packet" of gratitude, entirely "personal" for those who took part in the literary correspondence. As he emphatically explained in one letter, their recipients were under no circumstances to be compelled to share.[228]

These were the tokens of a long-term, long-distance friendship that Bertin developed with several of the missionaries. He grew particularly fond of Amiot, who was distinguished not only for scholarly productivity,

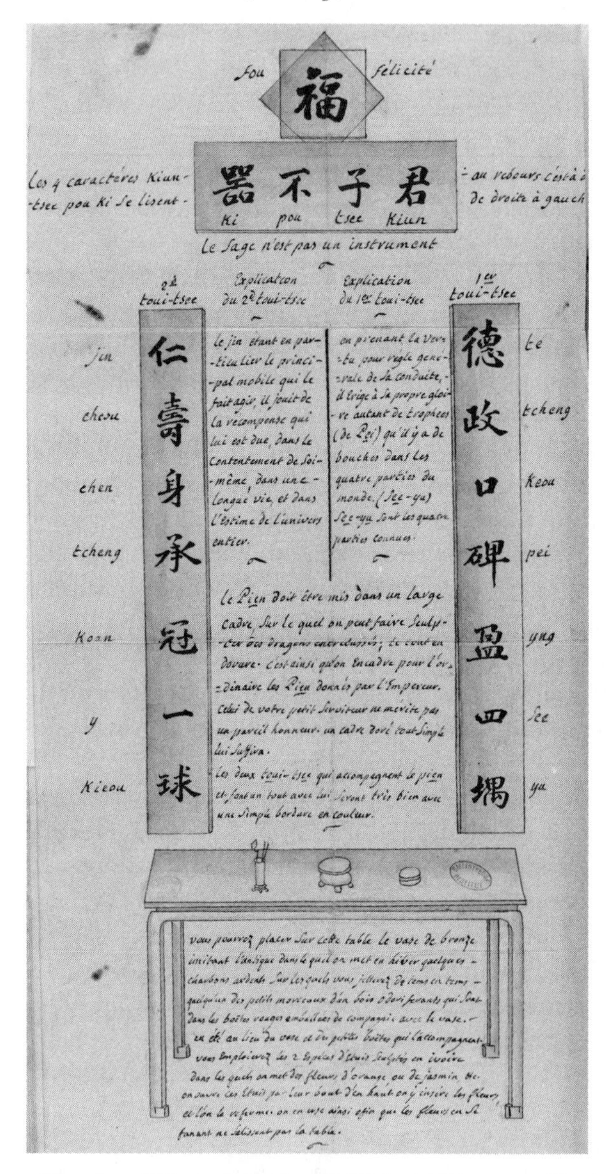

FIGURE 1.3 Plans for a Chinese study, Joseph-Marie Amiot, c. 1790. Courtesy of the Bibliothèque de l'Institut de France.

but also savvy politicking. There remained always a certain degree of distance in the exchange. Amiot referred to Bertin as "my lord" or "your greatness" and signed his own name "Amiot." For reasons that are quite unclear, Bertin persisted in spelling it "Amyot" for twenty-five years. He maintained the pomp befitting his station, as one signature, florid even by his standards, makes clear: literally, "Would that you be persuaded that I would like with all my heart to have occasions to convince you well of the sentiments with which I am, sir, your very humble and very obedient servant . . ."[229] Still, it seems that Bertin did come to regard Amiot with genuine affection. After Bertin's death, his former secretary wrote to inform Amiot of "the tender amity that he had for you, the tender amity that he often avowed."[230] In letters to others, too, the minister showered praise upon the missionary for his "indefatigable zeal" for "religion and his country," and, even more, his having "spared nothing to shed light on the arts and sciences."[231]

For, underlying all the exchanges of money and favors, books and equipment, presents and pleasantries, the ultimate purpose of the correspondence was to keep the missionaries writing about China. Bertin allowed them a great deal of freedom in this regard. Although he often gave them guidelines, they were always understood as suggestions rather than rules. In 1786, he wrote to Amiot: "I beg you in grace to abandon above all every kind of occupation that will not amuse you."[232] In practice, Amiot and the other missionaries had been already doing so for years. The entreaty involved little risk because both parties were generally amused by the same things. A somewhat surprising result of the suppression of the Society of Jesus was that with new state support, the missionaries became even more productive. Manuscripts poured into Paris, and Bertin went about securing their publication.

The Mémoires Concernant les Chinois

Historians almost uniformly agree—and have for two centuries—that the last generation of Jesuits in China did not live up to their missionary forbears.[233] The heroes of yesteryear already loomed large at the time: Giuseppe Castiglione the painter, Johann Schreck the mathematician, Johann Adam Schall von Bell the astronomer, and Matteo Ricci, a little bit of everything and the greatest of them all. Their late eighteenth-century successors lived through times of great duress for the Society of Jesus and died after it had ceased to exist. With China falling out of favor in Europe, they no longer corresponded with intellectuals like Leibniz and Voltaire. And with Europeans increasingly marginalized in China, they no longer spoke much with literati like Li Zhizao and Xu Guangqi. Unlike many of the earlier Jesuits, they left almost no literary legacy in Chinese. But despite their difficulties, and indeed in some sense because of them, they also achieved something that most of their predecessors had not: literary fame in Europe while they were still alive.

The final monument of the Jesuit mission to China was printed from start to finish only after the Jesuit mission to China was no more: the *Memoirs Concerning the History, the Sciences, the Arts, the Mœurs, the Customs, &c of the Chinese: By the Missionaries of Beijing,* frequently cited as the *Mémoires concernant les Chinois.*[234] Though largely forgotten by later scholars, it was hailed almost immediately as the most important European work on the topic to date.[235]As one contemporaneous critic wrote: "There is no one knowledgeable or curious who should not give a distinguished place in his library to such an interesting and useful collection."[236] From 1776 to 1791, fifteen volumes were published, each about five hundred pages quarto.[237] Subsequent reviews in major periodicals like the *Journal des sçavans* and the *Journal encyclopédique* showered them with praise: "It is always with a new pleasure that we announce the next of these *Mémoires,*" declared one; a "much more useful collection than any previously published by the missionaries," gushed another.[238] And yet, for many reasons—its unwieldy length, its haphazard organization, its diffuse style, and most importantly, its late publication with respect to the received story of China in the Enlightenment—the *Mémoires* has been the subject of relatively little historical study, with no European-language monograph to date on the collection or any of its major contributors.[239]

This collection was where European interest in China outlived, and eventually outgrew, the institutions that had fostered it. The publication of the *Mémoires* corresponded quite closely with the reign of Louis XVI, a time of momentous change in France. Most of the missionary contributors could still remember when China had served as a model for the Enlightenment, and they still saw themselves broadly as contributing to it. In a way, their ambition shows how successful the early philosophes had been. Amiot praised the "profound Bacon," quoted liberally from Fontenelle, and dubbed Voltaire the greatest poet of the age.[240] Yet he also realized that the audience for his work was shifting. The missionaries were no longer writing for philosophes, nor even for the society of Jesus, but for a rapidly evolving new constituency that was very far away, and whose fundamental interests they shared only to a point. If the *Mémoires* had an overarching argument, it was this: the culture of China was still worth exploring—"the America of people of letters, sages, and men of state"—though why and for whom were still being sorted out.[241]

The entire collection was edited and published in Paris under the supervision of Bertin. He was the one who decided which works were worth including, which should be rejected, and which required revisions. He also appointed the masthead, mostly with his friends. When the project's first editor, the philosopher Charles Batteux, died about halfway through,

Bertin paid for the erection of his tomb in the cemetery of Saint-André des Arts.[242] He then offered the position to the historian Louis-Georges de Bréquigny, whom he had employed years earlier in the hunt for archival documents.[243] Throughout the fifteen years during which it ran, other experts, including de Guignes and his supervisor at the Bibliothèque du roi, Jérôme-Frédéric Bignon, served occasionally as guest editors. These figures were both professional scholars and government employees. Each held multiple, overlapping positions at the Collège royal, the Académie des inscriptions, and the Académie française. [244] The project was backed by all the resources that the state could provide; and it, in turn, supported the state.

The *Mémoires* varied wildly in format and topic.[245] Generally, each year saw one volume, but some years saw two, and some years saw none. They included essays, letters, translations, and treatises, ranging in length from less than one page to more than four hundred. A few focused on a single subject, such as volume 8, featuring three missionaries' studies of natural history; or a single author, such as volume 4, which included works by Pierre-Martial Cibot on filial piety, smallpox inoculation, forensic examinations, and cabbages. Most, however, were diverse all around. Volume 3, appearing in 1778, was a fairly typical example. Among its contents was a 378-page "Biography of Famous Chinese," a fifteen-page essay on "Chinese Greenhouses," and an eleven-page letter on "The Conquest of the Land of the Miao," each by a different missionary author. Topics of particular interest were history, politics, and philosophy—not coincidentally, the same subjects that had most elicited Enlightenment appreciation for China—but the volumes also included studies of music, horticulture, military campaigns, famines, floods, cinnabar, meteorology, and much more. Taken together, there was virtually no subject related to China on which they did not touch.[246]

The diversity of the *Mémoires* reflected the diversity of its authors. Most of the collection was written by the French missionaries, and most of the French missionaries wrote for the collection. Six of the nine who were attached to the North Church at the time of the suppression ultimately contributed something. All of them had arrived in China with specialized training intended to make them useful at court. After they reached Beijing, their educations continued. Differences in language ability in particular opened up opportunities to some and closed them off to others. Amiot explained that his confrere François Bourgeois had arrived in China too old to master its language: "It is a lot for him to have learned to express himself passably in Chinese."[247] By contrast, the missionary artists and artisans were probably the most fluent, since their work involved speaking regularly with Chinese people; but on the other hand, this occupation left them little time to study ancient texts written in the classical language. A mix of aptitude, experience, opportunity, and interest all combined to determine what a missionary did or did not contribute.

Each of them accordingly developed something of a specialty. Louis-Antoine de Poirot and Giuseppe Panzi, artists by training, painted portraits of notable Chinese figures and also of the other missionaries. Jean-Paul Collas, who had studied mixed mathematics at university, wrote largely on chemistry, including borax, cinnabar, and indigo, and a little on biology and astronomy as well. Bourgeois, formerly a professor of rhetoric and the mission's chief administrator, reported on court ceremonies, demographic surveys, and military campaigns. But though most of the missionaries contributed, they did not contribute equally. Amiot and Cibot, who had both taught the humanities, demonstrated the widest range.[248] Cibot wrote historical and philosophical treatises, translations of the Four Books of Confucius, and shorter works on biology and agriculture as well. Amiot wrote largely on history, music, and philosophy—a subject he claimed to be particularly "within my reach"—but his studies of politics and science were also significant, as were his translations of classical and contemporary Chinese texts.[249] Cibot and Amiot were not only the most extensive contributors, but also the most prolific. Their works accounted for well over half of the *Mémoires* as reckoned by both number and length. With such diversity of interests and experience, disagreements were common. Yet there were several key areas on which the missionaries were both consistent and persistent, most of which reflected their shared Jesuit background.

The foremost source of authority for all the missionaries was always the ancient Confucian canon, and especially the Five Classics. These texts already had more than a century of European study behind them, making them potentially easier to understand. They also formed a discrete and convenient corpus to focus on. According to Amiot, the *Classic of Changes* was quite simply "the most ancient monument in the world"; after that, the *Classic of Documents* "unquestionably" held the highest rank.[250] The Classics were unequaled among all non-Christian writings, including even those of ancient Greece and Rome. Cibot expressed the common opinion when he wrote: "I see only the Bible above the Classics for authenticity, as for everything else."[251] The Classics were seen as both the oldest Chinese texts and the most congruent with Christianity. By contrast, some later pagan could usually be found to blame for anything in the Chinese tradition that the missionaries did not like: Buddhist idol-worshippers of the Han, Daoist libertines of the Tang, Neo-Confucian atheists of the Song, and so on.[252] There was nothing at all controversial about this emphasis on the Confucian canon. It was effectively the consensus among Chinese elites as well, and no European had yet contested its preeminence.

It is unlikely, however, that any of missionaries could simply sit down and read the *Classic of Changes* or the *Classic of Documents* unaided. Like their Chinese contemporaries, they relied on summaries, commentaries, and collectanea, which included the most important passages and explained them in clear, contemporary Chinese prose.[253] In one essay, Amiot included a fifty-page appendix of commentaries on each of the Classics, from ancient to contemporary times, a sheer list of Romanized Chinese names that could hardly have meant anything to any European reader except as a mark of erudition.[254] That the texts required and admitted such interpretation cut two ways for interpreters. On the one hand, the Classics could be coaxed into saying whatever a given author wanted them to say. The missionaries found support in them for a range of positions on a host of different issues. This flexibility worked to the advantage of Chinese scholars of the time as well, who couched even their most original arguments as commentaries on ancient texts. On the other hand, the ambiguity of the Classics also led to serious points of disagreement. For the missionaries, the meaning of the original texts was not always clear, and indeed it was difficult to say what the original texts even were. This was also a major problem of high Qing scholarship, which divided into "New Text" and "Old Text" schools over questions about which versions of the Classics were the most ancient, and therefore the most authoritative.[255]

Antiquity still provided the context for discussions of natural science. In the beginning, the *Mémoires* repeated the long-standing Jesuit position with new readings of Neo-Confucian natural philosophy. Noteworthy was an "Illustrated System of Chinese Knowledge," supposedly dating to the third millennium BCE: the eight trigrams of the *Yijing*, a very early European printing of a *yin-yang* diagram, and a tellingly editorialized version of a *taiji* diagram. At the top, instead of *li* or *taiji*, was *Shangdi*, the "Supreme Emperor"—a creative synthesis of Zhou, Song, and Christian cosmology that only the missionaries would have ever imagined.[256] Their emphasis on natural history continued with essays on worms, cotton trees, bamboo paper, and greenhouses, for example.[257] Conspicuously absent from the early volumes was any discussion of Chinese theoretical sciences construed as modern. One interesting exception, a liberal translation of desultory notes by the Kangxi emperor on topics of "physics and natural history," was presented only as a curiosity with the aim of casting him as an enlightened monarch.[258] But the natural sciences came into greater focus in the volumes starting around the middle of the 1780s, as noted by

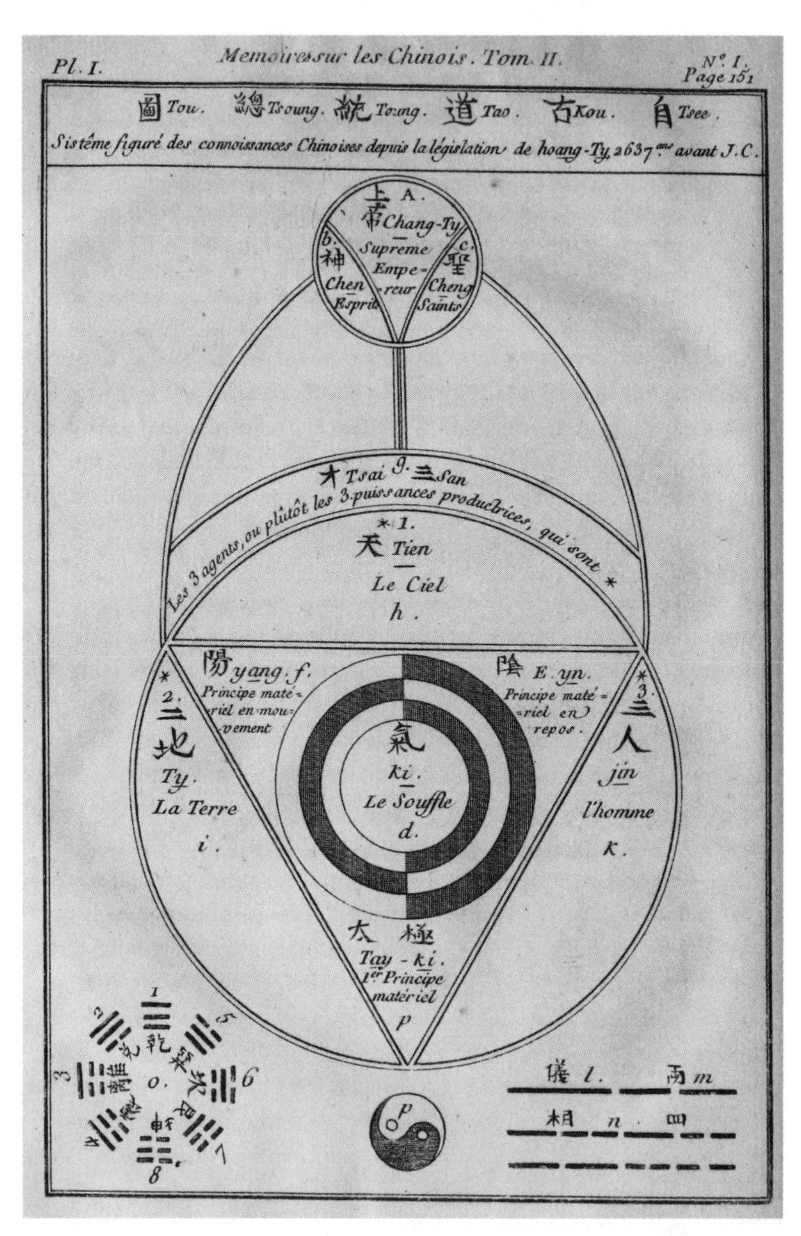

FIGURE 1.4 *Illustrated System of Chinese Knowledge*, Joseph-Marie Amiot, in *Mémoires concernant les Chinois*, volume 2 (1776). Courtesy of the Getty Research Institute.

a reviewer in the *Journal des sçavans* who singled out new letters on "subjects relative to electricity, magnetism, and Aerostatics." These were the cutting-edge topics in Europe at the time. According to the reviewer, the takeaway was that "the Chinese do not want to be below other Nations, and claim to find among themselves, and especially among their ancestors, everything that other nations discover."[259] In China, even the most modern science was now framed by the missionaries as ancient, a point that would animate exchanges in the decade to come.

The only contemporary Chinese authority that carried real weight for the missionaries was the Qing imperial imprimatur. The state organ responsible for scholarship, the Hanlin Academy, was often compared to the Académie des sciences.[260] Academicians were the "the Doctors of the Empire," according to one missionary; "the greatest literati of the Empire," in the words of another.[261] In the 1770s, the Qianlong emperor created a new bureau to compile a massive encyclopedia, the *Siku quanshu*, or *Complete Library in Four Sections*.[262] Bertin tried for years to obtain a copy for the Bibliothèque du roi; he only gave up when it was gently explained to him that this was "nothing other than a collection of all the books that have been made so far."[263] Reliance on imperial sanction also led to a notable Beijing bias. The Qing capital was the site of the metropolitan exams, so every successful scholar had to pass through it at some point. But at the time, economic opportunities and the heavy hand of the imperial censor were driving many intellectuals south to the Lower Yangtze region, where they constituted a somewhat separate scholarly community of equal or greater importance.[264] The French missionaries, who almost never left Beijing, were either were unaware of it or chose to ignore it. Indicative was Amiot's claim in 1773 that "the most able among the literati in China today" was Yu Minzhong, who, he later added, had "for more than forty years presided over or contributed to all the great works of science and literature in different genres that left the Chinese presses."[265] No historian today would likely agree. In reality, Yu Minzhong, who was then serving in the Qing government as head of the Grand Council, was more a statesman than a scholar.[266] Even in Beijing, others were considered far more original thinkers by their Chinese contemporaries: Dai Zhen, for example, who was making major contributions to history, philosophy, and science at just this time.[267] But for the missionaries, the best scholarship was that which most closely aligned with statecraft.

All the missionaries, without exception, were staunch monarchists, universalizing in their view of church and state. They drew the house of Bourbon and the clan of Aisin Gioro together for mutual praise: "One would be tempted to believe that there are in the world two Louis XVIs, or two Qianlongs, one in France and the other in China," read an unpublished letter.[268] A portrait of the Qianlong emperor adorned the first page of the first volume of the whole collection, right after the title announcing the *Mémoires*: variously described as the "*Grand-Priest* and *Sovereign Sacrificer* of his Nation," a magnanimous ruler, brilliant tactician, and poet second only to Voltaire, he was, in short, "among the greatest Princes who have ever governed their empire."[269] Praise for Louis XVI was no less extreme, though often left unprinted: according to one fairly typical encomium, "a descendant of Saint Louis and whose whole life is the most beautiful, the most triumphant, and the most invincible apology of our sacred Religion."[270] At the core of this recurring link between monarchy and religion remained the Jesuit reading of the Classics, which Catholic authorities had officially proscribed but never really expunged. In imperial ideology since the Han, the emperor as Son of Heaven kept both state and cosmos in good order.[271] The missionaries dated this doctrine to still more ancient times, when the emperors were supposed to have led sacrifices at the Temple of Heaven to the deity *Tian* or *Shangdi*.[272] In the present day, the Qianlong emperor continued to perform a similar ritual on the winter solstice, and "the Chinese people attribute to their sovereigns the benefits they receive from Heaven."[273] Church and state were united under a divinely endorsed ruler—which was just as it should be.

If the estimable ancient traditions of both ends of Eurasia were aligned, then so too were their disreputable modern enemies. Chinese Neo-Confucians and European freethinkers were disparaged together as *philosophistes*: The word, coined by an opponent of Voltaire, was a pun on *philosophe* and *Sophiste*.[274] In China, they were characterized by "their puerile taste for discoveries, innovations, and systems"; in Europe, they "dared to print against the creation, against providence, against the necessity of religion, against the immortality of the soul."[275] The rationalist Zhu Xi was compared to the skeptic Pierre Bayle: both had "opened the door to Pyrrhonism, the abuse of metaphysics, the distrust of the ancients, and the libertinage of heart and spirit."[276] Everywhere that it appeared, the "epidemic of philosophism" ravaged the body of "savant antiquity."[277] China had succumbed already during the Song, and France was now suffering the same disease. The missionaries' anxiety, still more evident in their unpublished letters, only increased over time. "O great Saint Louis," Cibot prayed in 1778, "save your France from the infernal deluge of impious books, in your honorable name, blasphemed and subjected to derision, as doomed to ridicule and defiance! Alas!"[278] In 1784, Amiot lambasted "the pride, the haughtiness, and the spirit of insubordination that your modern philosophers emphatically display in all their writings."[279] In one

FIGURE 1.5 *The Qianlong Emperor,* in Amiot et. al, *Mémoires concernant les Chinois,* volume 1 (1776). Courtesy of the Getty Research Institute.

of his last letters, he lamented "the number of these French whom modern philosophy infected with its venom."[280] As the philosophes turned against China, the missionaries turned China against the philosophes.

The Death of Voltaire's Confucius

Changing patterns of European engagement with China over the course of the Enlightenment at first had less to do with new knowledge of China than with new ideas in Europe. Voltaire portrayed Confucius as a French philosophe dressed in Chinese robes, an expositor of universal reason and a model for social reform. Condorcet saw Confucius as a boring pedant whose doctrine was a fundamental impediment to the improvement of society over time. Both, however, accepted him as broadly representative of Chinese thought. In fact, the image of China that the philosophes condemned in 1773 was quite similar to the one they had praised in 1733: an ancient and enduring polity grounded in a powerful moral and ethical philosophy. It was only that the early philosophes had judged these things positively, the middle ones ambivalently, and the later ones negatively. Crucially, no one yet proposed that China could not, at any rate, be judged. In this sense, China remained an analogue to Europe, rather than an alternative to it.

Throughout the early modern period, the almost-exclusive source of reliable information on China, for enthusiasts and critics alike, were the reports of the Catholic missionaries. By the late eighteenth century, many of those in circulation were more than a hundred years old and quite out of date. But no one in Europe had really been in a position to contest or augment them, and those who claimed to be were only marginally so. The Jesuits had consistently presented China as a land that was ripe for conversion to Catholicism. Their interpretations of its natural knowledge were relevant to this purpose in two ways: its pristine ancient science demonstrated that China had once been pious, while its impoverished modern science revealed an opportunity for the Jesuits to make it so again. There were some things the missionaries argued about, and some that they never discussed at all. But in general, the purported poverty of contemporary Chinese natural knowledge was something on which they could broadly agree.

French engagement with Chinese knowledge of all kinds continued nevertheless toward the end of the eighteenth century, though with notable changes becoming evident. Sinologue scholars and physiocrat economists shared a broad commitment to Church and State. Their enthusiasm for the absolutist government, pristine religion, and ancient philosophy of China remained strong as a way of defending these institutions in France during a period when all were felt to be coming under attack. They did not dispute that China had not developed far in the natural sciences, but the European consensus on this point bothered them less than it did many of their contemporaries, because they were not committed to the philosophes' progress program. In the final decades before the collapse of the ancien régime, the model of China became streamlined in its defense.

The suppression of the society of Jesus in 1773 accelerated this process, inaugurating a final wave of early modern European Chinese scholarship. Henri Bertin, already having promoted it for the better part of a decade, seized his opportunity to take control of the network connecting Paris and Beijing. The result was the publication of a monumental new collection, the *Mémoires concernant les Chinois*. Seeking to satisfy their powerful new patron and to vindicate their own life's work, the missionaries drew from ancient Chinese traditions to argue against modern European ones. Their new research pushed the model of China still further away from the philosophes, while promoting new kinds of interest among others. The reconfiguration of conditions of exchange that was taking place in Beijing at the same time allowed the missionaries to satisfy it.

✹ 2 ✹

The Ex-Jesuit Mission in China

Joseph-Marie Amiot, Qian Deming

The global suppression of the Society of Jesus was a defining moment in the history of Sino-European exchange. Announced in Rome on July 21, 1773, it brought a temporary end to the organization that had produced the great majority of important studies of China in Europe—and indeed, of Europe in China—reaching back almost two hundred years nearly to the beginning of sustained communications. Where the two worlds met at the North Church in Beijing, a dozen French missionaries were left reeling. They turned to other institutions in search of support, but the Catholic diocese, the French monarchy, and the Qing state could not replace the identity they had suddenly lost. The missionaries were now and "ex-Jesuits." What that would mean was still for them to decide. The suppression of the Society of Jesus has long been taken to mark the end of the story of European engagement with China; but in fact, it also signaled the beginning of a new one. The protagonist who brought them together was Joseph-Marie Amiot, the last great scholar of the early modern Jesuit mission to China, and also the first to come after it.

Amiot was born on February 8, 1718, to a family of civil servants in the Mediterranean port city of Toulon. His early life is somewhat obscure.[1] His grandfather had relocated from their ancestral region of Champagne, and his father worked in the government as a royal notary. He had nine siblings, two of whom he remained in touch with by mail for the duration of his long life. He received a classical education, focusing on philosophy and theology, then entered the Jesuit novitiate in Avignon at the age of nineteen. While training for the clergy, he also studied mathematics and astronomy at the Collège de la Marine. During his twenties, he taught grammar in Besançon, then rhetoric in Nîmes. In 1748, he was ordained a priest in the Society of Jesus. He immediately requested an overseas assignment and was selected for the prestigious China mission. He departed France from the port of Lorient the next year and never returned.[2]

FIGURE 2.1 *Joseph-Marie Amiot,* Giuseppe Panzi, c. 1790. Courtesy of the Bibliothèque Nationale de France.

In 1750, at the age of 32, Amiot disembarked in Portuguese Macau. He took a new Chinese name: Qian Deming, composed of the characters meaning "money," "virtue," and "bright."[3] It is not clear why he chose it. The name was perfectly conventional and, unlike many of those assumed by other missionaries over the years, would not have distinguished him in any way as a European. Word of his arrival soon reached the Qing court in distant Beijing, where two Jesuit courtiers informed the emperor of the arrival of a missionary "well-versed in music" who wished to "come to the capital to render his service."[4] Permission was granted. On March 28, 1751, Amiot set out with two Portuguese confreres for Canton and began the 1,500-mile journey north. Along the way, the impatient missionaries often complained to their Chinese escort about the slow pace. "Only the vulgar travel with haste," he retorted; "you are strangers . . . you do not know our customs."[5] On Sunday, August 22, 1751, Amiot finally arrived in the metropolis of Beijing. There is no indication that he ever left.

Amiot soon found part-time employment as a translator. With a particular talent for languages, he was one of the few missionaries to become proficient in both Chinese and Manchu.[6] His skill qualified him for limited work in the Grand Secretariat, or "Tribunal of Ministers" as the missionaries called it: the organ of the Qing state responsible for submitting memorials to the throne.[7] He was assigned to the "Mongolia bureau," where he dealt with the Russian delegations that occasionally arrived overland from Siberia. Since the Russians did not know Chinese or Manchu, they composed their messages in Latin. These messages were handed to Amiot, who translated them into Manchu for the emperor. The emperor's responses were written in Manchu, whereupon Amiot translated them back into Latin for the Russians. The whole process typically took three or four days. In some years, he was called upon five or six times; in others, none at all. The position was thus part-time, informal, and unranked. It was the only one that he ever had in the Qing government.[8]

Meanwhile, back in Europe, Amiot made a name for himself in the Republic of Letters. As one flattering reviewer put it in the periodical *Journal des sçavans*: "It is difficult to conceive how this missionary, in addition to his ordinary occupations, is able to manage such literary works in the different genres that he undertakes, and with which he never ceases to enrich us."[9] He began his literary career during the 1750s and 1760s with essays on ancient Chinese music and dance, earning such readers as the composer Jean-Phillipe Rameau and the philosophe Denis Diderot.[10] During the 1770s, he completed several landmark translations, including the first Western-language version of Sun Tzu's *Art of War* and a French rendition of a poem composed by the Qianlong emperor; "one always loves his translator," wrote Voltaire to Frederick the Great.[11] In the 1780s, his contributions to the *Mémoires* on topics ranging from Qing border affairs and military matters to Song philosophers and Chinese Jews secured his reputation. A set of instructions to the other French missionaries from around that time read: "Take for master and guide M. Amiot." They, in turn, duly acknowledged him as "our famous and venerable doyen."[12] A thin man with large brown eyes and a long white beard, he had become the most respected China missionary of his generation, and he looked every bit the part. This is the Amiot who survives most vividly in the record: not just a Jesuit, but a figure of the French Enlightenment.

Yet, for all his contemporary fame, Amiot occupies a relatively unsung place in the history of Sino-European exchange. Despite some attention in recent years, there is still no comprehensive biography of him in a European language.[13] As a scholar, Amiot has not seemed particularly important. Historians and Sinologists alike have broadly dismissed all the ex-Jesuits for their apparent lack of contributions to the accumulation of correct knowledge of China. In the words of one critic writing in the early twentieth century, "Amiot was the most notable of this group of missionaries, though they were far inferior to their predecessors; he himself was of lesser value as a historian than Father Gaubil, as a grammarian than Father Prémare, and he never had the high position at court of Verbiest or Gerbillon."[14] Furthermore, as a character, Amiot was not very charismatic. He was erudite, industrious, sycophantic, sententious, and, above all, self-satisfied. On occasion, he smoked a pipe, and he liked to drink champagne. But his favorite activities, at least according to his own testimony, were walking and working. He was not a good writer. As one nineteenth-century bibliographer remarked in the marginalia of an unsigned manuscript, "from the prattle that reigns there, it is not difficult to recognize the style of father Amiot."[15] In a word, Amiot was a little bit boring.

And yet, the very qualities that made Amiot uninteresting to generations of historians are precisely the ones that make him so revealing. While we know a great deal about the Jesuit mission to China, we still know relatively little about what happened after it came to an end.[16] Amiot was in a sense its last hero, and recent studies have mostly focused on his early writing, which was broadly continuous with the early modern Jesuit tradition. But it was only after he ceased to be a Jesuit that Amiot did his most original work—so original, in fact, that it has been difficult to make sense of it in the broader context of Sino-European intellectual exchange. This highly experimental corpus remained mostly unpublished during Amiot's lifetime, but it was very influential for his contemporaries, and in some ways even more so for his successors.[17] Amiot's somewhat unremarkable style and personality, too, come with advantages for the historian. If his writing was diffuse, it was also wide-ranging. And if his thinking was occasionally unsophisticated, it was also unusually legible. So too, it must be said, was his textbook-perfect cursive. Amiot's life and works open a vista onto a world of new ideas and rapid change. He lived in that world and operated by its logic. If we want to understand it, he is the right person to follow.

After the suppression of the Society of Jesus, Amiot became an orphan of the Enlightenment. He had grown up in a France that was fascinated by China and reached intellectual maturity in a China that was open to the French. His name had been known to Qing and Bourbon princes and he maintained far-flung correspondents from England to Russia and from Mongolia to Canton. He had inhabited a cross-cultural space that brought a certain amount of prestige on both sides. Until, slowly but all at once, it collapsed. Infighting tore through the factions of Catholic Beijing, while the Qianlong emperor became increasingly concerned about the role of Christians in his vast but fragile empire. Internal and external events compounded upon each other. The missionaries largely withdrew from public life. Nevertheless, cross-cultural conversations continued. Amiot's closest local friend in his final years was a cousin of the Qianlong emperor, the prince Hongwu.[18] Together, the Manchu and the Missionary experimented in secret with the newest Enlightenment science, from electrical medicine to gas balloons. Exchange continued in Beijing, but the conditions enabling it had not been so shaken for more than a hundred years. It was in this context that a new approach to the scientific knowledge of China began to emerge.

No Longer Jesuits, but Still French

The papal brief announcing the global suppression of the Society of Jesus was read aloud in Beijing on November 15, 1775. The events that followed were traumatic for everyone who lived through them.[19] The French missionaries took the news at first in disbelief, then confusion, and finally resignation. All the formerly Jesuit missionaries were now freed from their communal vows and became secular priests. Authority over the mission reverted to the Diocese of Beijing, which was under the supervision of the Archdiocese of Macau. Conflicts among the missionaries were as old as the China mission, but the Society of Jesus, organized explicitly on a model of military discipline, had usually maintained a certain degree of order. With the institution gone, Catholic Beijing broke out into open strife. The French missionaries faced an existential threat. From the outside, Portuguese and Italians clamored for control, while on the inside, long-brewing grudges bubbled up to the surface. In the new post-Jesuit reality, their status as Frenchmen and scholars would rise to the fore; even so, they never forgot that they had once been Jesuits.

The Society of Jesus had maintained a continuous and prominent presence in Beijing since the arrival of Matteo Ricci in 1601. There were as many as forty thousand Chinese Christians in the surrounding province of Zhili, out of perhaps 150,000 throughout the Qing empire.[20] The Jesuit Vice Province of China included about fifty European missionaries of at least five nationalities, plus a dozen or so Chinese-born priests, the majority of whom were based in the city.[21] Given Qing restrictions on foreigners in force under the Canton system that governed trade with European powers, this small population was still enough to make Beijing the second-largest community of Europeans on the Chinese mainland. Nearly all of them were affiliated with some religious organization, though not all were Jesuits; there a chapter of the Propaganda Fide and a Russian Orthodox mission as well. In the preceding century, four Catholic churches had been built in the city: the East Church, the South Church, the West Church, and the North Church, together forming a ring around the Forbidden City at the center.[22]

The North Church, or Beitang, was the headquarters of the semi-autonomous French mission and a gem of the Society of Jesus. Consecrated in 1703, it stood just a stone's throw from the imperial palace on land that had been personally donated by the Kangxi emperor. Its baroque chapel featured fleur-de-lis motifs alongside portraits of French princes.[23] Its library of several thousand volumes was the best collection of European books in China.[24] Ever since the days of Louis XIV, the French crown had maintained the right to appoint the superior by providing regular financial contributions.[25] Most of the money had been put into real estate. By the 1770s, the mission's properties were worth over 70,000 taels, generating yearly revenues of more than 6,000 taels in rent.[26] Considering that a skilled worker in Beijing made about fifty taels a year, this was a sizable fortune, and it afforded considerable luxury to the ten or so European Jesuits who called the North Church home.[27] Life there was generally pleasant and cosmopolitan, and it had been so as far back as anyone who was there at the time could recall.

The global events surrounding the suppression played out over the 1770s in a curious kind of double time. Daily life continued apace in Beijing while a slow-motion sequence began in parallel as the missionaries reached out to their various allies from Nanjing and Macau to Rome and Lisbon for support.[28] Communications between Paris and Beijing were sent on the yearly ships of the French East India Company. At a minimum, the return voyage took around twenty months. But since the ship set out only once a year, if someone wrote a letter right after departure, they might have to wait almost three years before hearing a response—and that was without any unusual delay, which storms both natural and political in fact made common. Correspondents on both sides frequently remarked on what a frustrating position it was to be in. A person might send out a letter, only to learn years later that its recipient was long dead.[29] At one point, Amiot joked that floating a miniature gas balloon over the entire Eurasian

FIGURE 2.2 *The North Church*, Anonymous, c. 1700. Courtesy of the Bibliothèque Nationale de France.

landmass might be an easier way to keep in touch.[30] Logistic impediments and communications delays were in fact a major reason that the fate of the French mission took more than a decade to fully sort out.

In principle, the extinction of the Society of Jesus meant that authority over all Jesuit property and operations should revert to the diocesan clergy; in practice, it was not at all clear what that was going to mean. Most of the Catholic priests in Beijing had always belonged to some religious order, so the infrastructure of the local diocese was critically underdeveloped. Complicating matters further, Beijing was at the moment without a bishop. No one seems to have cared much about the position until, all of a sudden, whoever occupied it stood to inherit from the Jesuits. Negotiations began in Europe to appoint a new bishop, while a struggle broke out in China over who was to be in charge *sede vacante*. On one side was the bishop of Nanjing, an Austrian former Jesuit taking his orders from Rome. On the other was the archbishop of Macau, a Portuguese Franciscan loyal to Lisbon. Each of them, not as remote as Europe though still very far away, sent his own vicar to represent his interests in the city. Factions formed around national and orderly lines. The "schism of Beijing," as it came to be called, lasted until a third candidate, an Italian Augustinian, was finally appointed under somewhat hazy circumstances in 1778, only to die there three years later.[31]

Seeking to preserve the independence of their mission through the turmoil, the French missionaries defined themselves increasingly in terms of their national background. Amiot put the matter succinctly in his first recorded response to the news of the suppression: "We are no longer Jesuits, but we are still French."[32] In another letter, jointly signed "in the name of all the good Frenchmen who live in Beijing," the missionaries pleaded to Bertin for help: "You are our father and our mother; it is by you that we live, it is by you that we are what we are."[33] Most of them believed that best way to preserve French autonomy in Beijing would be with French support from Versailles. "If the court of France does not take efficacious measures to support us here," Amiot wrote, "I pity the French who will come after us." [34] The twin objectives of the mission, to save souls and to advance knowledge, "would certainly not be achieved, if priests of different orders, Portuguese, Italians, or Germans, come to be our substitutes."[35] It was necessary to find or create a new affiliation soon, one that would not cede control to any foreign authority.

The first plan Amiot hatched to maintain their autonomy shows just how far the French missionaries were willing to go, and how contentious and bizarre their situation had quickly become. His idea was to establish a new ecclesiastical province, the "Diocese of Mukden." Its seat would be the original Manchu capital of Mukden, modern-day Shenyang, in Manchuria, and its jurisdiction would extend over all of "Chinese Tartary." Since the bishop of Beijing was traditionally appointed by the king of Portugal, the bishop of Mukden would be appointed by the king of France.[36] Now, to understand what this proposal was really about, one has to know something about the geography of Beijing. After the Qing had moved their capital there in 1644, they divided the city into three concentric districts. At the center was the Imperial City, which encompassed both the Forbidden City palace and the North Church. Surrounding it was the Inner City, or "Tartar City" as it was often called, where the Manchu garrison and the imperial bureaucracy were based. Outward from there sprawled the Han Chinese districts, which blended into suburbs and farmland. All told, Beijing was probably the largest city in the world at the time, with over a million inhabitants. The majority were ethnic Chinese, but, crucially, the Inner City was generally restricted to Manchus.[37] Herein lay the genius of Amiot's plan: if the Diocese of Mukden included all of "Chinese Tartary," that would include the very center of Beijing itself.

Amiot must have realized that the plan was somewhat ridiculous, because he concealed it even from his French confreres. François Bourgeois, the acting superior of the North Church, learned about it five years later by intercepting Amiot's mail. When another missionary, Jean-Paul Collas, asked Amiot about it, he refused to elaborate. Collas nevertheless managed to get the gist of the proposal, and he wrote Bertin independently to express his reservations.[38] He made the obvious objection that the main motivation behind the plan was not really to create a new diocese in Manchuria, but rather to carve out an isolated enclave within the existing Diocese of Beijing; a strange result was that "if the French bishop held the title of Bishop of Mukden, he would never be able to go to his seat." Furthermore, Christianity was technically forbidden to the Manchus who actually lived in the Inner City, while the Han Chinese who made up the great majority of local Christians were not even allowed to enter it. The bishop of Mukden would have no one in Beijing to minister to. And as for any nominal activity in Manchuria, there was already a small outpost there run by the Portuguese, who would surely not be pleased.[39]

Long before Bertin heard of these objections, he had already tried to move forward with the proposal; after all, no one would benefit more from the mission's independence more than he. First, he secured the consent of Louis XVI, who agreed that it was best not to let the French mission fall "under the spiritual jurisdiction of foreign bishops."[40] Next, he opened negotiations with the Vatican, assuring the pope that the king would make

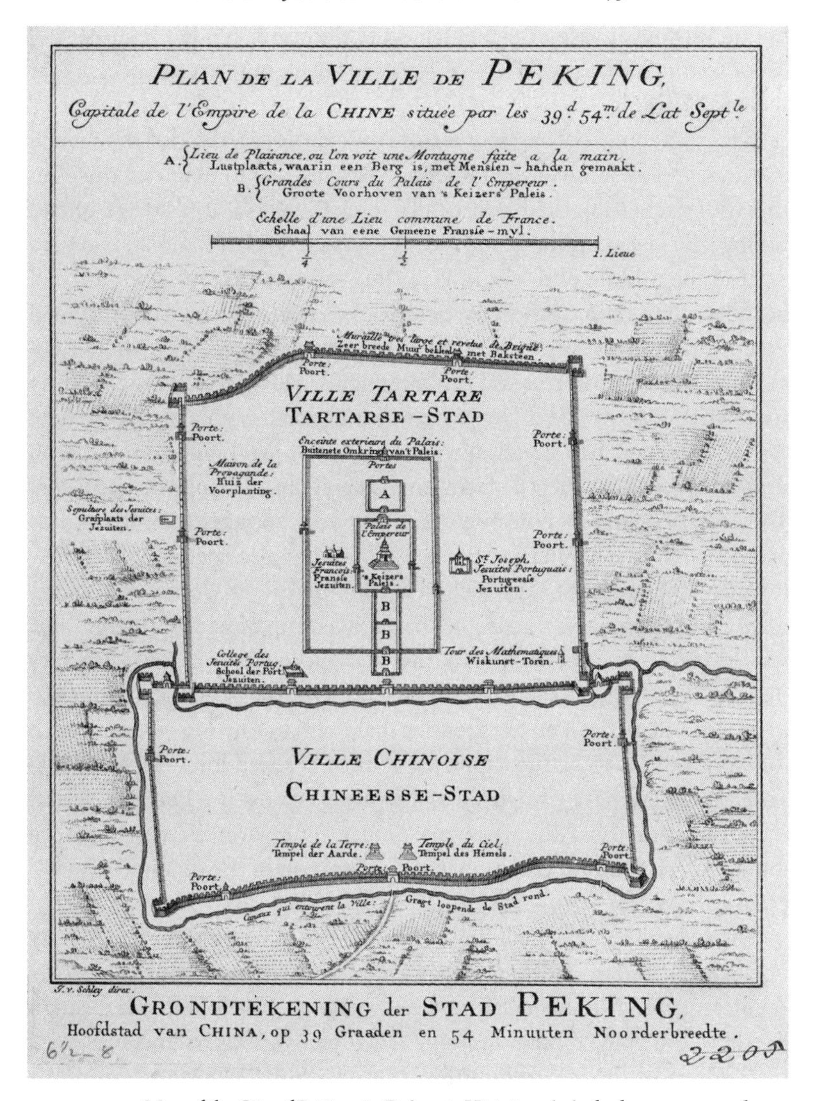

FIGURE 2.3 *Map of the City of Beijing*, in Prévost, *Histoire générale des voyages*, volume 7 (1749). Courtesy of Harvard Map Collection, Harvard Library.

a substantial "donation" to the new diocese and asking only the right to choose its bishop in return.[41] When the idea apparently met with resistance, he proposed downgrading the diocese to an apostolic prefecture—the prefect of which would be Amiot. Letters were exchanged between Bertin and the papal nuncio in Paris, the French ambassador in Rome, the minister of the marine, and the authorities of the Propaganda Fide, but he was never able to get a plan off the ground.[42] What is surprising is not so much that the farfetched scheme failed, but that its success ever seemed plausible.

Back in Beijing, new pressures from outside deeply divided the French mission, and as some of the missionaries struggled to maintain their authority, others took the opportunity to challenge it. In a stroke of bad timing, the last superior of the North Church, Michel Benoist, had died right in the middle of things in 1774; this was after the missionaries had heard about the suppression but before it was made official. His presumed successor, Bourgeois, tried to take over amid the chaos. But where Benoist had been a powerful and unifying figure, Bourgeois was apparently somewhat difficult and divisive. He was immediately challenged by three missionaries who had already formed a sort of clique: Louis Poirot, Joseph de Grammont, and their de facto ringleader, Jean-Matthieu de Ventavon. They were natural co-conspirators. They were younger than the others and had arrived in China more recently, and they also spent much of their time together as artists and technicians in the palace workshops, which might have also been a cause of shared resentment, since, though outranked inside the North Church, they had more clout in Beijing society outside of it.[43] The dissenters protested Bourgeois's ascension as superior of the French mission on the grounds that there was no longer any French mission for him to be the superior of. They accused him of sowing dissent among Chinese Christians and plotting against the Portuguese and Italian missionaries. Their primary demand, however, betrayed a more mundane motivation: communal administration of the mission's considerable property.

The power struggle that followed was all the more bitter because it was unclear who had the still greater power to resolve it. In response, the other French missionaries rallied around Bourgeois and pled their case through Bertin to the king, arguing that since their possessions had been granted by the king of France, to him they should revert.[44] In 1776, Louis XVI predictably responded in their favor, confirming Bourgeois as the strictly temporal head, or "administrator," of the mission and appointing Amiot as next in line.[45] "We are under the protection of the King, and we have nothing more to fear from foreigners," wrote Amiot; "Long live the King, long live great ministers."[46] This solution was satisfactory to Bourgeois and Amiot, but not to the three dissenters, for whom temporal authority was precisely what was at stake.[47] They contested the royal brevet on multiple counts: it was probably a forgery, they claimed, but even if not, the king of France had no jurisdiction in Beijing. Spiritual authority now rested with the Catholic diocese, and temporal authority with the Qing government. Predictably enough, the other powers in the city were universally sympathetic: Portuguese and Italian missionaries as well as Chinese and Manchu officials all stood to gain much from the dispersal of the French mission.[48] In fact, the only local opposition the dissenters faced came from their confreres within the North Church. Realizing his weakness, Bourgeois offered to concede the position of administrator to Amiot as a compromise, but the dissenters refused.[49] Realizing their strength, they took their case to the Qing authorities. The Manchu official in charge of the missionaries' affairs suggested that the matter should be resolved by the ranking European official at the Astronomical Bureau, but Bourgeois argued that delegating the authority to him would violate the precedent of French missionary self-government.[50] A special tribunal was therefore convened under the Imperial Household Department to settle the affair.[51]

A trial followed. Bourgeois described the proceedings with deep anxiety and evident fear. On December 22, 1780, the court considered the evidence against him, including a note attributed to Amiot that read: "Mr. Amiot rejects Mr. Bourgeois, and does not want him to be administrator," signed in the missionary's own hand. Bourgeois protested that the chief dissenter, Ventavon, "had extorted it from Mr. Amiot" four years earlier and was "preciously saving it" for an opportune moment—which this one was, since Amiot was sick and could not come in to denounce it in person. Two days later, the bishop of Beijing publicly insulted Bourgeois during the hearings, calling him "a poor man, and of a very mediocre merit." Bourgeois returned home beleaguered that night, only to find his confrere Jacques-François d'Ollières lying on his deathbed; a recent excommunication by the same bishop had left him broken in both body and spirit. The next morning, December 25, d'Ollières expired in Bourgeois's arms. No sooner had he finished performing the last rites than three Chinese commissioners showed up furious at the gate of the North Church and demanded to be let in. Assuming he had already been judged guilty, Bourgeois rushed to his office and burned all his papers. He then turned to the matter of laying d'Ollières to rest, which was scheduled for that night. As per custom, he invited all the missionaries in Beijing to attend the ceremony, but most of them snubbed him. "Christmas was a full day for me—full of fears and bitterness," he later recalled.[52]

A week later, the court issued its ruling, ostensibly in favor of compromise: the position of temporal administrator would be determined by lots and occupied in biyearly rotation. But when one of the dissenters drew the first straw, and another one drew the second, Bourgeois concluded that the whole thing had been rigged.[53] This outcome was in any case plainly at odds with the instructions of the king of France, but it spoke to the fact that the Qing authorities always carried more weight on the ground. Bourgeois and Amiot had no choice but to accept it. The ostensible resolution of the struggle in fact left the missionaries even more divided. As Amiot described it in the spring of 1781: "The demon of discord, which reigns more imperiously than ever in our mission of Beijing, was aroused among our *Messieurs* whom I would not dare call missionaries, still less Frenchmen, because they are not worthy of carrying either of these two names." For the next four years, the North Church was without permanent leadership.[54]

But Amiot and Bourgeois still had their allies back in Paris, where Bertin had been hard at work to keep the China mission under French—in practice, personal—control. From the perspective of the ecclesiastical authorities in France, it had apparently become something of a toxic commodity. Bertin first offered control to another French organization, the Paris Foreign Missions Society, which was already active in Vietnam.[55] The superior refused, citing the challenge of evangelization under an increasingly restrictive Qing regime. Just as well, Bertin confided, since they lacked the special intellectual caliber required for the charge.[56] He then reached out to yet another French organization, the Congregation of the Mission, more commonly called the Lazarists. Their superior general too turned him down at first, "because it seems to me beyond our abilities."[57] This time, Bertin was persistent, and in 1782 he finally prevailed upon the Congregation of the Mission to accept responsibility for the French mission to China.[58]

A Lazarist priest, Nicolas-Joseph Raux, was appointed in Paris as the new superior and set out with one confrere for Beijing. After an arduous three-year journey, they finally arrived to take over operations in April 1785.[59] By this time, things had calmed down a bit, thanks in part to the Qing government's renewed anti-Christian measures, which had united the missionaries against a common threat. The transfer of power went smoothly. Bourgeois wrote a glowing report of his "worthy successors" right upon their arrival: "These gentlemen are truly brave men, full of zeal, piety, and talent. . . . We live in common in the greatest intimacy; one would say that they have become ex-Jesuits, or that we have become Lazarists."[60] Even the dissenters seem to have become satisfied with the arrangement, perhaps because their leader Ventavon died the following year. Raux ruled gently and unobtrusively. He was generally respected and always obeyed.[61]

But it soon became clear that the Lazarists were not just Jesuits under a different name. The aging ex-Jesuits viewed their young superior with a degree of condescension. According to Amiot, Raux made only mediocre progress in the "profane sciences" that had once distinguished the Jesuit mission; "I would wish further that among the missionaries to come, there should yet be someone who might do something other than his St. Thomases and his St. Augustines," he complained.[62] He blamed the institution more than the man: "M. Raux understands very well all that has to do with the theology and ceremonies of the Church, but literature is a new world for him; the education of the seminaries is not the same as that of the colleges."[63] Another cause of concern was that the Lazarists, who had been born long after the Chinese Rites controversy was settled, seem to have finally internalized its resolution. Amiot lamented how they tried to institute "the ceremonies that took place in their House in Paris" and "worked to give the Chinese Christians a truly Roman form."[64] In particular, he was distressed that the syncretic ceremonial music, some of which he had himself composed, was replaced with plainchant.[65] Both the pillars of the Jesuit mission—scholarship and accommodationism—had crumbled.

Even so, in their hearts and minds, the French missionaries never really ceased to be Jesuits. For many years, they continued to harbor the hope that the Society of Jesus might someday be reestablished.[66] Only one of them lived to see it happen in 1815. None of them ever took new vows. Until their dying days, they referred to themselves as "*ex-Jésuite*." Two of them were still signing their names this way thirty years after the suppression.[67] In 1790, in one of his last letters, Amiot paid wistful homage to his "defunct society" with a line from Horace: "I have built a more lasting monument."[68] Even in his last will and testament, he recalled "the society of which I had the honor to be a member."[69] For Amiot, its dissolution had brought about what could truly be called a crisis of personal identity.

Jesuits and Ex-Jesuits at the Qianlong Court

The Qianlong emperor acceded to the throne in 1735 and abdicated it in 1796. Counting the subsequent three years, during which he retained power as "emperor emeritus," his was the longest reign in all of Chinese history.[70] Throughout this period, the imperial state, rather than independent literati, served as the most important patron of the missionaries in China.[71] Their influence at the Qianlong court reached a zenith in the 1760s, when they resided at the Qing palace in the capital and accompanied Qing armies in the provinces. In the early 1770s, the suppression of the Society of Jesus, along with internal unrest throughout the empire, set it on a precipitous decline. One thing, however, remained constant. The emperor appreciated European missionaries so long as they remained under his supervision; in this respect, they were like anyone else at the Qing court. They paid their service in whatever domains they were most capable: namely, arts, technology, and science.

The Frenchman who attained the highest rank, and one of the few Europeans who was ever on personal terms with the Qianlong emperor, was Michel Benoist, the last Jesuit superior of the North Church. Benoist rose to prominence in the 1740s, when he worked with the painter Giuseppe Castiglione at the Yuanmingyuan Palace, known as the "Versailles of China." Benoist helped design its mechanical fountains and hydraulic canals and continued to maintain them for the next two decades.[72] During this time, he reported, "the emperor seems to esteem us, and even to like us."[73] The two held frequent conversations, sometimes lasting an hour or more, about its lands and peoples, languages and religions, and politics and diplomacy. The emperor was particularly curious about the arts and sciences of Europe.[74] The missionary told the emperor about the exploration of the earth and the mapping of the heavens.[75] One of their conversations contains the first known discussion of the Copernican theory in China.[76] Others were more casual. Once, the emperor asked the missionary for a solution to the age-old question: "Between the egg and the chicken, which was created first?" Benoist had no clear answer.[77]

The emperor was especially interested in the technological and military applications of European science. He was delighted when Benoist presented him with a Newtonian telescope and encouraged demonstrations of an air pump at the Yuanmingyuan Palace.[78] In 1774, during the Second Jinchuan War, an engineer was needed to make calculations for the artillery to be used in an assault on a fortress occupied by indigenous rebels in western China.[79] The emperor ordered that an inquiry be conducted as to who was the better mathematician, Benoist or a Portuguese rival, Félix da Rocha.[80] The matter was decided in favor of da Rocha, who was promptly escorted to the Sichuanese front. European and Chinese accounts of the ensuing battle differ a little, but it seems that his technical expertise was considered a factor in the eventual Qing victory.[81]

Yet, even at the height of their influence, the missionaries were never a very high priority. According to Benoist, "the emperor here does not know Europeans under the name of secular priests, Carmelites, Augustinians, Jesuits; he knows only that they are Europeans."[82] Da Rocha and Benoist do not figure prominently in the Chinese record; virtually all that we know of their conversations with the emperor comes from the reports of the missionaries themselves. In 1772, Benoist recounted a revealing encounter that shows how insignificant they really were. The emperor wanted to appoint him to one of the official positions reserved for the missionaries and was considering vacancies. He asked Benoist about the German Father Gogeisl: "Surely, he must be very old now: How old is he?"—"He died last year," Benoist replied.—"Then there is a position open in the Astronomical Bureau," said the emperor.—"The position is currently filled by Father d'Espinha."—"I don't remember him."—"He is the one whom Your Majesty made a mandarin of the fourth order, when he went with da Rocha to make the map of the newly conquered lands."[83] Apparently, the emperor had forgotten the name of one high-ranking missionary official and did not even know that another was dead.

More serious problems for the missionaries were also beginning to accumulate. In 1774, a rebel named Wang Lun led a quasi-Buddhist millenarian uprising in Shandong Province.[84] Although it was put down with relative ease, the rebellion had a major impact upon the emperor's outlook: from then on, he believed religious sectarianism to be perhaps the biggest problem confronting his empire.[85] Meanwhile, the *Siku quanshu* project, a vast state-sponsored collection of Chinese texts, was just kicking into high gear: both a monumental editorial achievement and a thorough literary inquisition.[86] Amiot pointed out with pride that four Chinese works by European missionary authors were included in the collection; what he did not stress was that all of them had been written in the seventeenth century.[87] By the end of 1774, the emperor's two favorite missionaries, Benoist and Castiglione, were both dead. The fountains they had designed for his palace were turned on only when he planned a special visit, and even then, the water had to be hauled in manually because the hydraulic machinery no longer functioned.[88]

During the 1780s, things got much worse. The emperor's benevolence toward the missionaries had always been contingent in a way upon their not fulfilling their actual mission; active proselytization was more or less restricted in China, beginning with the crisis brought on by the Rites Controversy in 1724 and concluding only with the First Opium War in 1842.[89] While some, like Benoist and Castiglione, had been content to work toward this final goal indirectly, others were more hotheaded. In 1784, four missionaries sneaked into the southern province of Guangdong and began to preach without permission. They were soon apprehended, and the flagrant violation of long-standing laws set off a general inquest. In 1785, nine Christians were sent to prison. Two former Jesuits died in custody.[90] In Qing state archives of the Qianlong period, there is little mention of the missionaries dating from before 1785, after which there are many, and few of them are positive.[91] Representative of the bad air surrounding them at court is a memorial of 1786, in which the powerful minister Heshen referred to Christianity only with a new and derogative term, "western heterodoxy."[92]

The emperor became specifically worried about missionaries spreading Christianity in the provinces. He issued an edict declaring his intention to root them out: "Westerners sneaking into the interior to proselytize and delude the people is extremely harmful to minds and customs; of course, it is impermissible not to find and apprehend them."[93] During this period, the emperor was becoming increasingly concerned about Muslims, too, with the arrival of new Sufi groups into the northwest. He wondered whether "the Westerners and the Chinese Muslims belong to a single religion . . . one fears that they may not be without intent to collude and incite."[94] While it is hard to say whether this comment was truly ignorant or perhaps strategic, others at the time saw fit to disabuse him. One Manchu official who had regular dealings with the missionaries pointed out that unlike Muslims, Christians ate pork, drank alcohol, and did not believe in the Qur'an: "It seems believable that they and the Hui do not belong to the same religion," he gently suggested.[95] Popular comparisons between Muslims and Christians continued into the early twentieth century.[96] But despite the emperor's concerns about the missionaries' activities in the provinces, he allowed them to remain in Beijing, where they could be kept under supervision.

The state organ ultimately responsible for the missionaries was the Imperial Household Department, the director of which from 1769 to 1784 was the Manchu bannerman Fulungga.[97] He dealt with routine memorials concerning them, from the announcement of their arrival in China, to the publication of their obituaries.[98] He also managed them in more special circumstances; for example, he was the official who oversaw the compilation of cannon designs used by da Rocha in the Jinchuan wars.[99] The missionaries usually referred to him simply as the emperor's son-in-law.[100] But the missionaries were just one of the many responsibilities of the Imperial Household Department—and a very minor one at that—which was itself just one of the many responsibilities of Fulungga. Serving at various times as president of the Board of War, chamberlain of the Imperial Bodyguard, and director-general of the *Siku quanshu* project, he was a very busy man.[101] This was probably one reason why, when dissent tore apart the North Church, Fulungga was the one who on behalf of the Imperial Household Department first encouraged the missionaries to sort it out among themselves, and then appointed two Chinese officials to serve on an ad hoc tribunal in his stead.[102]

Some of the missionaries still held positions in the Qing bureaucracy.

As director of the Astronomical Bureau, da Rocha was ex officio a fifth-rank official.[103] For his particular service to the emperor, he was promoted to third-rank official, roughly equivalent to a prefect in the provinces or a vice-censor in the capital.[104] This made him by far the seniormost European in China from 1774 until his death in 1781.[105] He was the only one whom the French missionaries referred to by a Chinese name and title, Fu *Daren*, an honorific they generally reserved for very important people.[106] Da Rocha was thus also the one whom Fulungga had proposed to mediate between the French missionaries, on the thinking that "since he was the most distinguished of the Europeans, he should be in charge of them and govern them."[107] Following his death in 1781, his position at the Astronomical Bureau fell to its former vice president, Joseph d'Espinha, the missionary whose name the emperor had forgotten a decade earlier. In 1785, when five Europeans were invited to a ceremony, Amiot ruefully reported that d'Espinha's rank entitled him to march separately from the rest.[108] It is telling that something so trivial seemed by then like such a significant perk.

Missionaries were increasingly appreciated less for their mathematical and astronomical knowledge than for their technical expertise, particularly in the areas of painting, watchmaking, and machinery. These skills were considered by the missionaries to be the three main routes to court employment, confirmed by the fact that they are the subject of the majority of the surviving documents in the Qing central archives that mention the missionaries in a good light.[109] The French dissenters, for example, were particularly skilled in such fields: Ventavon was a watchmaker, Poirot was a painter, and Grammont was a musician. The three of them were invited to work and even sometimes live at a workshop in the Yuanmingyuan Palace complex called the Ruyiguan, or Wish-Fulfilling Studio.[110] According to Amiot, it was both a blessing and a curse: on the one hand, they there were graced "fairly often" with visits from the emperor; on the other, they had to be on their toes from dawn till dusk. Amiot also complained that the place was uncomfortable—though the cavil may have just been sour grapes.[111] But even roles such as these were becoming neglected. When Ventavon died in 1787, he was replaced by a lay brother who, according to his own superior, had the unhappy qualification of being "as capable in horlogery as he is bad in Chinese."[112] No one either in the North Church or the imperial palace seems to have cared any more that even if the emperor had visited the workshop, the missionary technician there would not have been able to converse with him.

Participation in public life had become mostly ceremonial. In the fiftieth year of the Qianlong reign, on February 14, 1785 (Valentine's Day,

FIGURE 2.4 *Illustration of the Feast of the Thousand Old Men, or Qiansouyan, Held by the Qianlong Emperor in Beijing, 1785*, in Helman, *Abrégé historique des principaux traits de la vie de Confucius* (1788). Courtesy of the Huntington Library.

as Amiot pointed out), the emperor hosted a celebration in honor of the elderly of the empire, the "Feast of the Thousand Old Men." Three thousand guests over the age of sixty were invited to attend, including Amiot, Bourgeois, and three other Catholic priests, "despite the reasons of state that would seemingly prescribe him to show his dissatisfaction with all the European missionaries." They arrived at the gates of the Forbidden City in time to greet the sun for a lengthy procession around the palace. After it was over, everyone sat down to a great banquet. The emperor's oldest living son, taking upon himself the position of "Maître d'Hôtel," personally greeted the French honorees. Each table was set with a quarter "lamb of Tartary" on a hot plate, nested among meats so exotic that Amiot could not recognize them. Feasting was followed by music, dancing, and a comedy show. Finally, presents were exchanged, and the guests were allowed to return home. Amiot claimed to find it all tiring, but for the rest of his life he jealously guarded the scepter he received as a memento. In virtue of his attendance, he was also granted the rank of fifth-rank official—the only title he ever held, bestowed for the sole merit of being old.[113]

To commemorate the event, officials compiled a poetry collection, the *Imperially Commissioned Poems of the Feast of the Thousand Old Men*. Amiot's short contribution, his only surviving Chinese language composition, presents a unique window into his self-fashioning after living more than half his life in China:

> Pharos Island, the Colossus, the Hanging Gardens of Babylon; from far away I recall the seven wonders, transmitting the ways of my country and reverentially wishing a long life to the Heavenly Emperor. Feasting and drinking, I respectfully raise a stone cup to the fortune of the whole world; his light near me like the Big Dipper; the constellation of the Black Dragon pours out the Heavenly Spoon, gazing upon the mansion of the North Star; in the center of the zodiac, the Heavenly Throne in the Purple Palace is luminescent.[114]

The poem spoke to Amiot's dual roles as European scholar and imperial servant. Astronomical metaphors for the emperor, drawn from the ancient classics, set it in a highly literary Chinese register, while at the same time calling attention to the mathematics and astronomy that were the signature contributions of the Jesuit mission. In China as in Europe, Amiot presented himself as a cross-cultural intermediary, and not without a little pride. After all, given the poem's obscure references to both Chinese and classical antiquity, there could have been no more than a handful of people anywhere in the world who would have understood it.

Mandarins, Manchus, Mongols, and Missionaries

As time wore on, the psychological distance between a missionary and the land of his birth increased. The normal age of departure for the mission field was around thirty, so a Jesuit could expect to spend most of his life in China. Few ever returned to Europe, so he could expect to die there, too.[115] The suppression added a further isolating effect. After the last Jesuits arrived in Beijing in 1771, it was thirteen years before the next infusion of French blood. By that time, Amiot had begun to refer to the French with second-person pronouns.[116] He contrasted "your European notions" with "our principles," and though he also continued to refer to "our France," he did so in order to express regret for what had been lost—the France he had left behind years ago, not the France to which he was writing now.[117] In his new home, he stuck out a little less than one might have expected. Beijing was the cosmopolitan capital of a multiethnic and polyglot empire. Many Qing subjects, like the Uighurs of recently conquered Xinjiang, probably looked more like Amiot than like the Qianlong emperor. Amiot could never become a Chinese person, but it might not be wrong to say that he became a Qing person.

Despite their growing isolation from political and intellectual life, the French missionaries still spoke with Chinese people every day. In 1779, they employed thirty-five regular Chinese personnel, divided into two teams, one at the North Church and the other at the Wish-Fulfilling Studio. Most were basic service staff: three porters, four rent collectors, a chauffeur, a shoemaker, a gardener, a baker, one cook for the missionaries, and another one for the staff. Some of these people, however, were quite learned. Of the Chinese people on the 1779 payroll, three were ordained priests and three were listed simply as "literati." In addition, nine were the missionaries' personal valets, who also acted as secretaries and research assistants.[118] Other more-educated associates, part-time contractors brought in at need, did not appear on the payroll at all. Amiot's Manchu teacher was one example.[119] His doctor was another.[120] The missionaries thus had plenty of local people to talk to, some of whom played an active part in their research.

Without question, the most important collaborator and the best friend Amiot ever had in China was his personal servant, a Chinese Christian named Jacob Yang. Yang was likely raised by the missionaries.[121] He was only about twenty years old when he was assigned to Amiot and worked faithfully for him for over thirty years. In effect, Yang was Amiot's research assistant. Amiot often referred to him as "my literatus," but that was not quite right: Yang had probably not received a classical education, and he had definitely not passed any civil-service exam.[122] Amiot insisted nevertheless that he was not inferior to the Chinese officials who had done so; in fact, he claimed, Yang was even "more universally and more solidly instructed." Amiot took no small credit for having trained his assistant "in our way of studying," instilled in him a "love of antiquity," and taught him "the art of reasonable critique."[123] The two men developed a distinctive way of working together. Amiot would assign Yang a research question, and Yang would present him with arguments for both sides. Amiot would then choose which to endorse. Sometimes Yang would disagree about which case was stronger, but he would help Amiot make his chosen one anyway. Amiot viewed this aid as invaluable, and in later years, he often complained of the difficulty of conducting research without it. Yang played the role in part because no one else would, suggesting again how isolated the missionaries had become.

Amiot's attachment to his servant was also deeply personal. None of his letters was more heartfelt than the eulogy he wrote upon his friend's death in 1783, when Yang peacefully succumbed to a "singular malady that is unknown in Europe" and had made it impossible for him to swallow. Amiot found it impossible to describe his grief: "If I converse with you here of the subject of my pain; the loss I have made is of a nature that renews my grief, even in conversing with you." Yang was not only "a veritable treasure" of a scholar, but a model Christian who approached death with a "courage approaching heroism."[124] Ten years later, when Amiot in his turn reached death's door, he had not forgotten his faithful servant. In his will, he ordered that a third of his money be given to the son of "my old Yang" for having served him with "a zeal, a fidelity, and an affection" that had contributed not a little to his own long life.[125] Amiot had never even tried to replace him.

The only Chinese scholar of any real significance who can be demonstrated to have had personal relations with the European missionaries during the second half of the Qianlong reign was the historian and poet Zhao Yi. During the 1750s and 1760s, he occasionally visited the South Church, home to the Portuguese missionaries, where he admired the musical instruments and telescopes. At the Astronomical Bureau, he conversed with d'Espinha and another missionary official, Ferdinand Augustin Hallerstein. In his writings, Zhao praised the famous missionaries of the late Ming for their intelligence and skill, and for having taught literati mathematicians and astronomers such as Xu Guangqi things that Chinese scholars had not then known.[126] Yet his own visits with the missionaries did not produce any collaborative projects, and they seem to have been discreet affairs. Some of his literati contemporaries who shared his interest in mathematics and astronomy undoubtedly knew about them, but there was no mention made in print until many years later.[127]

In private and unofficial settings, lesser figures continued to visit with the French missionaries on occasion. In 1779, Amiot thought he was close enough with a bureaucrat at the Tribunal of Rites to procure some semi-classified documents about the Jinchuan campaigns.[128] As it turned out, he was wrong; his friend advised him to give up the effort.[129] In 1783, he was still in touch with another official who was responsible for translating documents into Mongolian at the Grand Secretariat, but there is no indication that their relationship amounted to much more than the occasional lunch with a colleague from work.[130] Furthermore, these people were not major officials or famous scholars in the mold of a Xu Guangqi, or even a Zhao Yi. If they had been, Amiot would have likely mentioned them by name or rank.

Amiot eventually began to resent even the little contact with literati that he still maintained. In 1789, he regretted that the "literary career of the Chinese people" had been more open to him twenty or thirty years earlier.[131] That he found himself shut out of local circles is perhaps not surprising, given his own habits and attitudes. By this time, he rarely left the North Church. Most of the "mandarins" who visited him were led there by what struck him as frivolous curiosity—"to them, I am never at home."[132] With the others, scholarship was generally not on the agenda. In 1788, Amiot mentioned one frequent topic of conversation with his Chinese visitors: "The whiteness and prolixity of my beard impresses them." His pride on this point could hardly have been endearing: "None of them has a beard to which I could give without derision the elegies that they give to mine; for we must say; nature has endowed me very well in this regard, and I could, without much imposing on myself, draw from my rich stock enough to garnish the chin of fifty Chinese and equally as many Manchus."[133] By 1790, Amiot had, in his own words, "retired" from public life. He made no outside visits at all. Occasionally, some old Chinese acquaintance would still make the effort to see him at the North Church.[134] But most of his few visitors by then were not literati scholars or bureaucrats. Nor in fact were they even Chinese.

Close relationships between Manchus and missionaries dated back to the beginning of the Qing. The Kangxi emperor had taken a genuine interest in Western learning, and, for a time, his grandson the Qianlong emperor had, too. Both had employed Jesuits as advisers on Western studies and kept them close to the imperial court.[135] Other Manchu nobles had likewise engaged with the missionaries. In the early eighteenth century, some members of prominent Manchu families had even been practicing Christians.[136] The exchange of scientific knowledge had flourished under these conditions. As late as the 1760s, Benoist believed that the best locally trained mathematician in Beijing was not a Chinese literatus, but rather the Manchu prince Yinlu, whom he called the "patron" of the Astronomical Bureau.[137] While contact with Chinese scholars seems to have fallen, the missionaries' contact with Qing nobles continued. These relationships were easier to maintain, mostly because such figures had both reasons and opportunity to cultivate them.

One of the few people whom Amiot continued to call a friend was a certain prince of the Khorchin Mongols, who "had no greater pleasure . . . than that of coming to chat with me about the sciences or arts of Europe."[138] As a vassal prince, he was obliged to leave Mongolia to visit Beijing once or twice a year and pay homage to the emperor, whereupon he would usually take the opportunity to pay Amiot a visit. Both foreigners in the Qing capital, they chatted about their distant homelands. In 1777, Amiot asked for some information about Mongolia that he hoped might contribute to a debate about the population of the Qing Empire. During this time, travel between Mongolia and China proper was heavily restricted, but people were moving nevertheless.[139] The prince explained that Han Chinese immigration to his Mongolian domains had become a big problem, forcing him to forbid the planting of grains and fruits in order to maintain good pasturage for horses. "Wherever there is an inch of land, there is a Chinese to cultivate it," was Amiot's conclusion.[140] During another visit some ten years later, the Mongol prince asked Amiot to place a request for some coral-colored glass balls from France.[141] While Amiot presented this exchange as evidence of his friend's curiosity, it also betrayed the true basis of their friendship.

Local demand for European technology, particularly clocks and guns, was fed in part by the manufactured Manchu culture of the Qianlong court. Every year during the summer months, the emperor took his inner circle to Rehe near the Manchurian border to avoid the baking heat of Beijing and to encourage the Qing elite to reconnect with their seminomadic roots.[142] Activities included riding and hunting with the emperor; or, as Amiot described them, "those arduous exercises with which his ancestors amused themselves when they were still nothing but horde chieftains."[143] Indeed, by this time, many of the Manchu nobility likely felt little personal connection to that way of life. The Kangxi emperor and his lieutenants may have spoken Manchu at home, but their grandchildren did not.[144] Rehe must have seemed quite provincial to them. While it was an honor to accompany the emperor, it was also somewhat dull. Playing with European clocks and watches was a good way to pass the time, as was firing European guns, which Manchus exclusively were permitted to own.[145] Although some Han Chinese officials followed the court to Rehe, they were required to fulfill their regular government functions, leaving them with little time for such amusements. And, according to Amiot, tinkering and shooting would have been difficult for them anyway, on account of their long fingernails.[146]

As the only Western Europeans in residence in Beijing, the missionaries cornered a small-scale economy in luxury goods. Manchus had plenty of guns, but ammunition was limited; and although they received watches as patronage gifts from Canton, they lacked the materials to repair them.[147] The missionaries, with their technical training and access to European trade networks, offered supplies. Amiot signaled to Bertin that gun and watch paraphernalia were the most important items to be sent from France: "You would not believe how agreeable these sorts of things are to our Princes and other Manchu lords," he wrote. "It is a good thing, in a way, that one must wait three years to have from France what one desires. This long wait tempers the vivacity of the desires and turns to my advantage." Amiot quipped that it was also a good thing for Bertin, because if Beijing were any closer to Paris, he would have been pressured to ask for more materials every single day.[148] For Manchus and Mongols, a constant supply of bullets, powder, springs, and screws was a major benefit of having a missionary friend. For the missionaries, such toys were a cheap price to pay for their goodwill.

This pattern of exchange was what brought to the North Church its most illustrious visitor of the Qianlong period, the emperor's sixth and oldest living son, Yongrong.[149] He was exactly the kind of person who maintained relations with the missionaries, having grown up around them at court and developed a fondness for painting, mathematics, and music. He also held various positions that put him in contact with them in the course of his official duties.[150] He was attached to the Imperial Household Department, which oversaw the Astronomical Bureau.[151] He was also specifically responsible for the upkeep of the Yuanmingyuan, where the French missionaries worked. When an old Ming armillary sphere became loose in its swivel and a Jesuit astronomer requested that a new one be cast, Yongrong was the person whose name appeared on the approval form.[152] During the 1780s, when Amiot was working on a translation of a Ming musical treatise, Yongrong was preparing an edition of the same text for the *Siku quanshu*.[153] As one of its general editors—in effect, a censor—he was also free to explore sensitive topics that might have otherwise have aroused suspicion.

According to the missionaries, Yongrong developed a sentiment "in favor of Europeans, whom he esteems."[154] At the Feast of the Thousand Old Men in 1785, he was the prince who went out of his way to greet Bourgeois and Amiot. His favorite among the French missionaries, however, was the chief dissenter, Ventavon. As early as 1770, the Manchu prince had been charmed by the Jesuit watchmaker, who had "much to recommend him with his modest and polite air." Yongrong requested that Ventavon alone handle the repair of his European instruments. Whenever Yongrong wanted anything, Ventavon would "do his utmost to procure it for him."[155] The two spoke regularly and became close enough that Ventavon dared to speculate about Yongrong's chances of succeeding his father as emperor. In 1784, despite mounting state concern over Christianity—or, perhaps, because of it—Yongrong paid a personal visit to the North Church. Entourage in tow, he visited the chapel, examined the sacerdotal vestments, admired the observatory, and entered the private chambers of some of his "particular friends." Ventavon, who described the visit, was surely one of them.[156]

This friendship had real but hitherto overlooked consequences for the fate of the French mission. Historians have described the conflict in the North Church without noticing the role that Yongrong played in its resolution. Although Fulungga was theoretically in charge of the proceedings, Yongrong, who was also attached to the Imperial Household Department, soon became involved.[157] He was perceived by Bourgeois as "the great protector" of Ventavon, whom he allegedly treated with "particular favor." Apparently, Yongrong secretly promised Ventavon his support and put in a good word with the Chinese authorities. Without it, Ventavon might never have been able to move forward with his challenge to Bourgeois, given that the majority of the French missionaries and also the royal brevet were clearly against him; but with it, his superior was reluctant to fight the charges too strongly for fear of provoking the powerful Qing prince. When the tribunal's ruling came down for the nominal compromise, Bourgeois blamed Yongrong's interference; and when the first lot fell to the dissenters, he was sure that it was Yongrong who had rigged the draw.[158] While imperial politics kept most people out of the missionaries' affairs, personal relationships remained strong enough to draw some people in.

The Prince Hongwu

The Manchu prince Hongwu was a grandson of the Kangxi emperor, an artist, a poet, and a scholar; from the early 1780s on, he was also the French missionaries' closest local contact outside the North Church. He helped them with research, updated them on the affairs of court, and even secured them political favors. In return, they helped him learn about French culture and Enlightenment science. Hongwu and Amiot developed a special relationship—the only one between any of the ex-Jesuits and a non-European figure that can be reconstructed in any detail. Their story is unusual in many ways. Its characters do not fit neatly into the paradigm of Jesuit-Chinese exchange. Hongwu was not Chinese, and Amiot was no longer a Jesuit. Its setting was different as well. It took place not at institutional locations like the Yuanmingyuan and the Astronomical Bureau, but rather in private at the missionary's church and the prince's mansion. In fact, this story of cross-cultural exchange may have been the only one like it, and it was certainly the last.

Only a few historians have even noticed the existence of the "amiable prince" whom the missionaries called "*hong wu ye*," and not one has previously attempted to determine his identity.[159] The task presents two major challenges. The first is archival: His name appears with multiple spellings in just a handful of handwritten letters, scattered among several hundred in two different archival collections.[160] Most of those letters never went to press. In the few letters that did, the parts that mentioned him were usually edited out. And in the one or two letters where they were not, his name was changed to just "a prince."[161] No version of "*hong wu ye*" was ever printed. As a result, the published record provides little indication

FIGURE 2.5 *Hongwu Making Merry*, Hua Guan, 1785.

that there even existed a single person to whom the missionaries referred. The second challenge is historical: Even in their manuscript letters, the missionaries never gave Chinese characters his name.[162] Fortunately, the set of people to whom it may have referred is limited by the prince's identification as a grandson of the Kangxi emperor; unfortunately, this status alone does not narrow down the candidates very much, because the Kangxi emperor was one of the most fecund in Chinese history. The sons of his sons numbered about a hundred, and almost all of them had a two-character given name in which the first character was the generation name *hong*. What this convention suggests, though, is that the appellation "*hong wu ye*" did refer to someone with such name, to which the missionaries probably added the honorific *ye*, literally meaning "grandfather" but often used in Qing times as a form of address for powerful people. And there was only one grandson of the Kangxi emperor whose name had a second character that fits: Hongwu.[163]

Of all the paternal grandchildren of the Kangxi emperor who were active during the right period, what we know about Hongwu from the Chinese record matches well with the missionaries' descriptions. First, he was the right age. He was born in 1743, so when one missionary twenty years his senior first mentioned him in 1784, he could have been both the master of his own household but still plausibly "young," as described.[164] Second, he was of the right importance. If he had held a higher rank, such as "count" or "régulo," then the missionaries would have likely used that title, as they did for several of his cousins, instead of the generic term "prince."[165] Third, he was in the right place at the right time. From the mid-1770s on, he lived in the Imperial City, which would have made possible the regular visits that were reported. Fourth, and most importantly, we know from Chinese sources that Hongwu had the artistic, cultural, and scientific interests typical of those who maintained contact with the missionaries during this time, and that he and the missionaries shared several significant acquaintances.[166] All these facts suggest that "*hong wu ye*" was indeed the prince Hongwu.

Hongwu grew up in Beijing around the missionaries. His father, the Kangxi emperor's twenty-fourth son Yunbi, and his cousin, the Qianlong emperor né Hongli, had been childhood friends and remained close in adulthood.[167] Hongwu spent his youth during the 1750s and 1760s at the heart of the Qing court, when Jesuit missionaries were present and popular. He could have played in the European-style waterworks of the imperial palace, which he later painted, and marveled at the demonstrations of an air pump, which he later borrowed. The artist Castiglione may have encouraged the young prince's enthusiasm for painting; his early work dates from the time when the Italian missionary was working at the Yuanmingyuan.[168] Perhaps Hongwu was the "grandson of the Kangxi emperor" who once received instruction from Benoist on how to calculate eclipses.[169] Encounters with such famous and popular missionaries would have likely had some effect on a curious child.

As a young man, Hongwu held several commands of minor distinction, but he did not take to the martial way of life. He received a commission as second-rank general in 1763 at the age of twenty and eventually reached the rank of banner prince more than ten years later.[170] For a first cousin of the emperor, this ascent was not quick. In 1778, at the age of thirty-four, Hongwu's military career came to an end. His own brother accused him of corruption in conspiracy with a certain Manchu manor lord, "exploiting their power, unfairly seizing land, and so on."[171] The emperor was not pleased. If such crimes went unpunished, he reasoned, then they might become more common; even worse, what if his own sons followed their cousin's bad example? In a harshly scolding decree, he ordered Hongwu stripped of all his ranks and titles.[172]

Hongwu's career never really recovered. Though he remained a cousin of the emperor, he was marginalized in matters of war and government. His writings reveal a painful awareness of his disgrace: "My skill is unworthy to dedicate to my country, my lowly position a shame to my legacy," he lamented in one poem.[173] A year after the scandal, "having committed no great fault," he was given a new command in the Han military banners, and eventually received several minor promotions.[174] But these positions were no more than sinecures, signs of social rather than political rehabilitation. His obligations likely extended little beyond showing up for the occasional ceremony or display. Having effectively retired from public affairs, he spent most of his time in Beijing with little official business to attend to.

It was at around this time that Hongwu began to seriously dedicate himself to arts and letters. He took his freedom as an opportunity to paint watercolors, compose poetry, write calligraphy, collect books, and otherwise indulge his scholarly interests. The change in lifestyle seems to have suited him. According to the missionaries, he became "one of the most curious, intelligent, and learned of this court."[175] He achieved a modest reputation for his erudition, with a library of over two thousand volumes that impressed later Chinese scholars.[176] He also became one of the most respected Manchu artists of his generation, known especially for his calligraphy and landscape paintings.[177] He even won his way back into the good graces of the emperor, who ordered many of his scrolls to be hung in various palaces. The imperial catalogue included thirty-seven of Hongwu's works, and the Beijing Palace Museum today holds over a hundred.[178]

Hongwu immersed himself in a courtly circle of aspiring artists and intellectuals. Its other core members were his cousins-once-removed Yongrong and Yongzhong, both of whom were fairly close to him in age.[179] Together the three princes pursued artistic refinement and spiritual cultivation. They amused themselves with drinking games, poetry contests, and excursions to scenic sites. They discussed esoteric philosophy and recited curious tales. The famous scholar-official Ji Yun, general editor of the *Siku quanshu* and himself an advocate of foreign technology, later recalled one that Hongwu liked to tell. Once, a Mongol prince went out hunting and shot a fox. Upon examination he found something curious: on its hind legs were two tiny red shoes fitted for bound feet. It must have been a fox spirit—a notorious troublemaker in Chinese folklore and fable—just transformed from its disguise as a beautiful human temptress.[180] Receptive to the strange and wonderful, and with slightly more liberty to indulge it, the court culture of the Manchu aristocracy was one domain in which European oddities, from musical instruments to mustard seeds, seem to have retained some appeal.[181]

As a court artist working at the pleasure of the emperor, Hongwu lived not only in physical proximity to the missionaries, but also in somewhat similar circumstances. By the end of the 1770s, the ones most closely connected to the imperial court were the artists and technicians at the Wish-Fulfilling Studio. The most significant evidence of any European activities in Beijing during the entire post-Jesuit period to be found in Chinese sources comes from the archives of the imperial art studios. During the 1780s and 1790s, two of his Chinese friends, the artists Na Yancheng and Peng Yunmei, were collaborating with the ex-Jesuits Louis-Antoine de Poirot and Giuseppe Panzi at the Wish-Fulfilling Studio.[182] In 1793, the four men together signed their names on a painting commemorating the Khalkha Mongols' submission tribute of horses and elephants. In that same year, Hongwu painted the same scene and gave his piece an almost identical title.[183]

Hongwu may have also shared something in common with the missionaries in his pursuit of alternative religions. Buddhism and Daoism had always been a refuge for those like Hongwu who had found their political ambitions disappointed, thus failing to fulfill the highest duty of a Confucian scholar. The early death of several of his children seems to have intensified his sense of detachment, and perhaps encouraged his search for a more spiritual kind of fulfillment.[184] He described his worldly ambitions, or lack thereof, in a poem:

> Remote antiquity knew no fame
> A sincere heart keeps watch by itself
> The moon is bright, the wind and the dew are cold
> Relying on myself, I am quiet in middle age.[185]

His poetry invoked Daoist writings of the Huang-Lao tradition and elements of Buddhist iconography such as Amitābha statues and chrysanthemum flowers.[186] His artwork too expressed such themes. For example, in *Yu the Great Controlling the Waters,* he depicted a host of heavenly immortals observing the sage-king's labors—not unlike those undertaken by missionaries like Benoist at the imperial waterworks.[187] His chosen art name, Yaohua Daoren, "Daoist of Illustrious Jade," finally made his self-identification as a Daoist explicit.[188] Christianity was another matter. During the Qianlong period, the religion was forbidden to Manchus, all the more so to those close to the emperor.[189] Though some ignored this rule at great cost, there is no evidence that Hongwu ever considered conversion; if he had, the missionaries probably would have said so.[190] His general religious inclination might still have kindled an early interest in the missionaries.

But by the late 1770s, such an interest was difficult to pursue in public. The institutions that had previously sustained exchange between Catholic missionaries and local figures were decaying or defunct. The Society of Jesus was no more, and Jesuits who had once been closest to the emperor had been dead for many years. The palaces they had designed were falling into disrepair, and the machines they had demonstrated had been reclaimed. It was at this point that Hongwu began to frequent the North Church. From his princely mansion, still standing today in Big Matchstick Alley, it was a little over two kilometers through the Imperial City to the North Church.[191] There, he was well received by the increasingly isolated missionaries, who were delighted to entertain such an illustrious and well-disposed guest. By the early 1780s, his visits had become regular.

Hongwu and Amiot soon developed a singular friendship, forged in mutual curiosity and fired by genuine affection. Amiot was the spiritual and intellectual leader of the mission, having achieved a reputation in Europe as a preeminent living scholar of China. Likewise, Hongwu had earned his own reputation in China, at least among the missionaries, as "a great amateur of European things."[192] Building upon such shared interests, their friendship grew stronger over time. As Amiot's health declined toward the end of the decade, Hongwu was one of the only people he would still admit into his quarters. When the prince came to talk with the ailing missionary, it was what "one could call a veritable visit of a friend."[193]

Language exchange was a topic of perennial interest. Amiot taught Hongwu how to read and write French words "just for fun"; he reported that his student managed to pronounce them "better and more distinctly than our Germans and our Portuguese can," even though the Manchu had no idea what he was saying.[194] In turn, Hongwu taught the missionaries his own languages. He asked them to design a "machine that writes characters" to reproduce Chinese texts.[195] After the Paris publication of a landmark Manchu dictionary that Amiot had compiled, a copy was sent to Beijing for approval. Amiot showed it to Hongwu, whose compliments were so profuse he concluded that his friend had only "wanted to flatter me."[196] Amiot's short letter back to France to commemorate the release, composed in his own elegant Manchu script, reveals that he, too, had been a diligent student.[197]

Another specialty of Hongwu's, and one with an especially appreciative audience in Europe, was art. Just as Bertin sent objects for Amiot's friends in China, so Hongwu sent things for Amiot's friends in France. In 1788, after admiring some French hair braids, Hongwu insisted that Chinese artists could match French ones for skill. He gave Amiot and Bourgeois each a set of painted glass snuffboxes and asked them to send another pair for Bertin. One depicted a frog and a grasshopper, the other a goldfish, painted on the inside so that the minister might "compare difficulty with difficulty, patience with patience."[198] Amiot thought they were ugly, but he sent them along anyway, asking only that Bertin not take it as a sign of his own lack of taste. In the following years, the prince gave many more objets d'art to the minister: colored inks, Ming porcelains, an ivory flower box, an incense burner.[199] And, just as the French aristocrat maintained a cabinet of Chinese curiosities, so the Manchu nobleman with a "decided taste for all that comes from France" kept his own French one.[200] Hongwu thus became in effect a peripheral member of Bertin's "literary correspondence."

Hongwu lent what little political influence he retained to help facilitate it. Qing regulations mandated that all items shipped from Paris to Beijing had to pass a customs inspection at Canton. Amiot often complained about the customs officials' "roguery," by which he meant everything from casual disregard to outright theft. In reply, Bertin offered the implausible suggestion that Amiot ask the Manchu general and statesman Agui to intervene on his behalf.[201] Amiot had a better idea. When a shipment arrived at the North Church in 1788, he pointed out to Hongwu how the boxes had been ransacked. Hongwu promised to intercede with Agui's rival at court, his "good friend" the "prime minister" Heshen.[202] In this way, the missionaries managed to obtain a special order that their parcels be delivered from Canton unopened.[203] The double irony of the episode is that it was Agui, not Heshen, who was sympathetic to European interests; and, for all Amiot's accusations of the custom officials' corruption, it was Heshen himself who was then earning an ignominious reputation as one of the most corrupt officials in the dynasty's history.[204] But, through the efforts of a mutual friend, the missionaries were apparently able to turn his corruption to their favor.

By far the most valuable thing that Hongwu provided was intelligence. Information about the emperor was extremely useful, but it was carefully managed by a sophisticated censorship apparatus. The prince was a member of the imperial family and maintained a personal relationship with his cousin. He was an "eyewitness" source on the emperor's health and habits and thus also the privileged holder of what were in a sense state secrets.[205] On occasion, he shared them with the missionaries. In

1789, Hongwu assured Amiot that although the emperor was approaching eighty, he could still "ride a horse, hurl a lance, and shoot an arrow" like a man half his age.[206] The next year, Hongwu confided that while the emperor was otherwise in excellent shape, he had become a little hard of hearing. The misfortune, he suggested, could be turned into an opportunity. Hongwu had noticed on a table in Amiot's study an illustration of a novelty ear trumpet that could be inconspicuously attached to one's head under the hair. Perhaps the missionaries might offer one to the emperor, innocently presenting it as a mere curiosity, since "it is not unlikely that under the pretext of ascertaining the effect, the emperor will want to try it himself."[207] The record does not indicate what became of Hongwu and Amiot's scheme to fit the emperor with a hearing aid.

Advice about the emperor's personal preferences was particularly valuable at this time because the missionaries' participation in public life had become largely restricted to ceremonial gift-giving. When preparations began for his massive eightieth-birthday celebration several years in advance of the planned date in 1790, Amiot identified another opportunity. He consulted with Hongwu, who was "perfectly aware of his tastes," to help choose presents for the occasion. Hongwu suggested copies of a set of illustrations of Qing military victories that had been painted by Castiglione decades earlier and engraved in France.[208] Amiot requested the prints from Paris in 1787, and they arrived in Beijing three years later, just in the nick of time for the birthday celebration.[209] First, he showed them to Hongwu, who assured him that although there were a few minor mistakes, such as the improper coloring of certain uniforms, these would only serve as proof of foreign authenticity. The emperor would be delighted with this "monument to his glory" from "the first Kingdom of Europe"—"I take responsibility for the complete satisfaction on the part of the emperor," he promised.[210] Hongwu's own gift for the occasion, a commemorative battle scene that he painted himself, suggests his sincerity.[211]

Hongwu thus also helped the missionaries understand what mattered most in the imperial ideology: the Qianlong emperor's changing view of himself. Late in his reign, with corruption festering in the capital and unrest mounting in the provinces, major military campaigns remained a bulwark of Qing legitimacy. In the 1790s, the Qianlong emperor began to style himself as the "Old Man of the Ten Complete Victories."[212] Earlier Jesuits had burnished his image as a successful commander by painting his glorious victories; Hongwu advised their successors to remind him of it. The illustrations of Qing military victories were duly presented as birthday presents and recorded in the imperially commissioned *Grand Ceremony of the Eightieth Birthday*.[213] According to Amiot, the emperor ordered that they be placed as "ornaments" in one of his offices.[214] Hongwu's spot-on suggestion confirmed the value of his expertise. In return, he asked that the missionaries share theirs, too.

French Science in Beijing

If Hongwu had developed a "love of the sciences and of French people," then in all of China he could hardly have found a better interlocutor on either topic than Amiot.[215] Together, they studied the history, theory, and practice of late Enlightenment science. They investigated gas-balloon aviation and electrical medicine while performing cutting-edge experiments at the North Church and Hongwu's nearby mansion. Through his friendship with Amiot, Hongwu became "initiated in all the mysteries of physics."[216] And through his friendship with Hongwu, Amiot began to think of physics as a matter of mysteries and initiations. Their conversations also show how both European and Chinese scholars began to develop a truly remarkable idea: that the scientific discoveries of modern European might have been preceded by those of ancient China. Their cross-cultural investigation of natural philosophy thus played a significant though forgotten role in the history of science in both China and Europe alike.

This exceptional story calls into question much about the end of early modern Sino-European scientific exchange. Historians have suggested that the later missionaries were reluctant to share new breakthroughs in natural philosophy because of their commitment to Catholic theology, while local interest began to atrophy along with the Qing regime.[217] To the contrary, the ex-Jesuits were as eager as ever to discuss European science, and real enthusiasm for the subject in China remained. The fact that so little record of continued scientific pursuits survives reflects neither European zealotry nor Chinese xenophobia, but rather the rapidly changing social and political conditions in Beijing. The air had soured around the missionaries, and all their activities were increasingly suspect. At the same time, discoveries like electricity and aviation—then making natural science into public spectacle—were dangerous for the same reasons they were enchanting.[218] Hongwu had already learned the value of discretion, and he impressed it upon the missionaries. They considered publicizing their scientific investigations—and decided not to. The result was that their activities were very little known during their own time and entirely forgotten soon thereafter.

As the missionaries retreated from public life, the center of science in Beijing shifted from organs of the Qing state like the Astronomical Bureau and the Wish-Fulfilling Studio to their own private residences. Research at the North Church would have been impossible without Bertin's continued support. Believing that science remained crucial for the success of the mission, he invited the missionaries to ask for anything they might need to conduct research and never denied a request.[219] He sent custom-built instruments, instructions for how to use them, and supplies for their upkeep.[220] Just as important, he sent reading material. The North Church already possessed one of the best European-language libraries in all of Asia.[221] Bertin kept it updated with recent books on science by the likes of Jérôme Lalonde, William Herschel, and Benjamin Franklin, as well as periodicals such as the *Journal des sçavans* and the *Histoire de l'Académie royale des sciences*.[222] Amiot was proud of "our French books, so good, in such a great number, and in nearly every genre, which are the principal wealth of our library, in this extremity of the world."[223] The missionaries could perform experiments and conduct research in the comfort of their own home. All told, the holdings of the North Church were comparable to those of a decent French research institution.

The sight of it right in the center of Beijing was likely what inspired Hongwu to set up his own laboratory "as best he could in the French style."[224] It began with an air pump. Back in the early 1770s, Benoist had demonstrated an imported machine at the Wish-Fulfilling Studio, but after the emperor lost interest in it, the missionaries repossessed it for the North Church. They then lent it out to a certain Manchu count, but he died a few months later and his widow sent it back. Hongwu was apparently next on the waiting list. By 1784, he had set up the air pump in his own home.[225] Four years later, in 1788, he was still using it to conduct experiments. Seeking to encourage Hongwu, Amiot offered to lend him his own newer model as soon as he himself had finished with it.[226] Amiot promised to further augment Hongwu's laboratory with gifts of a cylindrical mirror and a barometer-thermometer. When the ship that was supposed to deliver them failed to arrive on schedule, Hongwu's disappointment was so great that Amiot vowed never again to mention a shipment from Europe until after it had arrived. Hongwu also found ways to avoid the wait. Upon seeing a new kind of water pump at the North Church, he had one built on specification locally for his own use.[227] In this way, he exploited his friendship with the missionaries to procure some of the most sophisticated equipment available not just in China, but anywhere.

By all reports, Hongwu became an accomplished experimentalist. Amiot was deeply impressed with his mechanical skill. When he showed Hongwu a French illustration of a certain machine, "he immediately understood the entire theory by himself, just by looking at it." Giving Hongwu an Argand oil lamp, he was certain that his friend would find a way to improve upon the French design.[228] Bourgeois confirmed that Hongwu's curiosity about technical machinery was matched by an exceptional talent: "He often asks for European books; looking at the plates and images inside them, he figures out the construction of the machines, their operation and their effects."[229] Hongwu's interest in machines extended to include the theory behind them. "Our *messieurs* are surprised by his penetration," wrote one missionary of his engagement with French science; "I do not know anyone but him in the empire who can do what he can."[230] Because of his unique friendship with the missionaries, there might have been no one else who had such opportunities.

The first piloted human flights took place to great fanfare in Paris in 1783, and balloon mania reached Beijing the very next year. The invention seemed to be most promising, both as proof of the power of European science and of France's preeminence in perfecting it. Amiot basked in Gallic pride: "The rays of all kinds of glory shining from our nation with such brightness before all the eyes of Europe seem to me in a way to reflect on me." Proposing improbable uses for air balloons became a favorite joke. In one letter to the French supercargo in Canton, he declared his desire to fly one down to chat with some fresh French faces.[231] In another letter to Bertin in Paris, he suggested that if the mail ship had already departed for China, "you could send me a little aerostat in which you will put your letter, in directing it toward Beijing, only paying attention to put the address in Chinese."[232] In his reply, Bertin played along: "one of our *aeronauts*" would fly to China, "but 1) he would have to be paid 2) he would have to be certain of returning to Europe."[233]

One day in 1784, Hongwu stopped by Amiot's study and noticed on a table a print depicting the sensational hydrogen-balloon flight performed by the Robert brothers outside the Tuileries Palace. When Amiot told him about the extraordinary achievement, Hongwu was at first incredulous: "Is this not just a pleasing story intended to amuse you?" In reply, Amiot explained that the balloon was lifted by a gas lighter than air. Hongwu immediately grasped the theory—but contested its significance.[234] Like many Chinese scholars of the late imperial period, he believed that much of the new science of Europe had already been known in ancient China. By the eighteenth century, this view had already become an important way to integrate Western learning with Chinese practices.[235] For Hongwu, French balloons were further confirmation. "We find in many fragments of ancient books examples of flying machines," he said. "We treated all this as fable, since we did not believe it possible; but what recently happened in your France proves to us the contrary." Modern French achievements, he proposed, could be understood as evidence of ancient Chinese ones.

FIGURE 2.6 *In Honor of Mr. Charles and Mr. Robert*, Rose Lenoir, 1783. Courtesy of the Bibliothèque Nationale de France.

Amiot took Hongwu's suggestion as a prompt for investigation. He found some evidence for the idea in an unnamed commentary on the *Classic of Documents.* One story recounted how the sage-king Shennong measured the earth with the aid of a spirit that was pulled through the sky by six flying dragons and further determined "that the diameter of the earth was greater from east to west than north to south." Amiot pointed out that "this has only been known in Europe for sixty years," referring presumably to the French geodesic missions of the 1730s.[236] Another story told of how the Yellow Emperor, a legendary magician in Chinese mythology, ascended a mountain in Hunan and flew to heaven on the back of a dragon to escape his impending death. These figures of deep antiquity were said to have been accomplished in "natural history" and "minerology"; was it not possible then that they had produced some gas lighter than air and used it to fill dragon-shaped balloons, thousands of years before the Robert brothers built their *globe aérostatique*? Amiot told Bertin that he still held the French as "the true inventors of the aerostatic machine"; but he might have said something different to Hongwu.[237]

This conversation in the fall of 1784 also marked an early step in Amiot's new approach to natural science. That very season, he wrote a striking letter about the natural philosophy of China, and particularly its cosmology and cosmogony. "Would you believe, *Monsieur*," he asked a French reader, "*that the central fire, the cooling of the planets,* and other similar systems were known to the ancient Chinese?"[238] Amiot had kept up with his reading from Europe: he was referring here to the cosmogony theory of Buffon, published in the *Epochs of Nature* of 1778 and sent to him by Bertin, in which the earth was formed from a piece of the sun that was blasted off by a comet.[239] But his interpretation was taking shape along the lines that had been argued in the abstract by Neo-Confucian philosophers and suggested to him directly by Hongwu: that new European science might be a rediscovery of things that Chinese thinkers had known already in a remote antiquity.

For his part, Hongwu did not doubt that he could learn something new about balloons from modern Europe; to the contrary, he "showed a great desire to have one of the balloons to try." Amiot protested that procuring him one for him would not be easy to arrange. In particular, there was a technical difficulty. The Robert brothers had filled their balloon with hydrogen. They had made it by dissolving iron in sulfuric acid, which Amiot did not know how to obtain. When Hongwu suggested that they try with nitric acid instead, Amiot had to tell him that the substitution would not work. But he placated Hongwu with the hope that the Lazarist missionaries, already on their way to take over the French mission, might have some sulfuric acid with them, or at least know how to make it. Hongwu asked to be informed upon their arrival.[240] Amiot wrote to Bertin, who was thrilled to hear about this opportunity to notch a new scientific win for the French mission. Unsure whether the Lazarists had in fact taken any sulfuric acid, he sent instructions for producing hydrogen without it.[241]

But by the time the Bertin's instructions arrived, Hongwu's enthusiasm had evaporated. When Amiot brought up balloons again in 1787, and again in 1788, Hongwu responded with only a few perfunctory words of caution; "he saw nothing but the danger to which they exposed themselves in flying to frequent an element that nature seems to have forbidden in refusing them wings."[242] Amiot continued to broach the subject from time to time, but Hongwu was "obstinate in not being willing to agree that this art could, in being perfected, become a great utility for men." Hongwu conceded only that the invention might be useful "exclusively for war, since then one has no regard for expenses, difficulties, or dangers."[243] he understood what kinds of knowledge would appeal to the local audience. Though he had wanted to experiment with balloons himself, he thought it imprudent to discuss them more widely. Amiot and Bertin were in the end disappointed to watch the opportunity float away.

The same thing happened with another hot topic in late Enlightenment science: electricity. The North Church had acquired an electrical machine as early as 1764, when Bertin commissioned the physicist Mathurin Jacques Brisson to make one on special order.[244] For many years, it sat in storage collecting dust, much to Bertin's frustration.[245] Finally in 1785, the Lazarist missionaries got one in working order to investigate the theories of electrical medicine that were then enjoying some popularity back in France. They arranged a private demonstration at the North Church and invited a select group of local figures to observe. When the machine was turned on, the Chinese and Manchu guests "began to wonder at the marvel" of the electrical phenomena. Some of them even volunteered to undergo electrical therapy for "conditions of the nerves"—a common malady in the busy capital city.[246] Hongwu was among those in attendance. He had been asking for years about the electrical machines, and when a demonstration finally took place, one missionary reported, "it gave him the greatest pleasure."[247] Another missionary was even more impressed: "Not only does this young prince understand all the theory of this machine; but he even gets it to produce new phenomena." Hongwu requested to borrow an electrical machine just like he had the air pump, but there was not an extra one on hand.[248]

Even so, just as he had with gas balloons, Hongwu advised the missionaries to conceal their wonderful invention, particularly from the emperor.[249] A misstep might have disastrous consequences; as one missionary put it, if a glass globe were to explode in the emperor's face, "all would be lost." Furthermore, demonstrations of electricity fed into a possible perception that the missionaries, with their religion already suspect, especially feared: "It is dangerous that the Chinese might attribute it to magic." Given the image of a strange-looking bearded man in robes with sparks flying all around, the fear may not have been unfounded. This time, the missionaries did not protest. The problem, as they explained it to Bertin, was that they did not understand the principles of the machine well enough to ensure its safety, nor to give a satisfactory account of its mysterious effects.[250] Natural science, which had once brought power and prestige, was becoming instead a source of secrecy and suspicion.

Bertin tried hard to assuage the concerns. He retorted that that "the example of the young prince" indicated that it was still possible that "Europeans and the French especially would have the merit of bringing this discovery to China and of having made the understanding of it easy." He exhorted the missionaries to demonstrate "these sorts of novelties," if not before the emperor, then "at least to the princes of his blood."[251] They should advertise their wonderful science, not hide it: "Instead of making them into a ridiculous secret, one performs experiments in explaining things; to the contrary, nothing can better cure all credulity in magic."[252] To further encourage them, Bertin sent another electrical machine, pointedly including a detailed set of handwritten instructions for how to use it.[253] This attempt, too, came to nothing. The Lazarists continued to tinker with their machine but abandoned electrical medicine demonstrations in favor of "purely physical experiments." Amiot claimed some excitement about the new machine, but it was last seen being held up in transit by one of the old dissenters, who was then causing new problems as the mission's procurator in Canton in 1789. It seems that all public demonstrations of electricity were conclusively discontinued.[254]

Just a few years later in 1793, when a British embassy arrived in Beijing and displayed a host of inventions including an aerostatic balloon and an electrical machine, the Chinese party to the meeting made a show of studied indifference.[255] Most historians agree that the reaction was due less to ignorant arrogance than to calculated politics: the Qianlong emperor understood the value of European science and technology and did not wish to concede anything that could be interpreted by the British as a weakness.[256] By this time, however, neither his knowledge of European science in China, nor indeed his control over it, was entirely firm. He may have understood its potential power, but he had also grown hostile to it, as judged by those who knew him personally, and as a result he had not been told about the experiments that continued within the very walls of the Imperial City. Those who still harbored an interest in scientific exchange were pursuing it strictly in secret. This explains in part why the British believed their interlocutors to be so ignorant in natural science; even if Hongwu had been at the negotiations, he would not have been likely to disabuse them.[257] It also explains why so little record of continued scientific research in Beijing survives. But the lack of documentation does not mean it was without impact.

In China, it is a puzzle that at the turn of the nineteenth century, while some literati remained as interested as ever in European astronomy and mathematics, they do not seem to have sought out contact with European people to discuss it.[258] Take, for example, the *Biographies of mathematicians and astronomers,* composed from about 1797 to 1799 and called by Needham "the nearest approach to a history of Chinese science ever written in China."[259] The book promoted selective aspects of Western studies while arguing as others had before that their origins were ultimately Chinese.[260] Its editor, the scholar and statesman Ruan Yuan, does not appear to have spoken with a single missionary—but perhaps he did so indirectly. In 1787, having passed the provincial-level civil-service examinations, Ruan Yuan arrived in Beijing, where he was taken in by a literati official who ran in Hongwu's circle.[261] In 1792, Hongwu invited six friends to a retreat at Wanshou Temple in the Beijing suburbs to paint the landscape, write poetry, and get drunk.[262] Two of the attendees were artists who were still working with the French painters at the Wish-Fulfilling Studio. Another was Ruan Yuan.[263] And after Hongwu died in 1811, it was Ruan Yuan who arranged the publication of his collected poems.[264] The Manchu prince must have made some impression on the young Chinese scholar.

In Europe, the conversations between Amiot and Hongwu set the stage for new engagement with Chinese natural philosophy. Back in the days of the Society of Jesus, pressures from both inside and outside the mission had promoted orthodoxies both Christian and Confucian. The Jesuits' research had been limited by corporate discipline and unity of purpose, and their Chinese collaborators' by the social expectations of the literatus scholar. Now, with Hongwu's help, Amiot was able to expand his research well beyond the traditional topics of collaboration. The other missionaries were aware that Hongwu might have had such an influence. Once, Bourgeois reported, "the prince gave me a book of secrets for making silver; but I did not need it. I know what I should think about that book and its secrets."[265] It was rejected by Bourgeois almost reflexively, as it would have been by most of the missionaries who had come before. Alchemy was, however, a topic that Hongwu's expertise in Daoism, art, and science would have prepared him to discuss.[266] This subject, rather than "natural history" or "mineralogy," was what the ancient sages Shennong and the Yellow Emperor were more accurately reputed to have understood. And it was just this kind of thing that Amiot was now beginning to search for in the Chinese record.

The Ex-Jesuit Mission in China

As a Jesuit, Amiot became a minor public figure of the global Enlightenment. In the land he had left behind, he had become known to artists and intellectuals such as Rameau and Voltaire, as well as generals, politicians, and monarchs including Frederick the Great and two French kings. Meanwhile, in his adopted home, he had learned the languages and customs, found work at the Qing court, and won the respect of the other missionaries. Amiot spent two decades contributing to a long-established tradition of Sino-European exchange. As a Jesuit scholar of China, he was part of a venerable institution that had secured the posterity of its illustrious members for nearly two centuries, and he had seemed to be on track to join their ranks. Amiot knew who he was and who he wanted to be.

After the suppression of the Society of Jesus, his identities began to collapse. Fundamentally, he was no longer a Jesuit. As a scholar of China, he found that his field was becoming irrelevant in Europe. And as an expert on Western studies, the door to public and intellectual life was closing in China. He tried several ways to address the situation. His Catholic solution, the Diocese of Mukden, came to nothing. His French solution, Bourbon control of the mission, was stymied. His Chinese solution, scientific proselytization, proved inadvisable. Amiot was still able to cobble together what he needed—financial support from France, religious affiliation from Church institutions, and local influence from the Qing nobility, among others—but he now lacked an outlet for his broader ambitions, which was an inevitable disappointment for a man who had once known fame. An orphan of the Enlightenment, Amiot was searching for a new home.

As Amiot began to cultivate a new scholarly identity, he drew upon the few relationships that could still flourish under these changed circumstances. The cross-cultural exchange of science in Beijing continued, though in a way that veered far from the course established under the missionary-mathematicians of the previous century. By the 1780s, a Manchu prince, Hongwu, had become the most enthusiastic local student of Western studies. As a failed soldier and aspiring intellectual, he brought with him new perspectives that had not been common among the Confucian scholar-bureaucrats who had once patronized the Jesuit missionaries.

His scientific collaborations with Amiot went beyond earlier literati studies of European mathematical and astronomical techniques to embrace the newest Enlightenment science. At the same time, they invoked previously overlooked aspects of the Chinese tradition in attempts to explain it.

The outcome of the serial crises brought on by the suppression of the Society of Jesus was thus a new research agenda. Changing sources of institutional and even emotional support made new kinds of investigations both possible and necessary to pursue. Frameworks that had once governed Jesuit scholarship—the compatibility of Confucianism with Catholicism and the superiority of European over Chinese natural science chief among them—were no longer very useful. For this reason, it became possible to challenge assumptions that had long gone unquestioned. Amiot's collaboration with Hongwu cast Enlightenment discoveries like electrical medicine and gas balloons as powerful, mysterious, dangerous, and secret. He would come to understand Chinese natural knowledge in just the same way.

✸ 3 ✸

The Origins of Esotericism

Ancient Wisdom in China

During the 1770s and 1780s, Chinese knowledge entered the French Enlightenment to inform an evolving debate about progress. At this time, European historiography was in a state of fracture and flux.[1] To play on a theme by Paul Hazard, Frenchmen did not go to sleep one day thinking like Voltaire and wake up the next morning thinking like Condorcet.[2] While some historians rejected Chinese culture because it seemed ancient, others embraced it for just the same reason. Many among the philosophes were coming to believe that knowledge would be perfected in a remote and indeterminate future. At the same time, other savants searched for it in an equally remote and indeterminate past. Orphans of the Enlightenment retained a once common but increasingly eccentric belief in ancient wisdom. In search of the origins of the arts and sciences, they developed core aspects of modern esotericism, from tarot-card fortune-telling to the legend of Atlantis. Their investigations in turn informed even the most future-oriented progress theories of the philosophes from Voltaire through Condorcet. In fact, modern ideas of progress and esotericism were empirically and conceptually intertwined: flip sides of the same coin, minted in part with Chinese evidence.

Through the early eighteenth century, many if not most Europeans who had thought about the matter believed that there existed a body of perfect knowledge that was commonly known to humanity in the first ages of the world and subsequently lost. In a way, the idea was an implication of the historical interpretation of Genesis: God had communicated His divine word directly with the antediluvian people, and it had been preserved by Noah and his descendants who repopulated the world after the great Flood. Historians of esotericism and occult science have long been attentive to such accounts, variously termed *prisca theologia*, or ancient theology; *prisca scientia*, or ancient knowledge; and *prisca sapientia*, or ancient wisdom.[3] More recently, historians of scholarship have critiqued these terms as vague or misleading. On the one hand, ideas that the ancients had possessed valuable knowledge unknown to the moderns were so diverse as to make it difficult to generalize about them. On the other hand, this protean quality strengthened rather than diminished their importance. In fact, a belief in ancient wisdom of some kind or another was not restricted to Italian Hermeticists of the sixteenth century and English Neoplatonists of the seventeenth but was remarkably widespread and persistent across early modern European intellectual culture.[4]

The search for ancient wisdom had always been particularly linked with orientalism.[5] Since ancient wisdom was based on a biblical pedigree, the most logical place to look for it was in the Holy Land and other places where those who had supposedly communicated with God had their dealings. Athanasius Kircher sought it in the apocryphal writings of the Egyptian Hermes Trismegistus, while John Spencer looked rather to the ancient texts of the Hebrew Bible.[6] Over the course of the eighteenth century, the search shifted gradually eastward, without losing its biblical connection. One of the founding members of the Asiatic Society, for example, speculated that Moses "stole" his description of the Garden of Eden from the Hindu Brahmins. Gradually, ancient wisdom grew more remote from the biblical story, but still its Eastern connection remained. As late as 1789, it was not considered totally implausible when another respected scholar suggested that ancient Indian sages had theorized about Newtonian gravitation, constructed optical telescopes, and discovered the moons of Saturn.[7]

And yet, during the Enlightenment, the search for ancient wisdom was beginning to seem somewhat quixotic. It was not essentially at odds with the new science; after all, Newton himself had believed in it to some degree.[8] It was, rather, a new view of history that more seriously called ancient wisdom into question. The Old Testament had provided a definitive account of humanity's ultimate origins, and the New Testament had laid out its final destination. Once that story was brought into dispute, it was no longer clear where human society had come from, or where it was going. It is hardly an original observation that the new ideas of progress that flourished during this time were in some sense a development of, or a replacement for, Christian millenarianism.[9] Yet there were significant differences that secularization introduced. Renaissance scholars had invoked ancient wisdom in the context of a classicizing historical idea of world cycles or a lost golden age, suggesting a view of history with no fixed direction. Enlightenment philosophes normally placed their golden age in a distant future, not a mythic past. But the transformation from one metahistorical narrative to another did not take place overnight.[10]

The idea of ancient wisdom persisted into the eighteenth century in a new form: as a secular explanation for the origins of the sciences. The Huguenot minister Antoine Court de Gébelin and the astronomer Jean-Sylvain Bailly were members of the social and intellectual world of the philosophes. Their research was motivated by a dominant question of the high Enlightenment: how had human culture and society begun? The answer they thought they found traced it all the way back to a single primordial people whose vestiges were preserved among the known traditions of the ancient world as a testament to their nearly forgotten grandeur. This was a solution with which most philosophes, including Voltaire and Condorcet, could not agree. But their skepticism about the story did not prevent Court de Gébelin and Bailly from trying to persuade them to embrace it. Orphaned by their commitment to ancient wisdom, some scholars struggled to remain members of the Enlightenment family.

United by their expertise on topics that were not obviously relevant to progress thinking, historians of the ancient world and specialists on China found each other. They associated the supposed wisdom of their own past with the real learning of ancient China, coming to see themselves as contributing to the same research project. Many of the missionaries, directly informed by their Renaissance predecessors, had already been searching for ancient wisdom in China. The reconfiguration of post-Jesuit networks encouraged them to look even harder. In Paris, Court de Gébelin and Bailly believed that the apogee of human achievement had existed in the past, not the future—and everyone agreed that China was the place where the past was best preserved. In Beijing, Amiot identified an opportunity. With the Society of Jesus gone and Qing support falling fast, he reinvented himself as an intelligencer and a go-between, aligning himself with French scholars because their work gave him an opportunity to valorize his own life's labor. Putting European research into conversation with Chinese scholarship, Amiot excavated evidence in confirmation of their theories that was to be deployed in French debates.

In this way, genuine though misunderstood Chinese evidence contributed to foundational components of modern esotericism. Where China had once informed the Enlightenment as a model for what Europe might triumphantly progress toward, it achieved new importance as a relic of what Europe had tragically left behind. Bailly and Court de Gébelin drew from conversations with Amiot, as well as the primary sources that he made available, in order to illustrate the supposedly advanced condition of human knowledge in ancient times. A Song-period stone inscription played a part in Court de Gébelin's invention of tarot-card fortune-telling, and a Qing travelogue provided new evidence in Bailly's search for the lost civilization of Atlantis. Amiot, who knew China too well, rejected some of their more venturesome ideas. But at the same time, he became convinced that his new interlocutors were essentially correct: that there had existed a lost body of advanced scientific knowledge that had been known to all the peoples of the distant past and might yet be recovered. It now became clear to him what China was good for in the late Enlightenment: it was a repository of ancient wisdom.

European Problems of Chinese History

If almost all eighteenth-century Europeans could agree on one thing about China, it was this: China was very old. In fact, this proposition remained stable and effectively unquestioned across the entire early modern period. In the words of an influential critic of China writing in 1773: "All that can be truthfully said is that the Chinese are an extremely ancient people."[11] This consensus in Europe, once suggestive of connections between Chinese culture and the Bible, became a source of increasingly secular concerns as the Enlightenment developed. In particular, European thinkers became interested in two questions about China's exceptional history: When had Chinese culture first appeared, and where had it come from? These two questions were always related, because if Chinese history was as ancient as some maintained, that might make it the best existing account of the origins of all the arts and sciences.

The 1770s stands out as a major turning point in European studies of Chinese history. The related setbacks of the suppression of the Society of Jesus and the philosophes' renunciation of China sparked renewed activity in what had become an almost dormant field. Two major new translations of Chinese histories became available: the *Chou-king*, based on the Confucian *Classic of Documents*, composed mostly before the common era, and the *General History of China*, based on a twelfth-century edition of the tenth-century text *Comprehensive Mirror for the Aid of Government*.[12] Both these French-language works were in fact publications of older missionary translations that had been edited under the supervision of the sinologue scholars attached to the Collège royal in the 1770s. For about a century thereafter, they remained the most thorough original Chinese histories available in any European language.[13] Meanwhile, two new book-length essays by missionary authors also appeared, printed in the *Mémoires concernant les Chinois*: Cibot's "Essay on Chinese Antiquity" and Amiot's "Chinese Antiquity, Attested to by Monuments," which were more syncretic accounts.[14] All four of these publications purported to tell the earliest history of China, and to use original Chinese sources to do so.

Although the questions asked in each work asked were broadly the same, the answers given in them were quite different. Reconstructions of ancient Chinese chronology varied by thousands of years, and European authors could not even agree on what continent the first people who inhabited the region had come from. Disagreement was significant enough that it was apparent to fairly casual readers.[15] There was even second-order disagreement about the disagreement: that is, whether it would buttress or impugn the credibility of Chinese history. De Guignes, the editor of the *Chou-king*, suggested that the conflicting accounts of the two missionary authors were a discredit to both; it would have been better for them to have presented a united front. On the other hand, the missionaries' patron, Bertin, thought that the discrepancy between them not only proved their good faith through lack of collusion, but was itself a sign of development in the field; open and public dispute was "so much the better for the truth, and by consequence for the good."[16] Neither the disagreements nor the meta-disagreements were ever really resolved.

Critical judgments on textual authority still had to be made. Credibility was always attached to the words of "the Chinese themselves." This standard reflected the immediately obvious fact that China had a written culture that was in many ways comparable to Europe's own.[17] As Voltaire put it: "Dare we speak of the Chinese without referring to their own annals?"[18] Bertin, who had some sense of what worked and what did not for French audiences, believed that translations, extracts, and anything based on a "Chinese original" were on the whole more successful than new compositions that purported to analyze them: "It is the Chinese themselves one wants to see," he insisted. Furthermore, originals were less open to scrutiny and criticism. For these reasons, he preferred Cibot's translation of the Four Books and Amiot's translation of the *Art of War* to their essays on Chinese history.[19] But even if European audiences evaluated histories of China as authoritative in proportion to how closely they represented original Chinese sources, there were still many different ways that the condition might be met.

In principle, the missionaries agreed on the primary importance of original Chinese sources. Amiot and Cibot conceded that their works were not pure translations, while at the same time insisting that they still represented the Chinese as they really were, since their essays were at core summaries of ancient texts.[20] Sometimes they rejected the interpretations of more modern scholars, and sometimes they embraced them; but they usually made plain their engagement with the robust indigenous critical apparatus.[21] They also consulted living Chinese people in the course of their studies. Cibot presented his essay on early Chinese history as primarily the work of the Chinese priest Aloys Ko, though most people at the time thought he had written it himself, and later historians have seen it as the product of collaboration.[22] By contrast, Amiot claimed full credit for his essay, though he too likely relied on the assistance of a Chinese Christian, his secretary Jacob Yang, to compose it. In sum, the missionaries were in conversation with living Chinese traditions and living Chinese people, which gave them a little extra leeway to depart from ancient Chinese texts.

By contrast, the sinologue scholars on the surface strove to more directly satisfy Bertin's demands. Both of them went so far as to include a transliteration of the original Chinese title in their own French publications. Deshauterayes's introductory essay to the *General History of China*, "translated from the Tong-kien-kang-mou," praised the missionary who had composed it, Joseph-Anne-Marie de Moyriac Mailla, as "a simple translator who narrates the facts as he finds them . . . in a word, he had no other aim than to represent the Chinese originals, which he submits to the judgment of his readers."[23] But in fact, the story of the text was anything but "simple." It seems that the French was not even translated from the Chinese at all, but from a more recent Manchu edition.[24] Once in France, it was further altered by both its "editor" and its "publisher."[25] Likewise, the title page of de Guignes's *Chou-king* announced that this "sacred book of the Chinese" had been "translated and enriched with notes" by the missionary Antoine Gaubil and "reviewed and corrected according to the Chinese text" by de Guignes.[26] But these revisions and corrections were not seen as such by everyone: it seems that the changes were significant enough that someone challenged in court de Guignes's publication rights to Gaubil's manuscript.[27] In sum, the *General History of China* was not the *Outline and Details of the Comprehensive Mirror for the Aid of Government*, and the *Chou-king* was not the *Classic of Documents*. No European-language history of China published in the 1770s was anything like a straightforward translation from an original Chinese source.

The complexity of translating, editing, and publishing works on Chinese history presented many challenges but also significant opportunities for commentators. Interpretation was an unavoidable and therefore justifiable necessity. Defending himself against Bertin's complaint that he wrote too much analysis and not enough translation, Amiot argued that Chinese histories were fundamentally inaccessible to European readers on account of their annalistic style. In a review of the *General History of China*, he pointed out that though it departed in places from the Chinese original, the changes had made the book easier to read for its target audience, and were fully excusable since the end result "pretty much contained what is most essential."[28] While this kind of approach to Chinese histories led some Europeans to take them wildly out of context, it also made it possible for their editors to streamline, or alternatively manipulate, them into giving clearer answers to the questions that European readers wanted to ask.

Those who took the time to edit or paraphrase Chinese histories generally thought that the originals were reliable. For Amiot, Chinese sources promised "the most ancient, as the most complete of the histories of the world."[29] Deshauterayes mostly agreed: "The authentic history of China is without question, of all the profane histories, that which gives us the most knowledge and certitude of the antiquity of time."[30] Cibot had certain reservations about the oldest Chinese records, but he still held them to be generally reliable, since no scholars in the world, he thought, were more scrupulous than the literati.[31] As far as trusting the Chinese historical record went, the one important exception among the experts was de Guignes. Well aware of his iconoclasm, he lambasted "the sentiment of all the missionaries and many other Savants" who placed too much faith in the antiquity of Chinese texts.[32] But even this criticism did not mean the Chinese texts were unimportant for de Guignes, since "despite these problems, these annals form a precious corpus for history"—just not for the history of ancient China, he would go on argue.[33]

The main debate, then, was not really whether Chinese history was trustworthy, but rather how far back in time it ceased to be. Following a partition of the classical past, ancient China was often divided into three eras: an age that was properly historical, an age of doubt or uncertainty, and an age of myth or fable.[34] The question was where to draw the lines. The most common position was that the records subsequent to the founding of the Xia near the end of the third millennium BCE corresponded to real events. Some missionaries argued that Chinese history could be pushed back still further to the end of the period known in Chinese historiography as the "Three Sovereigns and Five Emperors." Amiot gave the start of the reign of the emperor Huangdi in the twenty-seventh century BCE as the point where Chinese records became fully reliable.[35] According to Cibot, this position was popular in China because it reflected the opinion of the great Han historian Sima Qian.[36] Since it fit just barely within the chronology of the Septuagint, this chronology was broadly acceptable to Europeans, too. Going back much further than that, though, one encountered problems. The annals recorded few precise dates from earlier times and no eclipses that could be used to corroborate them. Even the missionaries therefore held them as accretions of myths and fables that the Chinese themselves did not fully believe.[37]

Nevertheless, it was often maintained that myths and fables should be studied anyway because of the deeper truths they might conceal. This interpretive strategy, called Euhemerism, had long been applied to the classical and Middle Eastern past.[38] Scholars in Europe believed that Chinese legends could clarify facts of history.[39] Missionaries in China encouraged the approach: as Amiot explained, "we strip them of all they contain that is fabulous, we explain what they contain that is allegorical, and we will be able approach the truth."[40] Some went so far as to discuss Euhemerism with Chinese interlocutors. In an interesting case of 1771, the Qianlong emperor asked Michel Benoist how the events of Genesis could be authentic if the Classics did not record them. Benoist replied that there doubtless had existed Chinese accounts of the creation, but they had been destroyed in the great book burning of the emperor Qin Shihuang. Efforts to recover them from peoples' memories in later years resulted in omissions and interpolations; nevertheless, "among these fables, we recognize facts conforming to the truth."[41] Perhaps the Qianlong emperor felt the same way about the Bible. But as European confidence in the historicity of the Bible too had begun to waver, the historicity of the Chinese record took on a new significance.

Somewhere in this morass of history and mythology was supposed to lie the answer to a distinctively Enlightenment question: where had the Chinese come from? Controversies surrounding this question were fierce, and recent historians have demonstrated their importance across the Enlightenment, from stadial theories of development to the comparative study of religion.[42] By the 1770s, three basic positions on the issue had been proposed. All of them assumed monogenism—the belief that all human beings shared a single origin—and each developed in conversation with the others. The first, held by virtually all the missionaries up until that point, traced the first Chinese to the immediate descendants of Noah. The second, associated during this period primarily with de Guignes, considered them to be a colony of ancient Egyptians. The third, which emerged essentially in response to the Sino-Egyptian theory, declared them to be an offshoot of the ancient "Tartars" or "Scyths" who had inhabited the lands of Central Asia. The Sino-Egyptian theory was long the most interesting to historians, probably because on the surface it appears to be the most bizarre. The thing to note, though, is that the other stories were just as fantastical; no one in eighteenth-century Europe seems to have considered a fully indigenous etiology.

The oldest European explanation of the origins of the Chinese traced them all the way back to the immediate descendants of Noah and the repopulation of the world in the first centuries after the Deluge.[43] This account was the most obvious in the context of early modern historiography, and it made the most sense in the context of Jesuit accommodationism. It explained the traces of monotheism that had been detected in the ancient religion of China and licensed the study of the Confucian Classics within a Christian paradigm. For example, the *Classic of Documents* referred to entities called *Tian*, "Heaven," and *Shangdi*, the "Highest Emperor," which seemed like the Christian God. If the ancestors of the Chinese had come directly from the Middle East, such traces of ancient monotheism were to be expected.[44] This position was not, strictly speaking, exclusive of other views; for example, the descendants of Noah might have arrived in China via a short stopover in Central Asia.[45] But in all its variations it granted a biblical pedigree to the people of ancient China and their purported beliefs.

There long existed one main alternative to the Noachic theory: that the first Chinese were rather a settler colony of ancient Egyptians. This argument had been put forth in the seventeenth century by Athanasius Kircher, whose works continued to be widely read.[46] But the theory had been essentially put to rest by the 1750s, when it was revived by de Guignes.[47] Originally, de Guignes had believed along with the rest of his contemporaries that the Chinese came directly from Mesopotamia; he changed his thinking rather suddenly, perhaps in response to the weaponization of Chinese history against the Bible by the likes of Voltaire.[48] His revised opinion was that all Chinese records of the times prior to about 1000 BCE were spurious. They referred rather to events that had taken place in Egypt, if anywhere at all. In his metaphor, it was as though the colonists of Quebec had appended the history of France to the beginning of their annals and labeled them as Canadian.[49] De Guignes deployed the theory to explain many remarked-upon commonalities in the two sets of ancient beliefs, from number mysticism to five-element cosmologies.[50] Most important were their systems of writing, which were supposed to have developed from a common ideographic origin. In the Middle East, it had evolved into the Phoenician script; in China, it had morphed into the radicals of which Chinese characters are composed.[51] A big payoff of the Sino-Egyptian theory, then, was that it also promised a key to the still-undeciphered Egyptian hieroglyphs—that is, the much better-known Chinese characters.[52]

De Guignes met with heavy resistance almost immediately and from all sides. If the missionaries thought Chinese records confirmed biblical history, while the philosophes thought they might provide a replacement for it, the Sino-Egyptian theory satisfied neither party. Amiot wrote a strongly worded letter on behalf of all the French Jesuits in Beijing. The gist of his response was that the ancient Chinese had been monotheistic, whereas the Egyptians had always been idolaters; since it was unlikely that a single people would abandon monotheism only to rediscover it, no genealogical connection was possible.[53] Voltaire was hardly more forgiving. He skewered de Guignes's speculations: who was he to say that the Chinese came from Egypt, when the Chinese themselves said no such thing? "The Chinese no more descended from a colony of Egypt than from a colony of Lower Brittany," he quipped.[54]

As the Sino-Egyptian thesis fell quickly in and then out of fashion, a third explanation more in line with the spirit of the age emerged. In his anti-Chinese diatribe of 1773, the Dutch scholar Cornelius de Pauw gave them a more proximate origin in Central Asia. He argued that their customs, including the sacrificial religion of "Scythism" and their writing with "runes," all matched the Greek and Roman descriptions of the peoples of Central Asia, who must therefore have been their predecessors. For de Pauw, the Scythian theory served a dual purpose. First, it condemned the Chinese by robbing them of their historical link to the wellspring of culture in the Middle East, associating them instead with the uncivilized inhabitants of contemporary Central Asia, whom he considered to be "coarse, unquiet, ambulatory or nomadic."[55] Second, de Pauw fancied himself a philosophe, as the title of his works made clear, and his was a more secular replacement for the other theories, infused as they were with broadly Christian and specifically Hermetic themes.

This message, too, proved satisfactory to no one, not least due to its messenger's polemicism. De Pauw framed the book overtly as a refutation of de Guignes, who he claimed was so ignorant about recent discoveries in Egyptology that "one must take all he wrote about this matter as a simple game of imagination."[56] Nor did he spare the missionaries from invective: "As for the French Jesuits of Beijing, the fragments that they send from time to their European correspondents are pieces of zero importance."[57], The indiscriminate barrage united all the China specialists of the day to unanimously dismiss de Pauw as doctrinaire and ill-informed. Cibot and Amiot each wrote their own hundred-plus-page refutation.[58] And even for many of the philosophes, de Pauw took an increasingly mainstream sinophobia a step too far. According to Voltaire, de Pauw's whole premise was misguided; the Chinese were a peace-loving people, unlike the warlike Tartars, "who knew only brigandage." The "moeurs, language, usages, religion, and government" of China could hardly be more different from those of Central Asia. Their origins therefore must have been somewhere else.[59]

By 1776, then, there were three major hypotheses for the origins of the Chinese—the sinophilic Noachic theory of the Jesuits, the sinophobic Scythian theory of de Pauw, and the idiosyncratic Egyptian theory of de

Guignes—and the Enlightenment's spokesman for China, Voltaire, had rejected all of them. But at around the same time, Voltaire was also joining a conversation that would radically reframe the issue over the coming decade. In 1778, the aged philosophe was formally inducted into the Masonic *Loge des Neuf Soeurs* in a ceremony that took place at the site of the old Jesuit novitiate in the Latin quarter. Presiding over the initiation was the savant antiquarian Court de Gébelin.[60] He had recently been working toward a new conception of the ancient past, one that suggested an alternative answer to the question of Chinese antiquity and origins, while taking the first steps to address Voltaire's original concern that Chinese people themselves should be asked about it.

Antoine Court de Gébelin and the Primeval World

Antoine Court de Gébelin was born in Nîmes, probably in 1725. Not long after, his father, a famous Protestant reformer, got into trouble for leading an underground network of Southern French Huguenots, and the family sought refuge in Switzerland. Court de Gébelin spent most of his first four decades in exile there, teaching oriental languages at Lausanne and eventually taking over his father's work. In 1762, the scandalous torture and execution of the Protestant Jean Calas in Toulouse drew him into pamphleteering. Filled with newfound zeal, he returned to the country of his birth and soon made his way to Paris. He began several projects of Protestant apologetics while on a pension from the synod of Languedoc, but when they ended in failure, the financial support dried up. By the early 1770s, Court de Gébelin had left behind his career in religious advocacy. He shifted his attention to scholarship and began work on his magnum opus.[61]

The eight-volume *Monde primitif*—fully titled "The Primeval World, Analyzed and Compared with the Modern World"—was a diffuse collection of essays inspired by Protestant, Masonic, and physiocrat themes.[62] Its historical vision was antiquarian, and its attitude was anti-philosophe. Its method was comparative linguistics, guided by the interpretive principle of Euhemerism: that true history underlies the myths of antiquity, but is shrouded behind a veil of allegory. Its overarching aim was to uncover "the antique Edifice that was the cradle of the human species and its development during the first centuries of the World." Indeed, the project was explicitly global from the outset: Court de Gébelin maintained that other savants had failed because "each was attached to one part of these materials disseminated upon our globe, and each regarded the portion that he was partial to as a kind of whole, which really formed a part of the

FIGURE 3.1 *Antoine Court de Gébelin*, François Huot, 1784. Courtesy of the Carnavalet Museum.

total Edifice."[63] Putting the materials from everywhere all together, the *Monde primitif* revealed an ancient people who had achieved unimaginable heights in the arts and sciences. Though they had long since disappeared, their traces remained in all the world's oldest traditions.

Court de Gébelin did not deny the possibility of the "advancement and progress of human Knowledge" per se; to the contrary, he believed that his own work contributed to it.[64] What he denied, rather, was that progress was unidirectional or metaphysically guaranteed.[65] He questioned the unique preeminence of his own times and the likely improvement of future ones. Advances were real, but not unprecedented. The ancients were not inferior to the moderns, nor even all that different from them: "The more we search through Antiquity, the more we find there numerous and surprising proofs that our most precious and rarest discoveries are nothing but a return to this Antiquity itself so surprising."[66] There was in the end no such thing as discovery; only recovery.

Court de Gébelin should nevertheless be understood as a full participant in the French Enlightenment. His most recent biographer has called the *Monde primitif* a "supplement" to the *Encyclopédie*; not because it contributed to the marquee project of the philosophes, but because it embraced the same questions, concepts, terminologies, and techniques.[67] Debts to ideas ranging from the empiricism of John Locke through to the sensationalism of the abbé de Condillac were evident throughout.[68] The general plan of the work, the first volume of which was published in 1773, began with a commitment to secular methods: there were only two foundations for all proof, "*Reason* and *Authority*."[69] A helpful acknowledgments section of those who had given comments or critiques on the manuscript prior to publication named many Enlightenment luminaries. Notably included were physiocrats such as Pierre Samuel Dupont de Nemours and Victor de Riqueti, marquis de Mirabeau, as well as establishment figures of academic erudition such as Charles de Brosses and Jean-Jacques Barthélemy.[70] Notably absent were all of the Encyclopedists; in fact, the listing reads rather like a roll call of ancien régime anti-philosophes.[71] What the diverse inclusions and exclusions seem to suggest is an active ambition to exclude just one section of the world of letters from a conversation in which they also had something to say.

And in the social world of the late Enlightenment, Court de Gébelin found himself very much at home. He was a particular fixture of the secret societies that dominated upper-crust Parisian life. Feelings of outsidership and yearning for ecumenicism had set him on the path toward Freemasonry, and his background in clandestine activities had taught him how to navigate it.[72] In 1778, Court de Gébelin became the secretary of the famous Loge des Neuf Soeurs, whose members in addition to Voltaire included Jérôme Lalande, Benjamin Franklin, and the abbé Emmanuel Joseph Sieyès.[73] He broke ranks with the Neuf Soeurs a few years later, but not because he had given up on secret societies. In 1780, he and a few apostates formed a splinter organization, called the Société Appolonienne, later the

Musée de Paris, which met down the street at the Rue Dauphine.[74] Court de Gébelin gave regular lectures there as the "Honorary President."[75]

It was likely through Masonic lore that Court de Gébelin developed a taste for esotericism. He promoted it in his new society, where one member presented a treatise arguing for the Egyptian origins of the arts and sciences, and another previewed an imagined utopia replete with dreams and Cabalism.[76] By the early 1780s, he was in regular contact with Louis-Claude de Saint-Martin, author of the frankly occult book *Des erreurs et de la vérité*.[77] Contemporaries regarded him as something of an authority on the subject.[78] In 1783, he became far more widely known as a staunch supporter of Franz Mesmer. As he sampled the many flavors of late Enlightenment occultism, he continued to incorporate new ingredients into the *Monde primitif*. The updates brought still greater success. The series went through at least six editions and earned over twelve hundred subscribers.[79] Their dedication to its author extended even beyond his death, when several of them began an informal subscription pool so that publication of the final volumes could continue.[80]

Taking into consideration Court de Gébelin's ties to physiocrats and *érudits* as well as his global vision of ancient history, it was only a matter of time before his interest turned toward China. He and Bertin had formed a relationship early on when the minister-savant was advising the savant-minister on the treatment of French Protestants. The two found they shared a common interest in history and antiquarianism.[81] Bertin particularly admired the passion and determination of the author of the *Monde primitif*. Seeking to aid the missionaries, he forwarded them manuscripts by the savant, sometimes years before they were published.[82] For his part, Court de Gébelin seems to have worked as a kind of sales agent for Bertin's *Mémoires concernant les Chinois*, procuring copies of it on behalf of his many friends.[83] He also had independent connections with the other scholars who associated with the publication. Bignon, addressee of many of the letters published in it and librarian at the Bibliothèque du roi, was his confrere at the Loge des Neuf Soeurs; while Bréquigny, its editor in chief, was his confrere at the Académie des inscriptions and among those whom he thanked in his own book's acknowledgments.[84]

Court de Gébelin drew upon his personal resources to obtain privileged access to the literary correspondence with Beijing. In particular, he believed that the missionaries' works promised to fill gaps in the historical record between biblical and classical sources on the development of language and confirm his thesis that alphabetic writing originated from pictorial forms. In 1774, he wrote to the missionary Cibot to inquire about two essays he had read "page by page," taking notes and extracts as he went along. One was about the origins of Chinese characters, and the other on parallels between a Song historical text and the Book of Esther. He wrote that many of his friends in Europe would be "delighted to know about these surprising reports." But the savant did have one revealing reprimand for the missionary: he was "quite upset that he did not have more courage" to be straightforward when Chinese sources seemed to contradict revelation; "one must never be afraid to tell the truth," he scolded. For Court de Gébelin, history and revelation were complementary, so scholarship could "shed light on origins and fortify the faith" at the same time.[85]

When issues of the *Monde primitif* reached Beijing, they were deposited in the library of the North Church. In 1777, Cibot and Amiot both had a look at the latest. Court de Gébelin's admiration for Cibot did not prove reciprocal. According to the missionary, for those Europeans who did not know Chinese, it was "a chimera to pretend entry into the domains of our republic of letters." Writing to Bertin, he suggested that "it will be giving the counsel of a friend, my lord, to tell him to give up on China in his work."[86] Amiot, on the other hand, was impressed with the scope and ambition of the project. He professed himself taken by a "great desire to be in correspondence" with its author, in the hope that he might help penetrate "the thickness of the shadows that hide from us the different routes that lead to knowledge of the *Monde primitif*." As proof of his sincerity, he compiled a package of materials on ancient China to be forwarded to the illustrious savant. One of the items he included was an album of a Chinese character written in a hundred different scripts: *shou*, meaning "long life," which Amiot wished to Court de Gébelin "so that he might accomplish the immense task that he has taken upon himself." The idea was that China might help him to do so.[87]

The starring item in the package was an ink rubbing of a stone monument called the Stele of King Yu the Great.[88] Amiot explained this artifact in an accompanying letter to Bertin. According to legend, the stele had been inscribed around 3000 BCE by the legendary sage-king Yu the Great and placed high in the Hengshan Mountains of Hunan.[89] Its mysterious inscription told the story of how King Yu had controlled a catastrophic flood and restored the people to prosperity before dividing China into nine provinces and founding the first dynasty, the Xia. Along with an ink rubbing taken from the stone monument, Amiot included a version of the inscription transcribed into modern Chinese characters as well as his own French translation of it. The stele, he believed, would be an invaluable resource for all those interested in mythology and the origins of language, but for Court de Gébelin in particular. Perhaps he might use the rubbing of the inscription to decipher the Egyptian hieroglyphs—as de Guignes's

FIGURE 3.2 Ink rubbing of the Stele of King Yu, Mao Huijian, 1666.

theory had promised but failed to do—or, better still, to learn more about the "primitive language" of the *Monde primitif*.[90]

Amiot must have been impressed indeed by the abilities of its author, for the Stele of King Yu had proven difficult even for contemporary Chinese interpreters to make sense of. The text was composed in an archaic style, with many rare and difficult characters, which furthermore were inscribed in an arcane script called *kedou wen*, or tadpole script, for its strokes' resemblance to tadpoles.[91] Several exempla of the stele from which the rubbing was taken existed in Qing times, each varying to some degree from a supposed ancient original. The inscription was therefore a subject of open debate for generations of literati scholars.[92] Only recently had real progress been made. The story of King Yu the Great, found originally in the "Tribute of Yu" chapter of the *Classic of Documents*, had become a fecund topic of research in the seventeenth century.[93] Originally, scholars had studied it to reconstruct the geography of ancient China, but with sustained contact with Europe the subject gradually expanded to include other subjects in cartography and cosmography more generally, and Renaissance world maps in particular.[94] Research on King Yu the Great had thus already become a part of Western studies in China, so it might have seemed like a natural choice for furthering Chinese studies in Europe.

At the time, a leading expert on ancient steles in Beijing was the scholar-official Wang Chang, an exponent of the evidential-studies movement in philology and textual criticism.[95] Wang was particularly noted for his work in epigraphy, and in this connection he had become something of a Stele of King Yu enthusiast. He traveled across the empire to collect rubbings from four different exempla at Changsha, Kunming, Chengdu, and Xi'an. His discoveries there led him to question the inscription's antiquity. How could the elaborate tadpole script have been developed so soon after the invention of writing, when people had recently been keeping records with only knotted cords? And if it was really so ancient, why did the legendary scholars of the past have nothing more to say about it? Although the existence of a certain stele related to King Yu the Great was mentioned during the Eastern Han, no record of the actual text predated the Song; "now still there is no definite evidence." Therefore, Wang concluded, the inscription was probably no more than five or six hundred years old. Those scholars who dated it to remote antiquity "deeply believed and did not doubt; the rest all rejected it as spurious." Still, he thought, the Stele of King Yu remained an interesting example of epigraphy that was worthy of study, mostly as a showcase for the critical perspective and sophisticated techniques of evidential scholars like himself.[96]

It is impossible to say exactly how Amiot learned about the Stele of King Yu, but if it was not through Wang, it was likely through someone who knew him. His friend Hongwu was a close associate of Wang: the two painted and wrote poetry together, and each appeared in the other's collected writings.[97] Circumstantial evidence is further suggestive. Amiot and Wang were close in age, lived in Beijing for many years, and performed similar functions at the imperial court. During the 1770s, Wang directed the compilation of an imperially commissioned dictionary of Buddhist terms in Central Asian languages while Amiot was working as a translator in the Mongolia bureau of the Grand Secretariat.[98] Amiot was also then following the Qing campaign against the Jinchuan rebels in Sichuan, for which Wang was an official chronicler.[99] And it was not long after Wang returned to Beijing with the victorious armies in 1776 that Amiot claimed a breakthrough in his investigation of the Stele of King Yu. Although he had already sent a copy of the inscription some years back, he decided to send another because he had only recently come to a better understanding of it through new Chinese commentaries.[100]

The only Chinese source on the stele that Amiot explicitly named was an obscure scholar from the Kangxi period, Mao Huijian, who had issued an interpretation of it in 1666.[101] Mao was one of the credulous scholars whom Wang Chang criticized; he had believed that the stele really did date from the Xia. It was true, he conceded, that earlier writers had not known about the monument, but their ignorance did not mean that the stele did not already exist. Amiot repeated Mao's own account of the modern rediscovery of the ancient stele. It began when a lumberjack stumbled upon it high in the Hengshan mountains, then led an unknown scholar on an arduous scramble over perilous cliffs to take a rubbing. When the lumberjack died, the location of the original was again forgotten. During the Kangxi reign, Mao himself went to investigate and concluded that it was more or less authentic. There were "savants now and to come" who doubted the antiquity of the stele; but Mao reasoned that even "if the monument did not truly belong to Yu the Great, it was nevertheless the most ancient that existed in all the Empire." After comparing different copies of the stele and all extant commentaries, he cut a new stele after the reconstructed original and placed it for the benefit of posterity in the far more accessible Xi'an Stele Forest.[102] This stone was the one from which was taken the rubbing that Amiot sent. He himself refused to pass judgment on its authenticity but deemed it in any case worthy of the attention of "savant antiquarians" like Court de Gébelin.[103]

There was just one caveat. Amiot wanted more than an explanation of the inscription; he also wanted proof of Court de Gébelin's interpretive system. He therefore proposed to Bertin that they conspire to put the savant to a test. Court de Gébelin should at first receive only the original tadpole script rubbing, without the modern Chinese version or Amiot's French translation. He would then be invited to attempt an interpretation of this possibly ancient monument based only on his relevant knowledge and interpretive technique. Amiot promised to be a fair judge: "If in the end he decodes the true sense, I declare myself from that moment the most ardent defender of his system and . . . count myself among his disciples."[104] It seems that Amiot really hoped he would succeed. Bertin agreed to the plan. He reasoned that if Court de Gébelin came to the same conclusions as Chinese scholars—"which is not probable," he conceded—then it would prove at once that both were correct. If, on the other hand, he came to a different one, that might be only because the tadpole-script inscription was not Chinese, but something older still. In any case, Bertin would wait until he heard back from Amiot before publishing anything about it.[105] He ordered his secretary to forward the rubbing to Court de Gébelin and "invite him to research and guess the sense, if this is possible by the principles of his system."[106] Court de Gébelin had no doubt that it was.

The Tarot of Yu the Great

At around this time, Court de Gébelin was just beginning research on the matter for which he is best known today: the invention of tarot-card fortune-telling. The tarot-card deck of the eighteenth century is essentially the same as the one of today. Much like an ordinary set of playing cards, it featured an additional fourth face card in each suit and an extra suit of twenty-one trumps, commonly known to modern enthusiasts as the "major arcana." The cards themselves were already well known all across Europe by the late eighteenth century, but they were mostly used for a trick-taking game a little like bridge that is still played in some places today. There are some indications that tarot cards were already featuring in some forms of popular fortune-telling. But there is no evidence that tarot cards were yet associated with any kind of esoteric or occult interpretation.[107]

Tarot cards as we now know them were really the invention of Court de Gébelin and his anonymous collaborator, "The Count of M."—in fact, the comte de Mellet, who also happened to be a nephew of Bertin. In the 1781 volume of the *Monde primitif*, they presented two essays which, for the first time in print anyway, ascribed to the tarot cards an ancient pedigree, esoteric meaning, and occult potency.[108] They proclaimed their findings not as discovery, but as recovery: "If one heard announced that there exists still in our day a Work of the ancient Egyptians, one of their Books escaped from the flames that devoured their superb Libraries, and which contained their purest doctrine on interesting subjects, everyone would doubtless rush to learn about a Book so precious, so extraordinary." And behold—"This Egyptian Book exists."[109] It was written in the trump cards of the tarot deck, which bespoke both a cosmological theory and a practice of divination. The knowledge it contained was a vestige of the "Hieroglyphic and Philosophic Science of the Sages," codified in the form of the tarot cards by the ancient Egyptians, but with origins stretching back further still—to the ancient wisdom of the *Monde primitif*.[110]

In this interpretation, the trump cards of the tarot deck were an alle-

FIGURE 3.3 Tarot cards, in Court de Gébelin, *Monde primitif . . . dissertations mêlée* (1781). Courtesy of the Wellcome Collection.

gorical representation of Egyptian cosmology, "containing in some fashion the entire Universe and the different States of which the life of Man is susceptible."[111] The twenty-one cards could be read backward to narrate a "history of the first times": the age of gold, the age of silver, and the age of iron. The first age was creation: so, for example, the twenty-first trump card depicted the goddess Isis ensconced in an egg, representing the birth of the universe. The second age was expulsion from paradise; accordingly, the twelfth trump card depicted a hanged man, representing earthly misfortunes. The third was civil society, beginning with the chariot, standing for the bloody battles of the bronze age, and concluding with the fool, for the Egyptian priests defiled the pristine knowledge.[112] The suit cards contained further information about the "political geography" of the ancient world; for example, the swords suit represented Asia, the land of great conquests, and the coins represent Europe, because gold was once plentiful there.[113] The tarot cards thus told a story of the development of the arts and sciences, of which they themselves were themselves supposed to be an early example.. The fact that ancient Egyptian priests had apparently written a kind of Enlightenment conjectural history does not seem to have struck their French spokesmen as at all strange.[114]

Court de Gébelin and his collaborator held that the tarot cards found their most important use in a practice of divination. They constituted the "Book of Destiny," which the Egyptian priests consulted to interpret dreams in a manner they had "reduced to a science."[115] Here is how it works. Two people begin by dividing the tarot deck into two stacks, the pips and the trumps. Each person takes one stack and cuts the other's. They begin to count together from one to fourteen. With each number recited, one person deals a trump card face down, and the other deals a pip card face up. Occasionally, the number of the count matches the number displayed on the pip, whereupon the trump that was dealt alongside it is revealed, and both cards are set aside as a pair. The trump stack is then reshuffled, and the count resumes. When all the pips have been turned up, the stack of those remaining is inverted and the procedure is repeated twice. The final result is a set of some number of pip-trump pairs that should be interpreted together and as a sequence.[116] As an example offered by the authors of the *Monde primitif*, suppose the Egyptian priests had wanted to read the Pharaoh's dream from Genesis, in which seven fat cows were eaten by seven thin ones. The Book of Destiny might produce as a first pip-trump pair the ace of batons and the sun, indicating agriculture from the pip suit and auspiciousness from the trump; but if the final pair were the five of batons and death, the message would be more ominous, and to a greater degree, since five is greater than one. The dream portended bountiful harvests, followed by famine.[117] This use of tarot cards should certainly be credited to the authors of the *Monde primitif* rather than to any Egyptian priest. It was, in fact, the first known tarot-card spread.

For the authors of the *Monde primitif*, the Egyptian practice of taro-card divination was essentially superstitious, but it had degenerated from something more pristine. As constituting the Book of Destiny, the tarot deck still contained "a great portion of the ancient Wisdom."[118] All the oldest records of the ancient world suggested that God really had given indication of His divine plan in the form of dreams, and "it would have seemed at least excusable to try to penetrate them."[119] But the Egyptian priests had grown ambitious and vain, adding new false cards to the tarot deck and inventing magic wands, talismans, and other sacrilegious elements. In modern times, only traces of the ancient wisdom remained, preserved in a sort of subconscious cultural memory among superstitious fortune-telling peasants from Spain to Romania. The historical transformation of the tarot deck was therefore an argument against the perfectibility of knowledge: "Such is the fate of human affairs: from such a sublime Science, which occupied the Greatest Men, the most learned Philosophers, the most respectable Saints, there remains nothing for us to draw on now but the games of children."[120] The story of the tarot cards was first told as not a triumphant proclamation of progress, but a sad rumination on decline.

At the same time, the fact that the story could be told at all was held up as a testament to the enlightened methods of the *Monde primitif*: tracing back origins and establishing etymologies. According to its authors, the tarot cards had survived in Egypt through to classical times and were brought by devotees of the Cult of Isis to Rome. From there, they later spread throughout the lands of the Holy Roman Empire and then into Southern France with the Avignon papacy.[121] In the present day, the cards had reportedly caught Court de Gébelin's attention when a certain countess invited him to a party in Paris where guests were playing with a deck that her husband had brought back from Germany "without instantly recognizing an Egyptian game modeled on the philosophical knowledge of the ancient Egyptians and arranged according to this famous number of seven."[122] That recognition was the singular accomplishment of Court de Gébelin. The most important clue was the word itself: he claimed that "'tarot' came from the Egyptian roots TAR, meaning 'road' or 'way,' and RO, meaning 'king' or 'royal'—thus TAR-RO, 'the royal road of life.'"[123] It should be noted again that historians today completely reject this account of the origins of the tarot cards.[124] It is nevertheless significant that the ancient wisdom they were supposed to contain was not presented in opposition to secular methods, but rather as among their most promising rewards.

It was just as he was preparing the essays on tarot cards for publication that Court de Gébelin received in 1779 the rubbing of the Stele of King Yu, without the benefit of accompanying translation or commentary, as Amiot and Bertin had planned. And, as they had predicted, he deemed it a fitting field of application for his theory and technique. His interpretation required some mental gymnastics, which he later explained. The key was in the esoteric significance of the mystical number seven. The number of cards in the tarot deck and the number of characters on the Stele of King Yu both totaled seventy-seven. Like the columns of the stele, the tarot deck was arranged in multiples of seven: four sets of fourteen suit cards (two times seven), and two sets of trump cards, one of fourteen, the other of seven. Now the relationship between the tarot deck and the Stele of King Yu was clear: "It is thus quite apparent that both the one and the other of these Monuments were formed according to the same theory and on the attachment to the sacred number of seven." After explaining the matching septenary organization, Court de Gébelin declared, "this monument is therefore perfectly similar, as far as its arrangement goes, to the game of Tarot." The characters corresponded with the cards, one to one, figure for figure.[125]

In an unpublished letter to Amiot penned in 1780, Court de Gébelin ventured a slightly more detailed reading of the tarot of Yu the Great. "I cannot stop myself from admiring the surprising rapport with a game known in Europe," he wrote; "It's the game called the *tarots*." The Stele of King Yu, like the tarot cards, must have presented a "tableau expressing human life from the point of existence all the way to death." It probably "had for its object the most useful knowledge for the entire nation of the Chinese," like tarot cards had for the Egyptians. Unfortunately, he confessed, the literal translation of the text was a mystery that he could not resolve: "I have never presented myself as a diviner," he wrote in his own defense, "but purely and simply as a comparer of words and monuments." He had found no other examples of the tadpole script, rendering his comparative method inapplicable. But he cleverly turned this ignorance into an argument: that the characters were so different from other Chinese characters only indicated that the inscription must be more ancient still. This suggestion was after all the one both Amiot and Bertin had anticipated. It was also the core idea of his whole project: "Antiquity was infinitely more savant and more enlightened than we imagine," he concluded.[126]

Herein lay the true significance of the Stele of King Yu for Court de Gébelin—not that it proved the authenticity of the tarot cards, but rather that it testified to the existence of ancient wisdom. In the tarot cards, Court de Gébelin had "rediscovered the philosophy and theology" of the Egyptians, and in the Stele of King Yu, he had rediscovered the tarot cards.[127] He believed that there must have been some historical relationship between the two relics. In earlier conversations with de Guignes and Deshauterayes, he had considered the various historical proposals for the origins of the Chinese people.[128] In the *Monde primitif*, he presented a new one: the Egyptians were not the fathers of the Chinese, but their brothers. As he explained to Amiot, the tarot cards and the Stele of King Yu were proof of a single shared inheritance, a tradition "born among the common mother of the Chinese, the Egyptians, and all the peoples." Scholars had "lost the traces of the origins of a multitude of things." Such things could be recovered through the painstaking empirical investigation of the shared global past.[129]

At this point, it was time for the reveal. Court de Gébelin gave Bertin his reply to pass along to Amiot. Perhaps from the desire to save Court de Gébelin some embarrassment, or maybe in consideration of the years-long wait required for a reply, Bertin decided that the gentlemanly thing to do was to allow Court de Gébelin an opportunity to revise his letter before submitting it for judgment. He showed him the rest of the contents of Amiot's package, the modern Chinese version of the inscription and its French translation, which made clear that the stele in fact recounted the accomplishments of Yu the Great and said nothing at all about tarot cards. Bertin had figured that "all this will probably make him change his letter"; but it seems that Court de Gébelin did not change a word.[130] He had ventured no guess as to the literal meaning of the inscription, only an explanation of its allegorical significance. The translation of the text itself made no difference. His interpretation was based on his broader knowledge of the ancient world, quite in keeping with the spirit of the test to which he had been put. Plus, he might have already figured out at some point what was really going on: "It was just sent to me from China to put me to the test," he wrote to another friend at around the same time.[131]

Amiot received Court de Gébelin's letter in 1781 and wrote a response from Beijing: the gist of it was that the proposed identification of the Stele of King Yu with the tarot cards was utter nonsense. The main issue was that the distribution of the characters into however many columns was arbitrary; there were in fact several extant versions of the steles with the same inscription but different layouts, and no Chinese scholar had ever said anything about its organization by groupings of seven.[132] Amiot thus considered the matter settled. As he wrote more candidly to Bertin, the interpretation "might well suffice among you; but it is not the same here." He gave a characteristically artless simile for the theory which at least got the point across: it was "a stroke of genius as wonderful as that which would ascribe purely fantastic objects with bodies that would be palpable even to those having the bluntest sense of touch." His main advice for Court de Gébelin was to go to the Bibliothèque du roi and ask de Guignes to show him the original of the *Éloge de la ville de Mukden*, which Amiot had translated some years back, to see another example of the tadpole script.[133] Court de Gébelin had failed the test, and Amiot never proposed another.

But by the time Amiot's letter reached Paris in 1782, it was too late—the first essays on the occult or esoteric tarot had already gone to press.

The interpretation of the tarot of Yu the Great was published in the 1781 volume of the *Monde primitif.* Toward the end of the essay entitled "Du jeu des Tarots" was a subsection called "The Relationship between This Game and a Chinese Monument"—that is, the Stele of King Yu. Court de Gébelin gave the identification of the matching septenary layouts almost exactly as he had written to Amiot. He further explained that he had attempted to print the tadpole script for the benefit of other savants but doing so proved too difficult due to the rubbing's cumbersome size. It was thus left for the reader to take it upon the author's word that the Stele of King Yu and the tarot-card deck indeed demonstrated the same layout. A genealogical connection could therefore be inferred. "It would be very strange," he declared, "if such a similarity was the simple effect of chance." The stele and the tarot cards reflected "a different application of a single formula, anterior perhaps to the existence of the Chinese and the Egyptians." Both, that is, were relics of the *Monde primitif.* In this way, the supposedly ancient Stele of King Yu became a real part of the modern invention of tarot-card fortune-telling.[134]

Orphans of the Enlightenment also learned something more general from the exchange: China possessed evidence that could be used in their investigations of the ancient world, and there was a particular person who would help them access it. In the following years, Court de Gébelin wrote again to Amiot, asking for more information on topics from the Jews of Kaifeng to the Daoist pursuit of alchemy, magic, and the philosopher's stone.[135] Meanwhile, his collaborator on the tarot cards, the comte de Mellet, soon began his own exchange with Amiot, which would last for the better part of a decade. As a site of ancient wisdom, China continued to grow its new following in France.

This is why the whole incident seems to have only galvanized Amiot's interest in the work of the "amiable and very estimable savant," even despite his interpretive failure. When he wrote that Court de Gébelin deserved "a most distinguished place among those who make up the small number of creative geniuses," it was ambiguous whether he meant it as a compliment.[136] Even though he rejected the interpretation of the Stele of King Yu, he continued to inquire about the tarot cards and "any other discovery of this kind."[137] Amiot requested new works by Court de Gébelin and received them "with pleasure."[138] The missionary came to see the scholar as "one of those profound and grave men, who by their application and their constancy digging in the almost arid den of antiquity, come finally in the end to bring forth a clear and brilliant water that all the world can drink in long draughts." Still further: "Their example gives me strength."[139] Court de Gébelin's minor failure belied a greater success. For, on the most important point, Amiot agreed with him: there was ancient wisdom in China, and it was the task of modern scholars to dig it up.

Jean-Sylvain Bailly and the History of Science

Jean-Sylvain Bailly was a great admirer of Court de Gébelin and an unacknowledged intellectual heir. He was born in Paris in 1736 and spent his whole life there, making his name as an astronomer and mathematician, then earning wider acclaim as an essayist and historian. As much a figure of the ancien régime establishment as it was possible to be, he was the first since Bernard le Bovier de Fontenelle to win membership in all three French academies. And yet, he also went on to become a leading figure of the French Revolution; it was he who was immortalized by Jacques-Louis David presiding over the Tennis Court Oath in 1789.[140] At once both *lumière* and *illuminé*, in the words of Condorcet, he shows that the roles were never really distinct.[141] Like the philosophes, he asked questions about progress and science; but like Court de Gébelin, he found answers in the "wisdom of the ancients."[142] In short, he was an orphan of the Enlightenment.

Bailly developed his ideas through the history of science, a discipline of which he has as much claim as anyone to be called a founder. In the late 1770s, having published extensively on classical and Middle Eastern astronomy, he expanded into a broader account of ancient science in two related books, *Letters on the Origin of the Sciences and the Peoples of Asia* and *Letters on Plato's Atlantis and the Ancient History of Asia*. The first was framed as an exchange with Voltaire and included authentic letters signed by "the old sick man"; the second continued the conceit, even after the philosophe's demise in 1778.[143] Like many before him, Bailly had noticed remarkable similarities between the astronomical traditions of all the most ancient peoples. Looking deeper, it seemed to him that the Chinese, Egyptians, Indians, and Chaldeans shared a whole cultural edifice in common: memories of a golden age and a universal flood, theories of metempsychosis and number mysticism, religious practices like wild Saturnalian parties, and more.[144] Human nature alone could not account for these similarities; it might have explained the independent development of a common astronomy, which was the product of reason alone, but the rest involved the exercise of imagination, which in humans was never the same.[145] Nor could the similarities have been the result of cross-

FIGURE 3.4 *Jean-Sylvain Bailly*, Jean-Laurent Mosnier, 1789. Courtesy of the Carnavalet Museum.

cultural communication or interaction; people were by nature factious and disputatious, more likely to argue than to agree, and not inclined to share their knowledge with their rivals.[146] The explanation therefore had to be genealogical.

Bailly concluded that in the expanse of time between the Creation and the Deluge, there must have existed a single ancestor people who had given birth to all the others. Comparing accounts collected from peoples across Eurasia about their own origins, he traced this ur-civilization to a Hyperborian homeland somewhere near Greenland, Spitzbergen, or

Nova Zembla. Citing Buffon's theory of planetary cooling, he argued that thousands of years ago, these lands had been warm and wet, and capable of sustaining thriving cities and powerful empires. Gradually, they became uninhabitable.[147] As their lands turned to frozen waste, the Hyperborean people were forced to flee. Setting out from the north in successive waves, they established their first colonies in the lands nearest to them, particularly on the Central Asian plateau between the 40th and 50th parallels.[148] From there, they dispersed throughout Eurasia to eventually become the known peoples of the ancient world, including the Persians, Indians, and Chinese.[149]

According to Bailly, this was the true origin of the legend of Atlantis. The name first appeared in Plato's cosmological dialogues, the *Timaeus* and the *Critias*, as an allegory told to the sage Solon during his travels in Egypt; but the story was "not a fiction."[150] The Atlantans had reached the pinnacle of human achievement: "This ancient people had sciences that were perfected, a philosophy sage and sublime."[151] The Atlantans had disappeared, and their science and philosophy were long forgotten. But still they lived, if not in history proper, then in myths and fables; and though their ancient wisdom was lost, its remnants survived in the science and philosophy of their successors. The most ancient human institutions were "the work of a people who disappeared from the face of the earth, a people whose name is lost, and of whom the histories make no mention; but the sciences vindicate it from this oblivion."[152] As with Court de Gébelin's invention of tarot-card fortune-telling, Bailly did not create the myth of Atlantis out of nothing. But he helped to give it a now-familiar form—as a real world-historical place, a lost techno-utopia—and his work provided the foundations for much later mythologizing.[153]

And if Bailly's Atlantis sounds a little like Court de Gébelin's primeval world, this was because the two thinkers had formulated their ideas in conversation with each other. Bailly may have been Court de Gébelin's confrere at the *Loge des Neuf Soeurs*; in any case, he was thought to be one by its enemies.[154] They also shared many mutual friends, including Bréquigny, editor of the China correspondence, and the Huguenot pastor and revolutionary Jean-Paul Rabaut-Saint-Étienne, who corresponded with both of them about orientalism and allegory.[155] Moreover, each author cited the other explicitly in his own works. Bailly especially recognized his debt to Court de Gébelin. The *Monde primitif*, Bailly held, was an "ingenious and profound work on comparative grammar" that proved the common roots of modern European languages in ancient Asian ones.[156] Its author was "more ingenious and more enlightened" than those who had previously written on mythology, and his study of languages had "advanced this science" more than anyone else.[157]

Bailly drew especially from these methods of comparative linguistics and mythological analysis. The original language of the primeval world was, Bailly opined, "a great discovery of our century."[158] Mythology, too, was always one of Bailly's dominant interests, guided likewise by a commitment to Euhemerism: "Fables sometimes contain debris of history."[159] Following Court de Gébelin, Bailly believed that the historical study of mythology and linguistics went hand in hand. For example, Chinese cosmogony was supposed to be resonant with that of the Greeks, Egyptians, and Indians. All of their myths featured sequential ages of gods, demigods, and men. The explanation for this commonality was that they had all descended from a single antecedent myth, and proof was to be found in their language. In Chinese, the rulers of the "age of men" were called the "*gin-hoang*"; this was etymologically connected to the Persian word *Gin*, or genie, beings who had been common at a particular time in the past; "It seems therefore that the name *Gin* was universal in Asia to designate the people who lived in the third Indian age."[160] Bailly thus followed Court de Gébelin in drawing diverse evidence from the language and mythology of the entire ancient world to be interpreted holistically and together.

Bailly was also responding to another interlocutor, one who worked on his own thinking less through influence than through opposition: namely, the marquis de Condorcet, the most ardent progress theorist of the late Enlightenment. The two men developed an intense professional rivalry, and, like many rivals, they had much in common. Condorcet was born in 1743, seven years after Bailly. He, too, spent most of his life in Paris, made his name early in the mathematical sciences, branched out into belles lettres and history, and eventually entered politics.[161] Bailly had already formed a grudge against Condorcet by 1776, when the younger man beat him to become permanent secretary of the Académie des sciences.[162] In the early 1780s, they fought over a position at the Académie française, which Condorcet also won. When Bailly was finally elected to join in 1783, Condorcet took the opportunity to give him a less-than-flattering reception address.[163] But Bailly's star rose later, during the early days of the French Revolution, when he and Condorcet served together in the Paris Commune. In 1789, Condorcet failed to win election to the National Assembly, and Bailly became its president. This turn of events does not seem to have settled their dispute, but the Reign of Terror soon would.[164]

In the intellectual world, their disagreement played out primarily over the history of science. At the heart of Condorcet's view of progress was a belief in the perfectibility of human society, ensured by cumulative developments in the understanding and manipulation of the human and the natural worlds.[165] Bailly, himself a respected astronomer, did not deny that progress in the sciences had occurred—in fact, he is largely credited with coining the phrase "scientific revolution" as it is now understood. Rather, he argued that history was not linear, but cyclical—as indeed the word "revolution" had originally implied.[166] Progress was neither regular nor inevitable. As he put it, "all that will happen in the future could have happened in the past."[167] It was myopic to think there was anything special about modern times: "Self-esteem tricks us; we believe ourselves at the top of the ladder; we are not there." Here, too, Bailly contributed to the terminology of progress discourse, with his early use of a ladder metaphor: "We likewise believe that none have climbed it before us, because time, which makes humans disappear, also erases their transient traces."[168] The course of history not only moved toward improvement; allowed to go on long enough, it would only lead to oblivion.

This alternative view of progress made Condorcet an ungenerous but careful reader of Bailly. In a 1775 book review, Condorcet contested Bailly's core idea that the ancient peoples of Asia owed their sciences to a Hyperborean ancestor as insufficiently evidenced. He pointed out that there was no direct written or archaeological record of any ancient civilization in Siberia that could be relied upon. As a result, "it was necessary to research the trace among fables and superstitious ideas" instead. Condorcet denied all but one kind of ancient wisdom: for him, only the Greeks had possessed the capacity for scientific thought that would come to distinguish the "modern nations of Europe."[169] In 1784, Condorcet returned to the subject of Bailly's works in a discourse at the Académie française. He complimented the author of the *History of the Origin of the Sciences* and the *Letters on Plato's Atlantis* in a way that was backhanded at best. These works were ingenious, exciting, and popular, for, "men need fables to support the truth." By explicitly praising Bailly's accomplishments in both sciences and the arts, he implicitly denigrated the former: "Posterity will pardon your Hyperborean people, as it pardoned the atoms of Lucretius."[170] To the extent that Condorcet is often seen as the most important Enlightenment proponent of progress, his thought emerged in conflict with Bailly's search for ancient wisdom.

Both Enlightenment scholars were working in the same intellectual world. This was not entirely lost on Condorcet himself: if the Hyperborean Atlantis was a kind of mythology, he conceded, it was nevertheless "the only mythology suitable for enlightened centuries."[171] In fact, Condorcet drew from Bailly both factual claims and conceptual tools. In his masterpiece, the *Esquisse d'un tableau historique des progrès de l'esprit humain,* he characterized Asia as not only static, but also as having declined. The conclusion implied elements of the idea of ancient wisdom that were particularly explicit in a short discussion of Chinese scholars: "They themselves forget a part of the truths hidden in their allegories . . . and they end up themselves the dupe of their own fables."[172] More broadly, one of Condorcet's key contributions to progress theory has been seen as the move from a discussion of the arts and sciences to a broader conception of civilization, analyzable in terms of relative states or degrees thereof.[173] In fact, the terms "state of Civilization," and even "progress of civilization," are to be found already in Bailly's histories of science, published more than a decade before Condorcet penned the *Esquisse.*[174]

Nor is it insignificant that one of Condorcet's last works was his own "Fragment on the *Atlantis.*" This utopian text continued the project of the *Esquisse* from the past into the future, describing the present state of the natural sciences and imagining the institutions that might perfect them. Long interpreted as a response to Bacon's *New Atlantis,* recent research has revealed the contributions of another *Atlantide.* Condorcet's utopian vision was not allegorical like Plato's, nor fantastical like Bacon's; it was set in a real-life present and a concretely imaginable future. Bailly's idea of Atlantis as an actual historical place had given to utopian thinking a certain realism that was lastingly influential. For Condorcet, the natural sciences were the primary marker of the state of civilization. Likewise, it was astronomy that had launched Bailly on his quest for ancient wisdom.[175] Condorcet set his Atlantis in present-day Paris amid the Revolution raging around him as he wrote, but in the final paragraph, he imagined something still greater, a global scientific organization "embracing all the peoples who have arrived at roughly the same degree of enlightenment and liberty."[176] For Condorcet, science was bound to diffuse beyond Europe; but for Bailly, Europe was the place to which it had already been diffused. Condorcet put science at the center of an emerging idea of European civilization by drawing from previous efforts that had been far more global in nature.[177]

In fact, nearly all of Bailly's work addressed Asia to some degree. The continent was widely acknowledged as the birthplace of the arts and sciences: "It is the spirit of Asia that animates Europe," he asserted. China was an especially interesting case study because it was close to the Hyperborean Atlantis, both geographically, due to its location, and temporally, due to its antiquity. To establish the ultimate origins of the Hyperboreans, it was necessary to begin in North Asia, "to see the closest men, to chat with the inhabitants."[178] The Chinese were "the most ancient people in the world, if one only relies on authentic monuments; the most jealous of their antiquity and the most careful to conserve its memory."[179] Their supposed conservatism, which for many of the philosophes was reason to ignore them, was for Bailly just what made them interesting. During the 1770s, he acquired both of the newly published Chinese historical works, the *Chou-king* and the *General History of China.*[180] By the 1780s, he, too, was plugged into the Paris-Beijing network and receiving manuscripts on loan directly from Bertin.[181]

Bailly did not deny the new consensus that the people of contemporary China were backward in the sciences. Although China was "the first enlightened," for forty centuries it had remained only at "the dawn of the beautiful day that enlightens Europe."[182] Like Voltaire, he sought an explanation for this paradox, and, also like Voltaire, he settled on conservatism. The people of China were by nature "incapable of this inquietude that makes change into a need."[183] This incapacity was a particular problem when it came to knowledge of nature: "Do you think, sir, that such a disposition might be favorable to the progress of the sciences?"[184] Certainly not—the Chinese "never at any time had the true spirit of the sciences."[185] They were obstinate even in the face of modern technology, rejecting it out of "respect for custom, long maintained by habit, and now defended by laziness."[186] Bailly's investigations were not flattering to the contemporary peoples of Asia; in fact, one of his last works established some of the racialized themes that would animate European orientalism for the next century. If the ancient wisdom of Asia had come from somewhere else, this observation was not meant to credit non-European peoples—to the contrary, it was only proof of the bankruptcy of their indigenous practices.[187]

Yet Bailly, also like Voltaire, was still able to appreciate China's historical achievements in a limited way, because he rejected the increasingly common view that scientific and moral progress were necessarily linked. He too believed they depended on different virtues, which were often as not opposed. The study of nature required "movement, genius and activity," which he claimed the Chinese lacked; but the study of ethics was "tranquil," and required only inward reflection. Bailly was thus able to resolve the other problem that had troubled Voltaire; that is, China's unequal achievements in natural and moral sciences. This inequality constituted no paradox at all but was perfectly expected: the same conservatism that held back the progress of the sciences had maintained the order of society, and as "men made for ethics, they are children in the sciences." Insofar as they lived in constant peace and virtue, the Chinese had "attained this final term of human sagacity." For Bailly, they were "the most enlightened of the people of Asia."[188]

Most important, Chinese conservatism, understood as a preference for stability and therefore a force against change, was precisely what proved the overarching point that the ancient wisdom of Asia could not have been original. According to Bailly, the Chinese qua Chinese were not only unscientific in modern times, but always had been so; the dual influence of "climate" and "national character" had seen to that. Yet, somehow, their ancestors had possessed scientific knowledge. Therefore, either those ancestors had gotten their knowledge from somewhere else, or they were not truly Chinese. The principles of ancient astronomy were said to be hidden in the riddles of the *Yijing*, which only the ancients had understood. The *Classic of Documents* in all its rich detail could not have been the work of one person, or even one century.[189] According to legend, it was Fuxi, some three thousand years before Christ, who had appeared suddenly and "brought the first lights to China," charted the movement of the stars, and described the progression of the equinox.[190] His knowledge was puzzlingly advanced, "which would indicate a science cultivated for a long time, and a people much more ancient."[191] He must have learned it from someone else: the Atlantans.

A Central Asian Atlantis

Amiot received Bailly's books in Beijing with great enthusiasm. Where he had found the *Monde primitif* suggestive, he found the *Letters on Plato's Atlantis* convincing. Here was a famous Enlightenment scholar who could benefit from his life's work on China; and here, too, was project on ancient wisdom that, in turn, could reshape his own. For Bailly, Amiot's claim that the people of China had "not taken one step beyond the limit placed by their elders of the most remote times" was now a reason not to ignore Chinese history, but to embrace it.[192] For Amiot, Bailly's work on the Hyperborean Atlantis finally made it clear who those elders of the most remote times actually were. In support of Bailly's research, Amiot marshaled contemporary Chinese scholarship and sent his studies back to France. New Enlightenment works in the history of science were confirmed by the recent research produced by the evidential-studies movement of the high Qing. European and Chinese historians alike believed that there had existed a body of ancient wisdom that had surpassed their own. Each could shed light on the other's history.

Along with nearly all early modern Europeans, Amiot had always believed that neither the people nor the knowledge of ancient China were truly home-grown. Early on, he had developed a story based on the biblical account that had always governed Jesuit studies, reinterpreted through an imaginative Enlightenment conceptual history. The first inhabitants, he believed, had come to China from the Fertile Crescent not long after Noah's flood. Leaving their brethren behind to place the last stones on the Tower of Babel, they wandered east and eventually settled along the banks of the Yellow River. As yet they were "without laws and without *moeurs*." These people who were the ones who had populated the region, but not the ones who civilized it. It was not until the later and separate arrival of the sage-king Fuxi that they became recognizably Chinese. Fuxi taught them agriculture, writing, and astronomy, first to the "most docile and least coarse" of the original inhabitants, then to the rest, who eventually made him "the general head of all the nation, solemnly recognized and proclaimed as such by common voice." Fuxi's arrival was thus the event that brought the period of myth or fable in Chinese history to an end.

Amiot's account of Chinese origins had given a clear answer to the question of the congruence between the Chinese records and those of diverse other peoples, especially the Phoenicians and the Chaldeans: "From where comes this accord between the three most ancient Peoples of the universe, between Peoples so different from each other?" The answer was that all of them shared a common origin in the Middle East not long after the Flood. All of them preserved in their records the traces of antediluvian knowledge, or "vestiges of the primordial tradition of the original state of the world." But their ancient wisdom had been gradually lost to the passage of time: "Such is the fate of the things of this world: everything passes, everything is insensibly destroyed, everything perishes, everything fades little by little from the memory of men." Some remnants remained buried in biblical and Chinese history, but digging them up would not be easy: "What light is bright enough to penetrate the depths of these shadows in order to dissipate them entirely? Where can we find among the ancients something that can settle our doubts and arm us against difficulties?"[193] When Amiot posed these questions in 1769, he meant them mostly meant to be rhetorical.

Two decades later, Amiot was immersed in the *Letters on the Origin of the Sciences and the Peoples of Asia* as well as the *Letters on Plato's Atlantis*, those "brilliant" works that had so "enriched the learned public."[194] These books seemed to promise answers to the questions that he had been asking all along. In his own words: "As the traces of this ancient people are entirely covered in the rust of time, I see from here no one but M. Bailly who, using a very powerful microscope, could discover something there with the point of his chisel."[195] A particularly important aspect of Bailly's work was its focus on the history of science. Amiot noted that ancient Chinese sources attested to a sophisticated calendar, "which could only have been the result of a science acquired long ago through experiments of more than one kind, or transmitted through the ages by the descendants or heirs of the inventor people." He therefore concluded, "along with the amiable and very learned author of the *Ancient Astronomy*," that the first Chinese astronomers could not possibly have developed their science "except with the debris of knowledge already acquired by an anterior people."[196] Amiot had once believed they had taken their sciences directly from the Holy Land; it was now "necessary to conclude with M. Bailly that they took them from an anterior people, a lost people, in a word an antediluvian people who were probably more advanced in the sciences than we are now."[197] His only complaint about Bailly's work was that it had not reached him thirty or forty years earlier, when he was younger and more energetic, and when "the literary arena of the Chinese people was more open to me," so that he might have followed up on it even more thoroughly.[198]

At the time, Chinese intellectual life was dominated by the *kaozheng*, or evidential-studies movement.[199] Evidential studies had begun in the late Ming as a rejection of Neo-Confucian metaphysics in favor of the "the investigation of evidence." Its major accomplishment was to historicize the Confucian cannon through sophisticated historical and philological techniques. The movement reached its height in Beijing during the Qianlong reign, when famous scholars such as Qian Daxin, Dai Zhen, and Ruan Yuan extended the techniques of evidential studies into the natural sciences, bringing it in contact with Western learning.[200] Modern historians have long have characterized evidential studies as a kind of parallel Enlightenment: Like their European contemporaries, evidential scholars advocated empiricism, took a critical approach to tradition, and valued the natural sciences.[201] More interesting still are the historical connections between the two projects and the appearance of each in the view of the other. In fact, some Enlightenment and evidential scholars were also united by an interest in a much more specific idea—that of ancient wisdom—which itself took form through cross-cultural conversations that were going on at the time.

Throughout the late imperial period, Chinese literati broadly concurred that science in Europe and in China had been genealogically connected in the ancient past. Just as the belief in ancient wisdom led European scholars to study China, so too did it lead Chinese scholars to study Europe—in both cases with the goal of claiming the others' scientific knowledge as their own. At the beginning of the Qing, mathematicians and astronomers such as Mei Wending pointed to supposed Chinese precedents for Euclidean geometry and the Tychonic system to justify their collaboration with the missionaries in "Western studies."[202] In the eighteenth century, the idea that "Western studies originated in China" became a basic assumption of the evidential scholars.[203] Ruan Yuan, for example, claimed that mechanical clocks had been invented in ancient China and "only failed to be transmitted" to posterity.[204] Nor did the missionaries really deny such claims, which many were sympathetic to anyway from preexisting commitments to Renaissance Hermeticism, and which only increased local appreciation for their skills and expertise.[205] The result was that China's Western-studies experts, the evidential scholars, and Europe's Chinese-studies experts, the missionaries, came to a kind of agreement on the common ancient origins of their natural sciences.

Furthermore, evidential scholars also believed that the state of the natural sciences in ancient times had not yet been and would not soon be surpassed.[206] At its most fundamental, evidential studies was about ancient Chinese texts. Its exponents developed their techniques in philology and textual criticism with the ultimate aim not of creating new knowledge per se, but of reviving the "extinguished learning" of the ancients. They were particularly interested in scientific subjects such as astronomy and geography, in which they explored extinguished learning through research on the story of Yu the Great and debates about the length of the tropical year.[207] Like orphans of the Enlightenment, evidential scholars believed that a coherent body of ancient wisdom that applied across the unified field of human knowledge had become corrupted with the passing of time. For example, Wang Chang cited Qian Daxin, another historian and epigrapher known also for his work in mathematics and astronomy, with a common formula that justified their efforts to recover lost knowledge: "The ancients are strong, the moderns are weak."[208] One major consequence of this belief was that the modern scholars should study ancient texts to learn about science. It is thus not surprising that many if not all of the major Qing mathematicians and astronomers were also practicing historians.

By the 1780s, orphans of the Enlightenment were in fact aware of their broad agreement with Chinese literati about the common source of their natural science and its advanced state in ancient times. Amiot was surely referring to evidential scholars like Wang Chang and Qian Daxin when he wrote in 1786 that Chinese scholars believed "that supposed modern inventions are really nothing but reminiscences."[209] Bertin was deeply interested in the notion but took a more moderate position: "I am in general of your opinion on the little attention given in Europe to the monuments of antiquity and the source of instruction that could result from them," he reassured the missionary; but as for the Chinese belief that all inventions were reminiscences, "maybe there is a little exaggeration in the opinion."[210] His doubts, however, did not stop him from discussing the idea with friends, or publishing about it in the *Mémoires concernant les Chinois*. In 1787, Bailly referred to the idea explicitly in his final book on the ancient science of Asia: "It was this science that was lost, and that in all the centuries the Chinese have strived to recover."[211] Not only did some Europeans believe that ancient China possessed scientific knowledge; they also believed that modern Chinese scholars agreed with them. The idea that was thought to unite Europe and China in the past demonstrably did unite them in the present.

Amiot thus sought to streamline European and Chinese scholarship. He hoped that each could guide the other, with himself acting as the go-between. Their shared historical vision encouraged a self-conscious kind of ecumenicism. Collaboration promised finally to reveal where the arts and sciences had originally come from: "M. Bailly would answer, without hesitation, the *peuple Atlantique*; I go further back than him, with the aid of my Chinese." Bailly had suggested that answers might be found on the ancient Central Asian plateau; not the original homeland of the Hyperborean people, he theorized, but rather the place where they had last flourished before their dispersal and collapse. And if research should start from there, Amiot believed his expertise on China made him well suited to contribute. China was nearest in time, since it was there that "the debris of ancient knowledge are in greatest number and best preserved"; and it was also nearest in space, since it now exerted political control over much of Central Asia.[212] Put into a conversation that only he would be able to facilitate, New Chinese and European scholarship could reveal much about a region that had until recent times been fairly marginal in both traditions.

Qing research on Central Asia had in fact made much headway in the previous century, thanks mostly to geopolitical events. The Manchus, who traced their own roots to the Jurchen people of the northern Chinese frontier, ruled over a vast empire stretching from modern-day Korea to Kazakhstan. Its borders reached their greatest extent only in Amiot's day, when the Qianlong emperor completed the conquest of Xinjiang. Conquest became an engine of scholarship. Understanding the lands and peoples now under Chinese hegemony was a political and diplomatic necessity.[213] The exchange of knowledge between the Qing, Ottoman, and Russian empires increased dramatically in the mid-eighteenth century as they came into contact along the great steppe. An enormous infrastructure developed in Beijing to make sense of the newly conquered domains.[214] Guided by the strong arm of Qing imperialism, scholars there, including not a few of the missionaries, did much to uncover the history, culture, and geography of the regions that had for millennia marked the boundaries of the Chinese world.[215]

And yet, the vast inland expanse beyond the confines of China proper was largely a mystery to Europe—the entire region from the Caspian Sea to the Sea of Japan was still lumped together under the toponym "Tartary"—and Amiot was in a unique position to demystify it.[216] Soon after his arrival in Beijing, he had mastered the Manchu language, finding its alphabetic orthography and systematic grammar to be a useful guide for understanding Chinese texts.[217] After the suppression of the Society of Jesus, Mongol and Manchu princes had become his closest local contacts. By 1767, he had begun translating a Chinese book on the Qing frontier.[218] Twenty years later, he completed the manuscript and gave it the title "Introduction to the knowledge of the people who were or are currently tributaries of China." In two accompanying letters dated 1786, he explained its value: "It seems to me that there are very interesting things about peoples who are not yet well known in your Europe, despite the universal histories that are written there."[219] Here then was "a host of objects very curious in themselves" that would be of interest to natural historians and philosophers, as well as information about the "mountains, rivers, etc. of Tartary and other lands" that would be useful for geographers. But the only person he mentioned by name was Bailly, who would "perhaps find there a station of his Atlantic people."[220]

The manuscript was in fact based on an obscure 1696 travelogue entitled *Investigations of the Translation Bureau,* which had just been reedited in Beijing for the imperially sponsored *Complete Library in Four Sections.*[221] Amiot's translation followed the original almost word for word. The preface explained how the book came to be written and placed it firmly in the ancient Chinese geographical tradition to legitimize the then-new Qing dynasty. It began with Yu the Great dividing China into the Nine Regions upon the founding of the first dynasty, making the peoples living beyond its borders subjects of the empire, and calling upon them to pay tribute. This story had long provided rhetorical precedents for imperial policies, but the Manchu conquests gave it new urgency. The preface then explained how the Qing emperors set out to learn more about the distant regions that were once again under Chinese hegemony. They established the Translation Bureau under the Board of Rites and sent agents to the far corners of the empire to investigate.[222] Their observations were published by the bureau chief Jiang Fan under orders of the Kangxi emperor, "to serve as a monument to the glory of his reign."[223] So ended the preface. The rest of Amiot's translation went on to describe China's western frontier, including the Muslim peoples of Xinjiang and the Silk Road city of Samarkand, as well as brief notices of places as far-flung as Thailand and Myanmar.

Bailly received the manuscript of the "Introduction to the tributaries of China" in 1787, two years before it was published in the *Mémoires concernant les Chinois*.[224] There is no direct evidence what he made of it, but his mature historical vision provides some clues. The preface might have drawn his attention to the myth of the Nine Regions, which he discussed in his final historical work.[225] The chapters might have confirmed his final opinion on the common source of ancient astronomy: "Its first seat, the place of its establishment, could have been in the West of China, to the North of India, between forty and fifty degrees latitude"—that is, in present-day Xinjiang.[226] In one late letter, Bailly explained how his system had resolved the old question of the historical relationship between the Chinese and the Egyptians: "I see beyond them a people more powerful, more enlightened than they, who had perfected sciences, a sublime and sage philosophy, the source of all the sciences that the Orient passed to our Europe."[227] In sum, Bailly agreed with Amiot about the history and value of ancient wisdom. And although no reply survives, it appears that Bailly did write one. It must have been somewhat encouraging, too, because it led Amiot to return to the text again for specific Chinese evidence that could further contribute to Bailly's project.

There was one story in the *Investigations of the Translation Bureau* that seemed particularly promising. In 1415, the Yongle emperor had dispatched an emissary to the city of Turfan in Xinjiang to negotiate terms of submission with the native peoples. The mission was a failure, but it did produce a curious report on the region. About 100 *li* northwest of the city was a mountain called Lingshan. Legend had it that at this site, 100,000 arhats had achieved their Buddhahood.[228] At the base of the mountain was a lake, in the middle of which arose an island of small rocks. Viewed from afar, the island cast a shadow upon the water shaped like a mass of human hair—in the very same spot where the arhats had shaved their heads to become monks. Nearby was a hillock covered with mounds of stones in the form of hands and feet, which were believed to be the arhats' mortal remains. This place was considered holy: "The locals say that it is there that Fo [the Buddha] himself became immortal." Amiot had always wanted to learn more about it, but the area was quite inaccessible, since it was at the time governed by a hostile Turkic vassal-state, in addition to being very far away.[229]

Originally, Amiot had thought that the mountain Lingshan was a monument of Noah's flood—what else could have left such a swath of destruction?—but Bailly's work suggested a new interpretation. Local tradition maintained that the fossils upon Lingshan, "one of the highest mountains of Tartary," were the remnants of the "Sages of ancient times."

It was no stretch, then, to think that the site might have been "one of the stations of the lost ancient people," its ruins a testament to their long-forgotten grandeur. Perhaps the stones of Lingshan were the archaeological evidence of Atlantis in Asia that Bailly had been critically missing. On June 26, 1789, Amiot wrote again to inquire as to his thoughts.[230] But by the time the letter arrived in Paris, Bailly had other things on his mind. Not ten days before it was signed, he had been elected as the first president of the National Assembly. The man who had dedicated himself to the ancient world was swept up in the currents that would create the modern one.

Bailly's influence on Amiot was nevertheless profound. In a kind of intellectual resume that he wrote toward the end of his life, he described what he thought to be three of his most important works—a translation of Sun Tzu's *Art of War*, an essay on the "*Antiquité des Chinois*," and a study of court music—as forming a unified account on the broader subject of ancient China. Looking back on it, he felt that he had only recently come to understand "the corner of remotest antiquity, which I could only so to speak glimpse with the aid of a torch, the light of which, too bright for my weak eyes, could not but dazzle me in the first moments, but which will not fail to enlighten me when I can withstand the bright day." That torch was the work of Bailly.[231] Amiot began to reconsider everything under its light. He became convinced that there had existed an antediluvian ancestor people with perfected ancient wisdom, and he began to see its traces everywhere. While Court de Gébelin had shown him what he could provide for the Europeans, Bailly had shown him what Europeans could do with what he provided. Together, the orphans of the Enlightenment constructed an account of ancient wisdom in which China was particularly, perhaps uniquely, important.

The Origins of Esotericism

After the epochal events of the 1770s, orphans of the Enlightenment on both ends of Eurasia found each other and began a global conversation. In Paris, Court de Gébelin and Bailly revived a theory of ancient wisdom that was increasingly discredited among the philosophes. In Beijing, Amiot searched for new ways to promote research on China even as many among the philosophes were openly disparaging it. The three scholars entered into communication about a subject on which their expertise overlapped and might still retain relevance for others among their contemporaries: the global origins of the arts and sciences.

The exchanges that ensued contributed to the invention of modern esotericism. Court de Gébelin identified Chinese knowledge with the lost wisdom of the primeval world, leading him to invoke the inscription of the Stele of King Yu in the invention of tarot-card fortune-telling. Amiot rejected Court de Gébelin's profoundly incorrect interpretation of Chinese evidence but saw in it a new way to advance his own agenda. Following in Court de Gébelin's footsteps, Bailly's research in the history of science led him to argue for the existence of a lost Hyperborean Atlantis, which the ancient astronomical records of Asia were supposed to confirm. This time, Amiot was persuaded, and he pointed to his own translation of a recent account of Xinjiang as providing further evidence. If the ancient world owed its scientific knowledge to a forgotten ancestor people, he was now in a special position to help recover it.

These discussions were part of the debates that shaped late Enlightenment ideas of progress. In their search for ancient wisdom, Court de Gébelin and Bailly suggested that remote places should be studied not despite their distance from modern Europe in time and place, but because of it. They presented a vision of global history that was in crucial ways opposed to the progress theories of the philosophes, while at the same time developing empirical narratives and conceptual tools that canonical progress theorists both contested and embraced. They explored techniques of comparative etymology and mythology, links between the development of science and the stages of society, a conception of perfectibility in secular time, and hopes for global collaboration in the future. Bailly and Court de Gébelin, as well as Voltaire and Condorcet, were members of a single community of thought, though a split was already beginning to emerge that would take sharper definition in the decade to follow.

The early attempts by orphans of the Enlightenment to deploy Chinese knowledge were halting and, in some respects, unsuccessful. Court de Gébelin's interpretation of the Stele of King Yu was plainly incorrect, while Bailly's engagement with the *Investigations of the Translation Bureau* was only impressionistic. Moreover, both of them used Chinese evidence mostly to confirm their preexisting theories. In their attempts to make their ideas legible to the philosophes, they said much about the achievements of an ancestor people, but they said little about the ancient wisdom that those people were supposed to have possessed. In the conversations that followed, that doctrine would be identified, bringing Chinese natural philosophy directly into the French Enlightenment.

✷ 4 ✷

The Yin-Yang Theory of Animal Magnetism

Chinese Science in Europe

During the 1780s and 1790s, the European search for ancient wisdom in China culminated in an unprecedented engagement with non-Western natural philosophy. Through a decade-long globe-spanning conversation, Amiot and the comte de Mellet developed an explanatory account of nature that united the Chinese cosmological principles of *yin* and *yang* with Franz Mesmer's new theory of animal magnetism. Mellet deployed Chinese sources in his study of ancient electrical medicine and presented his results to the Paris scientific community. Meanwhile, Amiot took mesmerism as an inspiration for groundbreaking new research on topics ranging from Neo-Confucian cosmology to religious Daoism. Orphans of the Enlightenment argued that ancient Chinese knowledge and modern European science were one and the same. But the result of their work was to associate Chinese natural philosophy instead with exactly those aspects of the European scientific tradition that their contemporaries rejected: mysticism, esotericism, and the occult. It has long been apparent that notions of science transformed ideas of progress; yet ideas of progress also transformed notions of science. The unique account of non-Western natural philosophy that the orphans of the Enlightenment created proved unrecognizable as science during their day, but soon found purchase elsewhere.

To understand how this conversation could have taken place, it is essential to recall that there had been a long history of European interest in Chinese natural philosophy and the grounds of engagement were already shifting. As late as 1700, Leibniz still believed that Europe could learn from China, not just in empirical fields such as natural history and descriptive geography, but even in theoretical ones like metaphysics and mathematics. By the 1780s, this position was not really tenable anymore. A consensus had emerged that the natural sciences had not developed in China past a certain point. Voltaire once considered it imperative to explain why this was so, precisely because it seemed to him surprising. For Diderot, the development of the natural sciences in China was less puzzling, though he found it worth addressing all the same. But crucially, Voltaire and Diderot had both relied upon an almost identical set of missionary sources that dated from the early eighteenth century and reflected the Jesuit position, which had changed little during the century before that. The upshot here is that shifting engagement with Chinese science did not essentially reflect a new understanding of the science of China; rather, it tracked with new ideas about the science of Europe.

The last years of the French Enlightenment produced not only a foundational narrative about science's past, but also a new program for its future. The historian Simon Schaffer has said that a key date for explaining European difference is 1775. In that year, the Académie des sciences prohibited discussion of perpetual motion, defining science thenceforth not by the inclusion of subjects that met some internal criteria, but by the exclusion of subjects that authorities could not accept.[1] True, astrology and alchemy had already been proscribed at the founding of the Académie in 1666. But its members continued to search for the philosopher's stone long after that, while the notion of celestial influence remained very strong in some fields.[2] It was only toward the end of the eighteenth century that the Enlightenment came to a stable agreement upon what science was—based on what science was not. Both occult science and non-Western science were excluded from the picture. The resulting conception of natural knowledge as necessarily both modern and European further fed claims to a monopoly on progress.[3] But just as alternative conceptions of history persisted into the end of the Enlightenment, so too did alternative conceptions of science.

The end of the "Age of Reason" also saw a turn toward esotericism, occultism, spiritualism, and mysticism. Historians have shown the persistence into this time of scientific thinking that was once thought to have become obsolete and demonstrated its importance even for the Enlightenment as most narrowly conceived.[4] First, major scientific figures drew from diverse intellectual traditions that have normally been placed outside the genealogical history of modern science. The vitalist Buffon revived Hermetic theories of analogy to reject mechanism, while the materialist Priestley brought back Aristotelian notions of organization to explain sensationalism.[5] Second, other figures once dismissed by historians as kooks or cranks also contributed to the Enlightenment's scientific legacy. The Neoplatonist Restif de la Bretonne may have inspired Pierre-Simon de Laplace's nebular hypothesis, while folk accounts of stones falling from the sky were discounted as popular superstition long before meteors were accepted as scientific facts.[6] Even in the most rarefied realms of the natural sciences, an Enlightenment epistemology according to which human reason was the only route to knowledge developed alongside an explosion of alternative ways of knowing.[7]

Likewise, notwithstanding the growing Eurocentrism of Enlightenment science in theory, it was also becoming ever more global in practice. One need only recall the Enlightenment's own scientific heroes to make the point: Joseph Banks observing the transit of Venus in Tahiti, Pierre Louis Moreau de Maupertuis measuring the arc of the meridian in Lapland, or Alexander von Humboldt botanizing up the mountains of Venezuela, to name just a few. This was a time of "go-betweens and global intelligence" circulating in an increasingly interconnected world.[8] The hallmark disciplines of the Scientific Revolution, such as physics and astronomy, had benefited from data and measurements taken far outside the confines of the European continent.[9] But the characteristic fields of the Enlightenment, empirical disciplines such as botany and geography, were open-air sciences that by their very nature had to be conducted outside and everywhere.[10] In sum, there has never been any question that Enlightenment science posed global questions that necessarily required global answers.

One question, however, remains vexing: what role did non-Western knowledge play in the global story of Enlightenment science? We know that Europeans learned *about* the wider world—but did they also learn *from* it? Many historians today would reject the distinction as a phantom of antiquated "internalist" / "externalist" debates, old-fashioned diffusion frameworks, and the positivist obsession with influence. In lieu of scientific ideas, they would point to social practices and embodied knowledge to evidence the globality of eighteenth-century science.[11] British cartographers relied on Mughal surveyors to map the Indian subcontinent, while German botanists studied the abortifacients of enslaved peoples in the West Indies.[12] But in examples such as these, non-Western people were often treated as a primary source, not a secondary source. That is, Europeans did not typically see themselves as engaging in conversation with other theorists whose activities were of a similar kind to their own. Nor did they often entertain the idea that their ideas might shift the theoretical underpinnings of their own knowledge, even in circumstances where it actually did. South Asian surveying techniques and West Indian botanical practices were thought to contain information about geography and botany, but they were not upheld as examples of the disciplines in their own right. The sciences remained essentially European in the eyes of Europeans, not least because an emerging orthodoxy held that they were supposed to.

The *yin-yang* theory of animal magnetism is quite exceptional in the global history of Enlightenment science, and almost entirely unknown until now.[13] This was a robust, explanatory, cross-cultural account of natural philosophy. It is thus the only example I am aware of in which Enlightenment Europeans not only believed that non-Western theory could explain new scientific ideas, but actually used original non-Western sources in order to do so. In 1781, the comte de Mellet wrote to Amiot to ask him about Chinese science and medicine because he believed that it might further European investigations in the new theory of animal magnetism. Amiot replied to the inquiry because he was convinced that it could do that—and also much more. Over the course of a decade, the missionary scholar and the gentleman aristocrat exchanged about a dozen letters putting European and Chinese theory directly into conversation, in which they explored topics ranging from alchemy and hypnotism to magic and extrasensory perception. They maintained that the ancient wisdom of China and the new knowledge of Europe were not only compatible, but actually identical. Each could be used to learn about the other.

The unintended result of their conversation was to recast the knowledge of ancient China as fundamentally different from that of modern Europe. For two hundred years, the Jesuits had accepted orthodox state Confucianism as the only legitimate intellectual tradition of China. Now, ex-Jesuits and other orphans of the Enlightenment explored the extraordinary range of ideas about nature, philosophy, medicine, and religion that permeated late imperial Chinese culture.[14] Amiot and Mellet interpreted Chinese ideas with the aid of concepts that were already familiar from the history of Europe: initiations and adepts, secrets and mysteries, magic and the elixir of immortality. They pointed to exactly those elements of their own historical traditions that the major figures of Enlightenment science were self-consciously trying to move past. For this reason, their theories proved roundly unacceptable to the scientific establishment. The union of Chinese and esoteric natural philosophy did not count as science. At the beginning of the Enlightenment, non-Western knowledge was already seen as ancient and unchanging; by its end, it was almost unrecognizable as knowledge as such.

Mesmerized by Kung-Fu

Louis-Raphaël-Lucrèce de Fayolle, comte de Mellet, was an aristocrat, soldier, and politician by profession.[15] He was a man of the world, most comfortable serving in the army, participating in provincial government, or managing his château at Neuvic in the Périgord. "When the weather is nice," he wrote to a friend in Paris, "I tend to my garden, I kill the ferns. That's all, and I would like only that what I do be yet further from any-

FIGURE 4.1 *Louis-Raphaël-Lucrèce de Fayolle, Comte de Mellet,* late eighteenth century. Courtesy of the Château de Neuvic.

thing useful."[16] He was neither an academician nor a philosophe, but a self-avowed amateur who, in his own words, had to "take up the rear when he comes to the sciences."[17] He described himself as merely "a soldier of the good old times, who knowing nothing but fighting, make all their glory consist in fighting well." Others rebuked him for this probably false modesty: "You have cultivated Minerva no less than Mars," wrote Amiot.[18] Still, it was perhaps on account of his martial métier that in his approach to the arts and sciences he was neither subtle nor timid, but always ready for war.

It is not clear when or how Mellet first entered the world of scholarship; indeed, the fact that he did so at all testifies to its importance in the cultural life of the late Enlightenment elite. All that can be said for sure is that the first indication of his interest reveals him to have been studying ancient wisdom under the tutelage of Court de Gébelin. In 1776, he was posted in the Loire Valley as the governor-general of the provinces of Maine and Perche and the county of Laval.[19] In that year's volume of the *Monde Primitif*, Court de Gébelin recounted a conversation with a certain "governor in the provinces" about the etymology of verbs.[20] By 1780, they were definitely collaborating on the occult interpretation of the tarot cards. In 1785, a mutual friend compared Mellet to the recently deceased savant; he demurred this "glory to which I am far from daring to aspire."[21] In fact, he does seem to have come to regard himself as a kind of successor, continuing the project that they had been working on together for some time already.

Mellet began his search for ancient wisdom in Egypt. In this sense, his project was not just generically esoteric, but out-and-out Hermetic.[22] Although he did not often mention the discredited *Corpus Hermeticum*, he accepted its central idea that the account of the New Testament was foreshadowed by Egyptian mythology, a common theme for magi from the Renaissance on down. He developed it further through Court de Gébelin's techniques of comparative etymology. For example, Osiris, both the son of the sun and the sun itself, was born on December 25, and his name was written with the hieroglyphic for "sun"; thus, he was Jesus. Isis, both mother and virgin, was written with the hieroglyph depicting the constellation Virgo; thus, she was Mary.[23] Returning to the defining problem of Renaissance Hermeticism, he wondered: How had the Egyptians known these Christian truths thousands of years before the birth of Christ?[24] His answer was that the hieroglyphs had descended from a "sacred language" which expressed divine knowledge: "The ancients, confusing the future with the past, would thus have seen events prophesied in a symbolic writing as having already happened."[25] Like the tarot cards, the hieroglyphs were used in a practice that scrambled the relationship between the future and the past, while themselves starring in a nonprogressive story of degeneration, corruption, and decline.

Mellet took a further position on ancient wisdom that was unusual even among orphans of the Enlightenment: he explicitly denied progress in the natural sciences. The philosophe icon Francis Bacon had said that the moderns were the true ancients, since the world had grown older with the passage of time.[26] According to Mellet, the ancients were rather the first moderns, because they were already inheritors of a still more ancient tradition. They, too, had struggled "to decipher with pain and labor the writings of the first ages." Progress was inverted; scientific knowledge increased the further back one looked. The Greeks had only "weak notions" of the speculative sciences, because they were "feeble imitators" of the Phoenicians and the Egyptians. In the "first ages," men had possessed science that was "a pre-sensation so fine, that it approached divination."

They had even mastered electricity and magnetism, which had only been "recovered" in recent times. Ancient wisdom was thus the same thing as modern science, but even better.[27]

It was this realization, Mellet later recalled, that led him to China. Since the late 1770s, he had been working to reconstitute a supposedly antediluvian theory of electromagnetism, which he believed had been used especially for medical purposes. It was an "aerial, celestial medicine," related in some way to fire or electricity, that had been practiced by the "primitive antediluvian people." Some knowledge of it survived the flood among the Israelites, who called it *Rapha-el,* meaning "medicine of God"—one wonders if Mellet was not pleased that "Raphael" was also one of his own three given names. In the late 1770s, he was scouring the historical and scientific literature for clues, trying to figure out how exactly this medicine had been controlled and administered. In 1779, he come across two new leads: the European theory of animal magnetism, and the Chinese theory of kung-fu.[28]

The German physician Franz Anton Mesmer was at the very point of intersection between science and occult science.[29] In his *Mémoire sur la découverte du magnétisme animal* of 1779, he proclaimed his discovery of animal magnetism, uniting a cosmology of microcosm-macrocosm correspondence with a vitalist medicine based on unseen forces and subtle fluids. Animal magnetism is a "universal agent" extended through all of nature that produces a "universal harmony" between the heavens, the earth, and living things. It flows as an infinitely subtle fluid and functions according to mechanistic laws. By understanding this fluid, "the art of healing will come thus to its final perfection" and produce hitherto miraculous cures.[30] The extravagant doctor claimed to do just that. Enrobed in lilac taffeta, he delivered stirring lectures and produced medical wonders that enchanted the salons of Paris.[31]

In some ways, animal magnetism was quite consistent with Enlightenment science, but in others it manifestly was not. For Mesmer himself, the theory was essentially materialist; furthermore, subtle fluids, vital spirits, and analogies to electricity and magnetism were common across a broad range of establishment physics and medicine.[32] But much of the scaffolding for animal magnetism was plainly drawn from the history of esotericism and occult philosophy, especially the central importance of microcosm-macrocosm cosmology and the charismatic figure of Mesmer himself. Many of his early acolytes, and even more of his later ones, were drawn to these aspects of his doctrine.[33] One of the most prominent was none other than Court de Gébelin, who published a letter addressed to the subscribers of the *Monde Primitif* in which he praised the

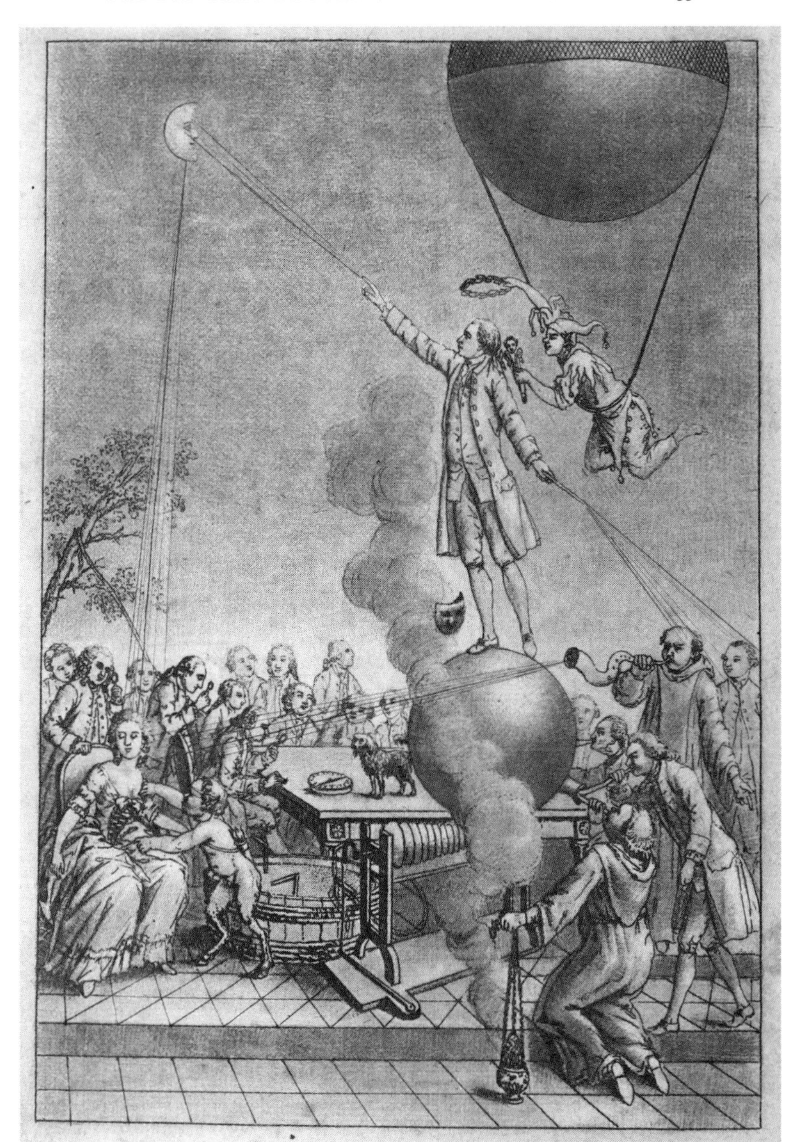

FIGURE 4.2 *Franz Anton Mesmer*, in Paulet, *L'antimagnétisme* (1784). Courtesy of the Wellcome Collection.

"sublime theory" of animal magnetism and claimed that "the discovery of M. Mesmer stems from primitive times."[34] On the other hand, famous opponents of Mesmer included none other than Court de Gébelin's admirer, Bailly. And yet, in a report he delivered to the Académie des sciences, Bailly proposed that while animal magnetism did not exist, Mesmer's cures were based on something that did: the power of imagination.[35] Thus even if it was never embraced as establishment science, the subject remained somewhat open for a while, and it was especially appealing to orphans of the Enlightenment.

Right as Mesmer's theory was exploding onto the scene, an anonymous essay entitled "Notice du cong-fou des bonzes Tao-sée" appeared in the fourth volume of the *Mémoires concernant les Chinois,* in 1779. The report was one of the most extensive discussions of Chinese medicine published in Europe to date, and another example of the new research directions that had opened up after the suppression of the Society of Jesus. Its missionary author explained Chinese "cong-fou," or kung-fu, as a theory and a practice in terms that would be familiar to his French readers.[36] Kung-fu is founded on the principle that "the human body is entirely hydraulic"; it is a machine, maintained by the circulation of bodily fluids, including the blood as well as more exotic ones such as "spirits" and "humors," and regulated by the pulse, which aerates the fluids and keeps them in equilibrium.[37] The hydraulic machine gains energy from the pumping of the heart and loses it due to gravity and friction. Based on this theory, the practice of kung-fu can affect the body in two ways: by controlling respiration through the mouth, the nose, or both, so as to affect the aeration of the humors; and by positioning oneself in standing, sitting, or prone postures, so as to counteract the effects of gravity and friction. The result of these breathing and kinesthetic exercises is to "operate a physical change in the hydraulic mechanism which either facilitates or impedes it," such that different posture and breathing exercises could cure particular ailments according to the cause of dysfunction.[38]

The essay on kung-fu also explored another topic that had been mentioned only in passing in previous European publications: the mysterious doctrine of the "*Tao-sée,*" or *daoshi,* literally "masters of the Dao," a term used in late imperial China to refer to Daoist priests.[39] According to the missionary author, kung-fu not only heals the body but also prepares the soul "to enter into commerce with the spirits and open the door of I-don't-know-what immortality, where one arrives without passing through the tomb." Adepts guard their secrets jealously, like "our alchemists of the Great Work." Still, even the scrupulous literati grudgingly acknowledge their wondrous feats, which they attribute to "an ancient practice of medicine founded on principles and quite independent of the absurd doctrine of the Daoists."[40] Here, too, was the suggestion that ancient wisdom had degenerated into a modern superstition. The reader was finally encouraged to withhold judgment: "The point of this notice is not to teach kung-fu, but to propose that physicians and doctors examine what to think about

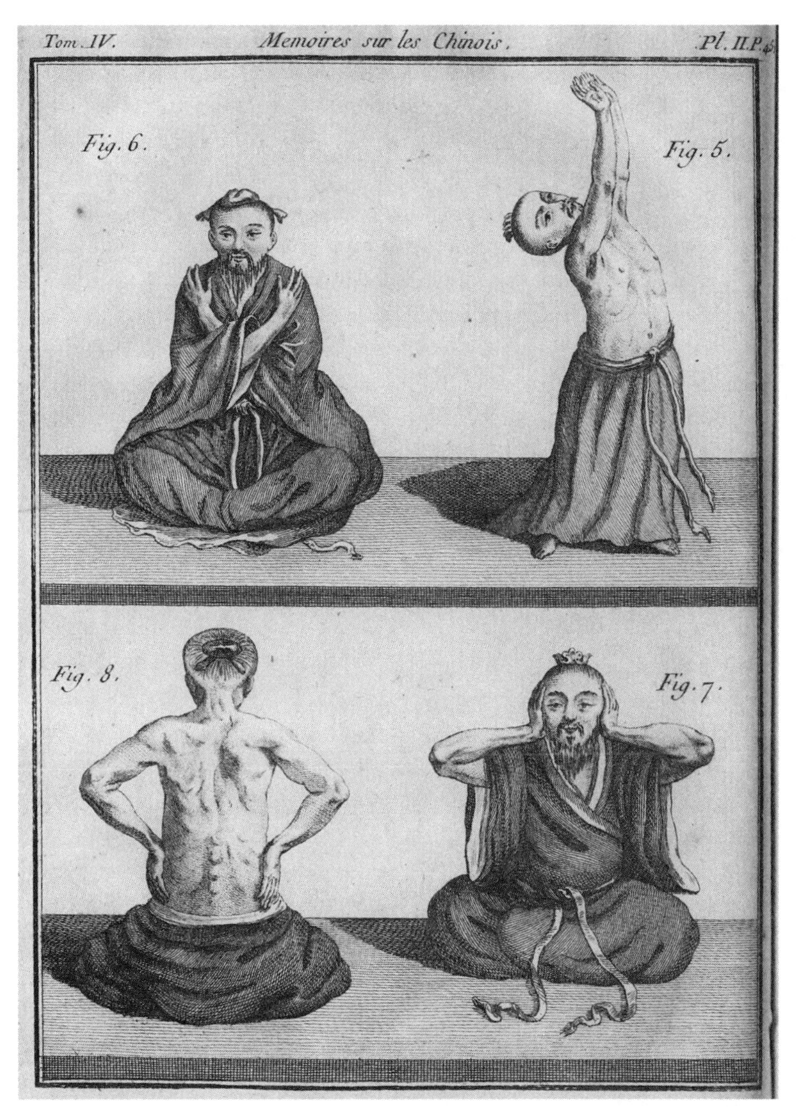

FIGURE 4.3 *Notice of the* Gongfu *of the Daoist Masters,* Pierre-Martial Cibot, in Amiot et al., *Mémoires concernant les Chinois,* volume 4 (1779). Courtesy of the Getty Research Institute.

it without prejudice. If the system upon which it is based should be false, it could lead them to find one truer."[41] On balance, this conclusion was as generous to Daoism as any thitherto published in Europe had been.[42]

With their coincidental appearance in the same year of 1779, animal magnetism and kung-fu came together at just the right time for Mellet.

The similarities must have been striking. After all, what was animal magnetism, according to Mesmer, if not a subtle fluid that runs through the body? And what was the body, according to the Daoist masters of kung-fu, if not something through which fluids flow? Both practices involved strange postures and odd gestures. Both claimed the rediscovery of an esoteric medicine that had maybe been known to the ancients, and both were condemned by their more orthodox contemporaries. Mellet also had personal connections at the forefront of research in both fields. His friend Court de Gébelin, a prominent mesmerist, was already in correspondence with Beijing. He was close with the Puységur family, one of whom was a major theorist of animal magnetism, and another of whom had recently written a book on Chinese military history. Most important, Henri-Léonard Bertin, initially curious about animal magnetism and by then the French gatekeeper to China, was his uncle. As a result, his opportunity to pursue both interests was substantially unique

In 1781, Mellet wrote to the missionaries in China with the explicit intention of putting kung-fu and animal magnetism into conversation. He compiled a "little collection of works on mesmerism" and drafted an accompanying letter for the missionary author of the article published in the *Mémoires concernant les Chinois* to see "if *kung-fu* did not have something to do with all that." He addressed his letter to Amiot, "missionary of very great learning, whose truly apostolic beneficence for the progress of the sciences and the well-being of humanity" he already knew.[43] What Mellet did not know was that the author of the article on kung-fu was in fact a different missionary, Pierre-Martial Cibot.[44] But this mistake was fortunate, because Cibot had died the year before, so any letter written to him would not have elicited a reply, and the one to Amiot did anyway.

The Yin-Yang *Theory of Animal Magnetism*

Amiot's first response to Mellet consisted of two long letters that were conceived of together and written between 1783 and 1784; impassioned and fervent, he articulated a theory of natural philosophy with far-reaching implications for the study of China's past and possibilities for Europe's future: the *yin-yang* theory of animal magnetism.[45] The main point was that Mesmer's modern theory was perfectly consistent with, and indeed a rediscovery of, the ancient wisdom of China. The realization revolutionized Amiot's approach to Chinese natural philosophy: "What you tell me about the miracles Mesmer performs among you opens my eyes," he explained to Mellet; "all that M. Mesmer said about animal magnetism, the

Chinese have said of *yin-yang* for almost four thousand years." The subtle fluid of animal magnetism was another name for *taiji*, and its positive and negative poles were *yin* and *yang*. In China, *yin* and *yang* were "the key to all the sciences. There is nothing that cannot be explained using them properly."[46] The discussion of Chinese medicine, cosmology, physics, and metaphysics that followed was extremely idiosyncratic and distorted. It was also among the most enthusiastic of the entire early modern period.[47]

Amiot's interpretation of Chinese natural philosophy was essentially based on the orthodox Neo-Confucianism canonized during the Song and maintained throughout the Qing. Though he cited no specific authors or sources, he likely drew from recent textbooks and encyclopedias such as the *Great Compendium on Nature and Principle* of 1515, which had been used by generations of Jesuits before him. According to Amiot, "the physical principles that the Chinese admit in nature" could be essentially reduced to *taiji* and *yin-yang*, united to "form a totality, which is in every thing and everywhere." *Taiji* is the "universal agent" and "material principle," which "in French you might call 'nature.'" It is constituted by *yin* and *yang* and acts through them. *Yang* is the active, perfect, or complete: movement, visible things, force, light, the sun, and the male. *Yin* is the passive, imperfect incomplete: rest, water, shadows, weakness, the earth, the moon, and the female. The essence of every being derives from the relative proportions of *yin* and *yang* within it, and "the order that reigns in nature" is nothing more than "the effect of the concourse of *yin* and *yang* combined with each other, following the rules of harmony."[48]

To illustrate, Amiot included an image that he called the "Universal Figure" (see fig. 4.8). In fact, it was based on the "Diagram of the Supreme Polarity," attributed to the eleventh-century philosopher Zhou Dunyi and fundamental in later Chinese natural philosophy.[49] The particular version Amiot sent appears to have been extracted from another Ming compendium, the *Foundational Meaning of the Six Books*.[50] As he explained it, the circle represents the universal agent *taiji*. The light half labeled A is *yang*, and the dark half labeled B is *yin*. In nature, *yin* and *yang* are always found in agglomeration, represented by the small circles of *yin* within *yang* and *yang* within *yin*. Their mixture is the manifestation of *li*, the "first mover," or the "fuel" of creation, which excites *yin* and *yang* and causes their mutual attraction.[51] Though it appears that *yin* and *yang* are drawn to each other, each actually seeks to return to its own kind, having been separated by the activity of *li*. Through this mechanism, they "produce, by the continual explosions of their rays, all the corporeal beings that exist, make them take a form, insensibly destroy them and them renew them to destroy them again, &c. &c." The "Universal Figure" thus presented both a physics and a metaphysics, constituting a complete system of natural philosophy.[52]

This was the ancient wisdom that the orphans of the Enlightenment had been searching for. It should be noted here especially that Amiot distorted the Chinese record. The physics of *yin* and *yang* was developed during the Warring States period, and the metaphysics of *li* and *taiji* did not predate the Song.[53] Amiot knew that the theory he was describing essentially dated back less than a thousand years, but he believed that it was consistent with the natural philosophical principles of the time of Confucius, which "fell little by little into discredit, and were then almost entirely forgotten, to reappear in the writings of the self-styled sages after more than ten centuries of oblivion." Looking back further, he contended, the tradition of the classical period could be traced back all the way to the Three Sovereigns and Five Emperors, who lived not long after the Flood. Their knowledge was "an encyclopedia so vast, and profound, and at the same time simple" that it must have been yet more ancient still; a "system of the world, the knowledge of which it is very probably that men were not lacking before the deluge." At its ultimate origin, Amiot identified the lost Atlantan people described by Bailly or an antediluvian people who had lived even before them.[54] He was finally in a position to say not just where ancient wisdom might be found, but to describe what it looked like.

Amiot now believed that Chinese natural philosophy could show a way forward for European science. He saw himself as a kind of spokesperson on its behalf: *yin-yang* was "the only key that can open for us the sanctuary of nature; it is only in it and through it that we can make sense of all the phenomena that are encountered in the tenebrous regions of physics, and that we can come to form a clear idea of the true theory of the world."[55] Chinese philosophers were just as clear as "your Newtonian philosophers and others, when they speak of principles of attraction, of electricity, of the movement of magnetism, of the cooling of the planets, etc. etc." Indeed, Amiot took the remarkable step of suggesting that Chinese philosophers had already made the same discoveries: "The Newtonians and all attractionists, both ancient and modern, could without scruple adopt *li* for the *unknown principle* of their various attractions; all natural philosophers, of whatever sect they are or may be, could make it the base of their systems: and all those who search the truth, could regard it as the solid foundation upon which rest all the truths that they want to discover."[56] Not before had a missionary had so much faith in Chinese natural philosophy as to believe that contemporary Europeans should adopt it in their science.

To help them do so, Amiot sent for Mellet a certain "Chinese book" along with a brief translation of an extract that he believed was especially relevant for the medical applications of mesmerism.[57] As he explained its theory, *yin* and *yang* operate in the human body through a host of "particular agents," each of which acts as "a kind of guardian" entrusted with the care of a specific organ. For a given malady, treatment involves the manipulation of the particular agent associated with the site of its etiology. For example, the liver is governed by a particular agent called "smoke of the dragon," also known by the courtesy name "reservoir of clarity" and represented with the symbolic image of a dragon. It resides at the bottom of the heart, extending its influence through three capillary nerves to the thumb of the left hand.[58] Suppose that a superabundance of *yin* sticks to a patient's liver, preventing it from functioning properly. A doctor should take his own left thumb and press it against that of the patient, allowing *yin* and *yang* to flow between their two particular agents. Eventually, the *yin* and *yang* surrounding the patient's liver will be restored to balance, and the patient will be healed.[59] Amiot's conclusion was that "Chinese doctors possess the knowledge of the precise quantity of *yin* and *yang* necessary to create equilibrium in this or that subject"; and, as he put it later, "it is precisely this [knowledge] of which Mr. Mesmer is in possession, to the exclusion of other doctors"—or, at least, to the exclusion of other doctors in Europe.[60]

The "Chinese book" in question was in fact a chapter called the *Zangfu bu*, or "Organs Section," taken from yet another recent compendium, the *Complete Collection of Ancient and Modern Figures and Texts*, which was begun under the Kangxi emperor and completed in 1726.[61] The "Organs Section" is a miscellany of classical, medieval, and modern sources. Some of the material came from the *Compendium of Materia Medica*, composed during the Ming, and some from the *Inner Canon of the Yellow Emperor*, traditionally attributed to deep antiquity, though likely composed no earlier than the Warring States period.[62] Also included were passages from the syncretic *Huainanzi* and the Daoist *Zihuazi*, which did not predate the Han.[63] Taken together, the "Organs Section" thus put forth a theory that reflected the universal ambitions of that and later periods, and had almost nothing to do with any real pre-classical medical theory or practice.[64] The original text reveals that it is primarily concerned with the *zangfu*, or internal organs of Chinese medicine. The *zang* organs, mostly related to the cardiovascular system, are *yin*, and the *fu* organs, dealing generally with digestion, are *yang*. For each organ, the chapter includes two illustrations. The first depicts the organ itself, with a description of its size and position in the body. The second depicts a *shen*, or "spirit," that presides over the organ's functions—this was what Amiot termed the "particular agent."[65]

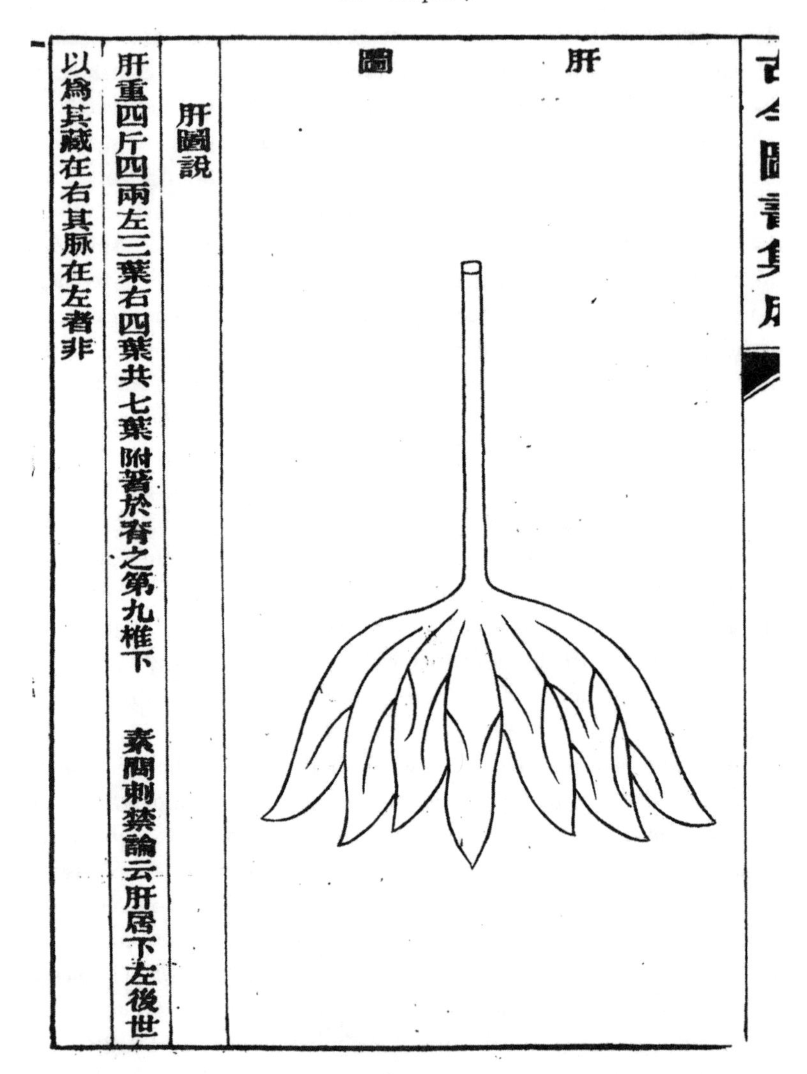

FIGURES 4.4, 4.5 *Liver* and *Spirit of the Liver,* in "Zangfu bu" (1726). Courtesy of the University of California Library.

Based on the original, it is not difficult to see why Amiot identified conceptual similarities between late imperial Chinese medicine and animal magnetism. First was the idea of microcosm-macrocosm correspondence. The "Organs Section" referred to the three powers or realms of heaven, earth, and man, an organizing concept in Chinese natural philosophy since the Ming.[66] The first proposition in Mesmer's *Mémoire* was: "There exists a mutual influence between celestial bodies, the earth, and animal

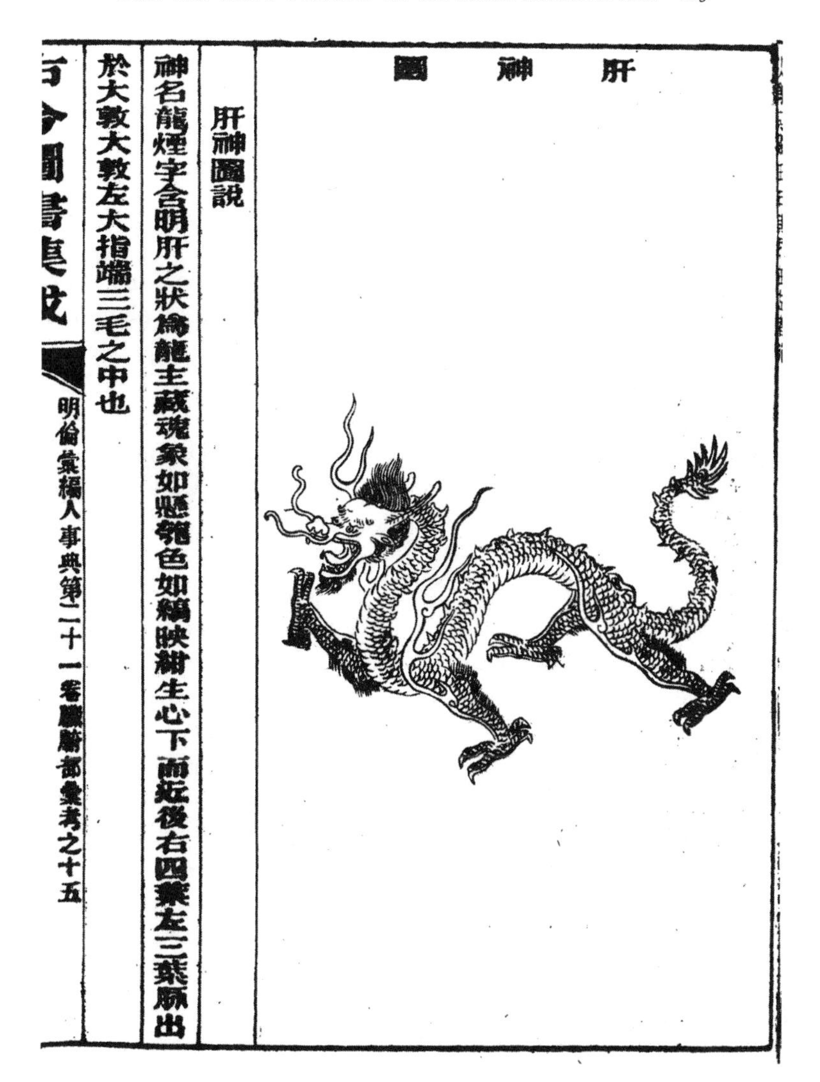

肝神圖

肝神圖說

神名龍煙字含明肝之狀爲龍主藏魂象如懸匏色如縞映紺生心下而近後右四葉左三葉脈出於大敦大敦左大指端三毛之中也

今圖書集成

明倫彙編人事典第二十一卷臟腑部彙考之十五

bodies."[67] Second was the importance of paired polar forces. Just as the dual principles of *yin* and *yang* explained the relationship between the seasons and the human body, animal magnetism was supposed to flow between heavenly and human bodies.[68] Even the technical vocabularies had some overlap. The Chinese word *shen* is a pretty good translation of the French *ésprit,* which featured not only in mesmerism but more broadly in the vitalist medicine of the time.[69] Both philosophies emphasized universal correspondence; terms meaning something like "harmony" appeared frequently in the "Organs Section" as the measure of successful treatments and proper bodily functions, while the major mesmerist club in Paris at the time called itself the "Society of Universal Harmony."[70] Referring no doubt to such similarities, Amiot wrote: "The *yin* and *yang* of the Chinese are the base on which they have built the system of universal harmony, that is to say, the existence of a mutual influence between the celestial bodies, the earth, and animal bodies." The observation was at least somewhat perceptive, as far as it went.

Still, there were also components of the "Organs Section" that Amiot either overlooked or elided. The passages about spirits that might have struck earlier commentators as superstitious did not much trouble him; the bigger problem was those that suggested too much naturalism. Notably absent was the fundamental concept of *qi*, meaning *air, breath, energy* or *force*.[71] *Qi* was invoked ubiquitously throughout the "Organs Section": bowel-*qi* clears the stomach and ghost-*qi* causes disease, while earth-*qi* is produced in the womb and water-*qi* irrigates the veins.[72] Amiot once misleadingly defined it as "the all-powerful breath of God," but seems to have also understood it more correctly as relating to matter.[73] Likewise absent was another crucial concept, the five phases: fire, water, wood, metal, and earth.[74] For example, the "Organs Section" held that "heaven and earth are completely full and replete with things; all things are the five phases; the five phases exist; nothing can be without them."[75] In his discussion of the "Organs Section," Amiot focused exclusively on *yin-yang*; he did not even mention *qi* or the five phases. He thus muted some of the materialist overtones that had made much of late imperial Chinese natural philosophy unacceptable to missionaries since the days of Matteo Ricci.

Amiot did not deny that some elements of the text were dubious, nor that the theory was controversial even in a Chinese context. Justifying his decision to send it anyway, he recounted a conversation that had supposedly taken place between the Kangxi emperor and the scholars of the Hanlin Academy who were responsible for editing the *Collection of Ancient and Modern Figures and Texts*. Some of the academicians protested "that it is necessary to reject from the new collection all these traditions and supposedly antique monuments on literature and the arts and sciences that contain manifest absurdities." The emperor replied that the "Organs Section" should be included nevertheless: "It may happen that what you regard today as absurdities will be viewed by those who come after you as so many luminous and incontestable truths," he replied. Amiot exhorted Mellet to accept the volume, "supposing with the emperor . . . that that which appears against reason is not always absurd."[76] Not all readers would be so generous.

Amiot knew that selling Chinese natural philosophy in Enlightenment

Europe would be no easy task, though its promise was almost without limit. "Admit, sir," he entreated in one letter, "that your presumptuous savants, who regard the Chinese as such bad natural philosophers, would be well surprised if they had to adopt the same system of these same Chinese, as being, if not true, at least the most satisfying and most close to true."[77] The value of Chinese natural philosophy was proved most of all by the efficacy of Chinese medicine: "You say, this is the language of a charlatan. I would agree, if you like, but at the same time I ask that you accept that my doctor is by no means a charlatan. I am living proof of it, for without his aid, I would have died long ago."[78] The recent rejection of Chinese natural philosophy in Europe was lamentable indeed: "Oh! If our philosophes of Europe, with their minds closed in such different ways from those of the Chinese, wanted to give themselves to this study, what a crowd of knowledge would they not be able to enrich us with!"[79] But now, there was new hope, presented in the person of Mesmer. If he could only overcome the obstacles of prejudice and false science, he would reveal "the knowledge of the operations of nature, and of the true system of the world."[80] Better still, his work might bring Europeans to study the natural philosophy of China, and "every year would be marked by the appearance of some phenomenon that would delight with just cause the admiration of the public."[81] This was the ultimate promise of the *yin-yang* theory of animal magnetism.

Amiot outlined a plan to bring about this philosophical union; he called it "the Order of the Universal Figure." [82] A secret society half Salomon's house, half *Loge des Neuf Soeurs*, it would lead the united nations of Europe in plumbing the secret depths of nature. "Knights" of the order would be chosen for their works in "abstruse philosophy," or alternatively for "having a taste for this kind of work." Each of them would carry a medallion of white ivory and black ebony, with the "universal figure," the *taiji* diagram, engraved on one side, and the "sacred quaternary," the trigrams of Fuxi and suggestive also of the Trinity, depicted on the other. There would be a "grand master" chosen by election, although Amiot already had a candidate in mind: the "Unknown Philosopher," author of *Des erreurs et de la vérité, ou les hommes rappelés au principe universel de la science*, whose thought the universal figure "contained in its entirety."[83] The grand master would be aided by two assistants, representing *taiji* and its agents, *yin-yang*. Other officials would be chosen too, one for each of the sixty-four hexagrams of the *Book of Changes*. If not enough worthies were to be found in France, then England, Germany, Italy, "and even Switzerland" could fill out its ranks.[84]

As Amiot's proposals ran wild, it became clear what the Order of the

Universal Figure was really all about. It would meet four times a year, once for each season. Knights would be seated according to age, showing Confucian respect for elders. Normal meetings would address philosophical topics, with no limits on what could be said and no fear of causing offense. Their other main activity was a strange but revealing funerary rite: knights would assemble before the corpse, prostrate themselves, and perform a solemn kowtow in recognition of the deceased. The Jesuits had long defended ancestor worship as compatible with Christianity. Pursuant to the Chinese Rites controversy, sacrificing to the dead was forbidden to all Christians in China. Nevertheless, Amiot offered, "I believe it might be permitted to the Christians of Europe, since there is not for them the same danger of perversion." It was as if the Society of Jesus, extinct in China, could be revived in Europe with its Sino-European syncretism fully intact. Being so far away, Amiot left the details for Mellet to hash out on his own.[85] No one ever seems to have mentioned the Order of the Universal Figure again, but its imagined charter set the real stage for the *yin-yang* theory of animal magnetism.

The Comte de Mellet's Treatise on Chinese Magnetism

Over the course of the 1780s, Mellet concluded that ancient Chinese natural philosophy could indeed contribute to modern European science. None of his longer works survive; nevertheless, by piecing together Amiot's responses to his inquiries, scattered letters he exchanged with other French savants, and one or two obscure published pieces, it is possible to partially reconstruct his thought and to understand how China fit into it. In conversation with other orphans of the Enlightenment, Mellet had already outlined his own system of ancient wisdom, which focused on a mesmerist interpretation of electricity and magnetism. After reading Amiot's first letters, he concluded that China offered a way to prove and expand upon his theory. He actively looked for Chinese facts and theories and planned the publication of a *Treatise on Chinese Magnetism*. He sought support for his work from among the most respected authorities of the French scientific establishment. In the end, he failed to win it. What he presented may have been some kind of knowledge, but by the end of the Enlightenment, it was not what counted as science.

In 1784, the sensation stirred up by Mesmer prompted Louis XVI to appoint not one but two royal committees to look into animal magnetism. One was composed of leading members of the Paris Faculté de médecine; the other was conducted under the auspices of the Royal Society of Medicine. Their investigations have been much studied as a textbook case of scientific validation: they conducted a series of blind trials to determine the empirical effects of Mesmer's treatments and considered a variety of naturalistic explanations to account for them. After several months, the Faculté de médecine commission, headed by Benjamin Franklin and including Antoine Lavoisier, Jean-Baptiste Le Roy, and Jean-Sylvain Bailly, among others, issued a scathing report. Mesmer's treatments did seem to have some healing power, but it did not seem to matter which one he applied in any given situation; animal magnetism, therefore, was not the explanation for their effects. Mesmerism was thoroughly discredited and excised from the French medical and scientific community. Mesmer left Paris in disgrace the next year.[86] Yet his story exposes the importance of Enlightenment esotericism in two historiographically significant ways. First, Mesmer was taken seriously by many of his contemporaries: the rejection of animal magnetism was not foreordained.[87] Second, modern psychology drew much inspiration from this work: the establishment rejection of animal magnetism did not make it disappear.[88]

Mellet's particular spin on the idea of ancient wisdom was to take it in an openly mesmerist direction. He placed his own theory among other "systems" of the day: "Here is mine. There was a people who possessed the arts, all the arts, better than us, and more than us." His influences were quite explicit: "I was friend and servant of Mr. de Gébelin, and am still incorrigible in my amity," he wrote; as for Bailly, "I am his servant, as any man who seeks to cultivate letters must be of he whom he regards as a master."[89] Mellet's version began with the hypothesis of what he called the "Adamic Medicine," which he believed could be recovered through modern science. It was supposedly a treatment administered remotely by control of a fluid with electrical and magnetic properties. According to his explanation, just as the electrical fluid charges a battery, so too can it charge a human body. And just as the magnetic fluid passes between magnets in proximity to each other, so too can it pass between one human body and another. In sum, he maintained: "Man is the most perfect magnet and electrical machine."[90] This theorizing was all quite consistent with Mesmer's animal magnetism, which Mellet saw as an expression of his own discoveries. He even claimed to have been a mesmerist avant la lettre, since he had suspected the existence of an ancient medicine based on electricity and magnetism "a long time, sir, and a very long time" before anyone had heard of Franz Mesmer.[91] Such claims too show both how quickly Mesmer's theories had spread and how slow they were to dissipate.

Even after Mesmer's disgrace, Mellet remained a true believer. While the royal committee's disparaging report pushed animal magnetism further away from the scientific establishment, Mellet seems to have only grown more interested in the idea on that account: "That which seems absurd, extravagant not always is," as he put it.[92] He went on from there to accept the whole parade of mesmerist theories that appeared in the following years, from Barthélémy Bléton's use of the divining rod to Armand-Marie-Jacques de Chastenet de Puységur's practice of magnetic somnambulism. By his own estimation in 1787, he had read everything available on related topics.[93] Every morning, he spent four hours studying animal magnetism. In the afternoon, he would venture out on his domains, curing at least one sick person and easing the suffering of thirty or forty more. Where more-sensitive savants might fear contagion, Mellet proudly got his hands dirty. Mesmerism withstood the scrutiny of his own experience: "I have denied, I have doubted, I have struggled, I have seen, I have done, I believe." It was practice, not theory, he claimed, that convinced him that the mesmerist phenomena were real—not that he seems ever to have doubted it.[94]

Nor did the rejection of mesmerism discourage Mellet from seeking approval from the very people who had been responsible for it. Probably in 1784, with the major advantage of an introduction from his powerful uncle Bertin, Mellet sent a now-lost "Mémoire" to the prominent physicist and former director of the Académie des sciences, Jean-Baptiste Le Roy.[95] Le Roy was a leading expert on electricity and magnetism and an early proponent of Benjamin Franklin's work (and, incidentally, the recipient of the 1789 letter in which Franklin wrote that "nothing can be said to be certain, but death and taxes").[96] He was also a prominent critic of animal magnetism. After expressing some early enthusiasm for Mesmer's theories, Le Roy had been personally embarrassed when Mesmer performed a demonstration before the Académie des sciences at his invitation and was laughed at by the other academicians.[97] He also served as one of the commissioners who found against Mesmer's medical and scientific claims. So it is a little surprising that Mellet chose him as a reviewer for his mesmerist memoir. More surprising still is that Le Roy seems to have read it with genuine care.

His response contains the fullest surviving account of Mellet's theory of ancient electricity and magnetism. According to Mellet, the oldest records, from the Greek myths to the Hebrew scriptures, were replete with accounts of abnormal or paranormal events. Commonly attributed to supernatural powers, these phenomena were explicable as effects of electricity and magnetism. The ancients had not only understood the forces that had recently been uncovered by modern science but had further possessed the capacity to manipulate them. The Eleusinian Jupiter called down lightning from the heavens. The prophet Ezekiel experienced a vision of God as an explosion spewing plumes of flame. Lesser examples could be found all the way through to the times of Livy and Ovid, when, for unclear reasons, the ancient knowledge was lost. Mellet explained these accounts in the language of contemporary science, referring for example to experiments with the Leyden jar and studies of lightning's effects on the compass. The thrust, however, was not to explain away the ancient miracles, but rather to show that they had really taken place.

Le Roy found the whole thing ridiculous. He thought Mellet was wrong on ancient wisdom and even more wrong on modern science. The Hebrews and the Greeks had not understood electrical or magnetic phenomena, much less how to control them. No translation of the Book of Ezekiel suggested electrical or magnetic effects, and no report on the Eleusinian Jupiter came from a qualified witness. Mellet's discussion of electricity and magnetism assumed properties that were not known, or, in one case, that Le Roy himself had already disproved. Most disturbing of all to Le Roy was Mellet's open embrace of occult science: "I cannot help but observe on the subject of Kabala and magic that it seems from your explanations that you want to show the existence of sciences that it seems to me today are generally agreed to have no foundation in nature." The problem with such practices was that they were not predictive: "Even if all events were enchained such that we are subject to a fatality without limits," occult sciences cannot "give us the power to tear the veil in which the future is enveloped for us." Le Roy again invoked the philosophe champion: "There is no magic besides philosophy, as Bacon said." Mellet's memoir was essentially unscientific: "It seems to me that a *bon esprit* such as yours, monsieur le comte, cannot undertake to explain that which does not exist." Le Roy suggested revising all the parts that touched on electricity and magnetism. Since that was apparently most of the manuscript, one might have assumed that the comment was not entirely sincere.[98]

Disappointed but not discouraged, Mellet took his memoir to someone whom he had good reason to think would be more sympathetic: "I am delighted that M. Bailly has my memoir; I report myself certainly ready to delete, correct, or throw in the fire."[99] Mellet thought his theory would "find defenders among the Atlantans," that ancient people whom his system credited with the invention of telescopes and electric machines.[100] But a brief reply to Mellet, probably written in 1785, shows how much Bailly's own thoughts on the matter had changed: "There was certainly an ancient state of sciences, which was followed by barbarism; this is the ancient science the debris of which I collect; but whatever was the perfection of these sciences, they cannot be compared to those of our modern times."[101] To believe in ancient wisdom had been one thing; to equate it with modern science was quite another. For Mellet, the rebuke stung: "I had some reason to count on the favors of M. Bailly; it was possible that in traversing the unknown lands of the *monde primitif*, we would meet each other, but perhaps I travelled in the land of chimeras."[102] Nor was Mellet the only person to wrongly make such assumptions. Bailly had by this time become something of a cult figure in mesmerist circles. For example, in 1784, another orphan of the Enlightenment had published a letter addressed to him—"my first master in Animal Magnetism"—explaining it as consistent with both Newtonianism and astrology. [103] The author insisted that Bailly's historical and scientific theories had already said as much, so he should change his opinion on Mesmer's.

But Bailly always had one foot in the door of establishment science—he indeed served as one of its doormen during the Mesmer affair—and it was closing fast on both esoteric and non-Western alternatives. In 1787, he published his last major work on the history of science, the *Treatise on Indian and Eastern Astronomy*, in which he claimed to clarify but in some ways walked back his earlier position on the ancient scientific knowledge of Asia: "The Orientals were never what we are. Whatever good opinion the study of their astronomy might suggest in their favor"—a clear consequence that many had drawn from his own previous studies—they lacked "this genius of discoveries, which seems to belong now to Europe alone, and which, substituting for time, causes the sciences and the human mind to make rapid progress."[104] In sum, a high opinion of ancient eastern knowledge was not consistent with progress as defined by the development of the arts and sciences over time. This was the consensus position among the late philosophes, and even their late Enlightenment critic seems to have come around. Thus, in 1784, when Bailly joined Le Roy in rejecting the modern theory of animal magnetism, his enthusiasm for the knowledge of Asia and the ancient world alike had faded. For orphans of the Enlightenment, the effort to unite them would be a last-ditch effort.

At some point in the following year or two, Mellet responded to the two respected academicians' revise-and-resubmit reports. He repackaged his memoir with a new title, the *Argonautide*, presenting it as both an "antique encyclopedia" and an adventure story narrated through the voyage of Jason and the Argonauts.[105] Mellet envisioned it as a primer on science and language that would be less boring than the "*contes bleus*" that schoolboys were forced to learn and immediately forgot. After forwarding copies to Bréquigny and Bertin, he optimistically sent it back to the original reviewers as well. Despite Le Roy's pointed criticism, he claimed, "I nevertheless have much respect for his opinion." Bailly he was still more optimistic about bringing on board: "It seemed to me that M. Bailly had judged me with less severity; maybe now that I have developed my ideas more, he will approve them."[106] Six months later, he was still waiting to hear back from either of them and inquired with another about the resubmission: "If it should be thrown it to the fire, I assure you that I hardly care, and if these *messieurs* think I am wrong, it is pretty much all the same to me. This would not stop me if I had a young man to educate."[107] The author of the *Argonautide* had himself come to believe that his theory of ancient wisdom would find its greatest success as a children's story. It is unclear if any of those figures whose approval he sought granted even that much.

By this time, however, Mellet's correspondence with Amiot was finally beginning to bear fruit. He had avidly kept up with the missionaries' publications: as he wrote to the editor, "I know the Chinese *Mémoires* too well not to be sure in advance of the pleasure" the upcoming volumes would bring.[108] He had studied Amiot's first two letters on the *yin-yang* theory of animal magnetism so carefully that he was able to recall them from memory, and mostly without error.[109] He had used already them to supplement his own theories: for example, in one study, he explained that the *yin-yang* diagram, like the Egyptian sign of Isis and also the twenty-first trump card in the tarot deck, confirmed the ancient allegory of a universal egg.[110] Amiot came to consider Mellet an excellent pupil, "very learned in different cosmologies and mythologies of ancient Eastern nations."[111] For his part, Mellet declared: "I would very much like to be a disciple of Confucius."[112] After a decade of mutual learning and conversation with other orphans of the Enlightenment, their interests were now aligned.

Mellet wrote to Amiot again. This time, he inquired about another essay that had been written by Cibot about another medicinal practice with Daoist associations: the uses of deer blood.[113] Said to contain the "soul and the life" of the animal, the blood was to be harvested in autumn, bled warm from the jugular vein, mixed in a wine-based concoction, and dried to a powder that could restore vigor and combat smallpox. As with kung-fu, there was a veil of secrecy surrounding such practices, related again to Daoist alchemy. Cibot had acknowledged that deer blood probably did possess some anti-aging qualities, but he lamented its superstitious association with the elixir of immortality.[114] For Mellet, however, this association was the most interesting part, because it presented an opportunity to link ancient allegories with natural effects. Throughout the primeval world, he believed, certain animals had served as emblems for representing things "in an occult manner." One of his questions for Amiot, then, was whether Chinese myths or legends discussed a "deer of the immortals," which he thought might explain the global use of its blood in alchemical decoctions.[115]

Amiot was once again happy to answer. Once again, he turned for research to a more modern work, the *Compendium of Materia Medica*, completed by the naturalist Li Shizhen in 1578.[116] There, he found still-more-grotesque instructions for the creation of deer-blood elixirs: the deer could not be brought down by dogs and its blood must be extracted through a tube while still breathing. Given these conditions, Amiot thought, the medicine might perhaps be effective: "It is better to believe even the impossible which has not yet been demonstrated as such than to dare to place limits on nature." To encourage further study, he sent Mellet an assortment of medical curios, including a "fulminating needle"—a mixture of animal, vegetable, and mineral substances that produced a tonic fume for arthritic pains—and a wind instrument used as a deer call, so that he could capture a stag and make the medicine himself.[117] Amiot further approved of Mellet's idea regarding "the rapprochement one might make between Chinese emblems and those of the other peoples of antiquity." Together, he imagined, "we would gather our lights in the same foyer, allowing us to put out a work that would astonish" their less enlightened contemporaries.[118]

Mellet already had an idea of where such a project might begin: with his "Chinese manuscript," by which he meant the "Organs Section" of the *Collection of Ancient and Modern Figures and Texts* that Amiot had sent him. As he put it to Bréquigny, the academician responsible for editing the missionaries' letters in the *Mémoires concernant les Chinois*: "You see how much I wish that the Chinese, well translated, will succeed in enlightening us on an agent that, according to me, relates to everything."[119] He therefore asked Bréquigny to help him find a translator for the manuscript "who would not be fooled by European ideas." At the time, there were only two good candidates, the aged sinologue scholars at the Collège royal. Mellet steered the project away from de Guignes, whose disregard for Chinese antiquity aroused his concern that he would "see nothing there but the conquest of disasters and I would like that one sees there what is there." Instead, he proposed Deshauterayes, who he likely knew had written a Euhemerist interpretation of Chinese cosmogony decades earlier and was then planning a magnum opus about the apocalypse, so could have seemed more likely to support Mellet's own project. An unaided translation of the Chinese original was to be preferred since it would be the truest to the original source. If Deshauterayes really needed some assistance, he could be given the brief excerpt already translated by Amiot.[120]

Mellet put together a plan for publication under the title *Traité du magnétisme chinois*. In addition to the "Organs Section," it would also include Amiot's letters as well as additional critical commentary by Mellet himself. The overarching thesis was that "magnetism, conserved and studied for four thousand years in China, should be much better known there." In particular, Mellet thought that Chinese sources would explain the "causes and effects" of the divining rod, which had recently been revived in France. He cited the physician Pierre Thouvenel, whose work combined rhabdomancy, or dowsing, with electricity and magnetism as interpreted along mesmerist lines.[121] The theories struck Mellet as highly resonant with the story that Amiot had recounted to explain how Chinese physicians manipulate particular agents. During the Tang or Sui period, a doctor had cured the ailing wife of an official by inserting a bamboo tube through a hole in a wall and touching the site of the patient's malady, facilitating the flow of *yin* and *yang* between their two bodies and restoring the sufferer to equilibrium and health. Probably, Mellet thought, Thouvenel's divining rod controlled the flow of animal magnetism in the same way that the bamboo tube directed *yin* and *yang*. He sent Amiot a recent overview of research on the topic that had been published in the *Journal de Paris* for further comment, knowing he would have to wait a few years for a response. In the meantime, he gathered together everything that he had been sent already. By the end of 1787, Mellet had completed his own original preface for the *Traité du magnétisme chinois*.[122]

But in the end, publication was delayed, for two reasons. First, Mellet was missing the most important part: the translation of the "Organs Section." Unfortunately, the original Chinese manuscript seems to have gotten lost somewhere, and when asked about it years later, Deshauterayes denied having ever received anything from Bréquigny on the subject of animal magnetism.[123] It seems Mellet had already realized that he was unlikely to ever get his translation from the Chinese. Instead, he had a new idea: perhaps Amiot could send a Manchu version of the book, in which case Louis Langlès, a young scholar who was then working with the support of Bertin to publish Amiot's Manchu dictionary, might be more forthcoming with assistance. This expectation was reasonable enough, since Langlès was not likely to neglect such a request from his patron's nephew. Yet, there was also a second reason to delay publication: Mellet worried that his "extraordinary ideas might distress Father Amiot." This concern turned out to be entirely unfounded, but, wanting to secure the missionary's approval before moving forward with the project, he decided that it was best to wait.[124] Once again, by the time Amiot's responses finally arrived, it was too late. The delay in publication of the *Traité du magnétisme chinois* turned into a cancellation with the beginning of the French Revolution.[125]

Magnetic Somnambulism and Daoist Magic

Mellet and Amiot agreed that ancient Chinese natural philosophy pointed the way toward future scientific discoveries. But where Mellet was restricted by skill and circumstance in pursuing this shared ambition, Amiot could bring to bear on it the full weight of over three decades living and thinking in Qing Beijing. During the last five years of his life, he began to reconsider aspects of Chinese history, religion, and culture that almost all Europeans before him had either disparaged or ignored. Historians today often distinguish philosophical Daoism, as expressed by Spring and Autumn and Warring States period thinkers such as Laozi and Zhuangzi, from religious Daoism, a syncretic popular movement that began to take form hundreds of years later.[126] Amiot was primarily concerned with aspects of the latter, although his picture of it would hardly be recognizable to a modern scholar of Chinese religion. His studies of magic and alchemy through the lens of Daoism ultimately served less to shed any real light on ancient Chinese traditions than to associate them with new narratives about European ones.

The transformation in Amiot's thought on the subject he understood as Daoism was marked. It took place entirely during the period after the suppression of the Society of Jesus, reflecting both new interest from Paris and new opportunities to pursue it in Beijing. For most of his life, he had agreed with almost everyone else in Europe that Daoism was "a vile sect, given over to a kind of magic art founded on the most absurd principles."[127] In the mid-1780s, his evaluation was undergoing a shift: since it represented the first phase of the corruption of China's most ancient traditions, Daoism was wrong, but still worth studying.[128] By 1790, his evaluation of Daoism had substantially reversed: "This sect of the Dao, one of the most ancient of all those that have existed among men, subsists still such as it was in its state of origin, while all the others have been absorbed in the gulf of time." It was his great regret was that he had only come around so late to realize that Daoism was the doctrine that preserved the ancient wisdom of China the best.[129]

This transformation was catalyzed by Amiot's deepening commitment to mesmerism. By the end of his life, he possessed over a dozen works on the subject.[130] In 1784, he received a copy of Court de Gébelin's letter of support to the subscribers of the *Monde primitif*. If the Faculté de médecine and all their "henchmen" were against Mesmer, Amiot thought, very well: Court de Gébelin "should have no need for them."[131] The next year, Amiot heard from Bertin about Court de Gébelin's death in Mesmer's arms after a series of animal-magnetism treatments: "I do not know if it was right to think that he owed his first healing to Mesmer, but in his relapse the same method sadly unfortunately not have the same success."[132] Amiot assumed the voice of a Chinese doctor to respond. He invoked the concept of *yuanqi*, or innate *qi* energy: "When the *yuanqi* begins to exhale and to evaporate, it is impossible to retain it within its old limits and death necessarily follows."[133] Therefore, he concluded, no human art, however wonderful, could have saved Court de Gébelin. Mesmer's cures were undeniable, and the explanation for them had to be animal magnetism since the alternative incorrectly attributed "to the imagination and to imitation effects far beyond what either can produce." The royal committee's discrediting of Mesmer was further evidence of the depths to which Europeans had fallen; by contrast, the ancient Romans "would have erected temples in his honor."[134] Like Mellet, Amiot only grew more devoted to animal magnetism as it diverged from mainstream science.

Amiot was particularly interested in a new mesmerist theory known as "magnetic somnambulism," which was developed by Armand-Jacques-Marie de Chastenet, marquis de Puységur. He first learned about it even before it was widely known in France, thanks to some pre-publication manuscripts that were sent to him in 1784.[135] Here again, social and intellectual ties had paved the way for a confluence between esoteric science and Chinese studies. Back in 1773, Puységur's father had written a popular book about Chinese military matters, based largely on Amiot's translation of the *Art of War*. It began quite simply, and against the spirit of that ignominious year in Chinese studies: "The more one delves into the history of China, the more one is convinced that the Chinese are the best governed people, and consequently the happiest."[136] Puységur fils was a close friend of Mellet. He was one of the few people Mellet inquired about after fleeing France around 1791, and in the early 1800s he was responsible for publishing the first item to appear in print under Mellet's name.[137]

To learn about Puységur's theory, Amiot relied on a short overview that was published in 1786, the *Essai sur la théorie du somnambulisme magnétique*, which he read "with the greatest pleasure."[138] There had been some early question in mesmerist circles about whether or not the theory of animal magnetism was materialist; according to the précis, it was not, and claims to the contrary had been viciously contrived by opponents in order to malign it.[139] In this interpretation, animal magnetism, properly understood, acts as material in that it is extended through space and penetrates all objects; but it is essentially spirit in that it is also "the principle and the cause of movement and life."[140] "Magnetic somnambulism," then, referred to a sleeplike state that humans enter when animal magnetism flows through the body in a particular way, and following Mesmer's rediscovery of animal magnetism, doctors could learn to induce it artificially. In this state, one gains access to a mysterious "sixth sense," finer and more sensitive than the other five and comparable to instinct or the conscience, which explained such mysterious phenomena as sympathy and antipathy, platonic love, and even clairvoyance.[141] The ancients had possessed some knowledge of all this and now it could be studied once again with modern science.[142]

It was through magnetic somnambulism that Amiot made the specific identification of animal magnetism with Daoism. In his second letter of 1784, he had already suggested a mesmerist interpretation of Chinese magic: by controlling *yin* and *yang*, he wrote, adepts claimed to "go from one end of the world to the other, and ascend even above the stars." But at that time, he attributed such stories to the "overactive imaginations" of the "bonzes," ambiguously referring to both Buddhist monks and Daoist priests.[143] In his last two letters to Mellet, written in 1789 and 1790, Amiot specifically referred to Daoist masters and used the Chinese term *Tao-sée*, or *Daoshi*, to do so. His new attention to the topic was made possible by changed circumstances in both Beijing and Paris. At the North Church, it was clear that the old Jesuit rules about non-Confucian traditions no longer applied, and Amiot's friendship with Hongwu, a self-identified Daoist, was flourishing. Meanwhile, he was receiving specific inquiries about the Daoists from Mellet, to whom he wrote: "Since you spoke to me on their account, I started to study them." But the final stage in the evolution of his thinking seems to have begun only with Puységur's theory: "This somnambulism communicated by magnetism seems to me not to differ from the state of ecstasy experienced by the *Tao-sée*."[144] Both phenomena could be explained by natural philosophy—which meant that both could be real.

According to Amiot, the Daoists claimed extraordinary powers: "I candidly report not what I have seen, but facts attested by thousands of others who claim to have seen," he explained.[145] Here is what he described. The Daoist master prepares himself by collecting himself and assuming a particular posture. He then begins to see with what he calls the "third eye" or the "intellectual eye," located at the solar plexus; this is an eye "only allegorically," in the sense that it allows one to see. Entering a state of ecstasy, he acquires "knowledge far superior to that of ordinary men." He can remember the future as he remembers the past and see "from one end of the earth to the other, in the void of the airs, in the regions of the planets and the stars." All humans could master these powers "if the superior part of their being knew how to conserve its prerogatives and its superiority over the inferior and terrestrial part." It is "an entirely simple

FIGURE 4.6 *Grand Lama and Daoist Master,* in Breton de la Martinière, *China: Its Costume, Arts, Manufactures, &c.,* volume 1 (1812). Courtesy of the University of California Library.

and natural art, developed by application and study and perfected by a long experiment," just like any other science.[146]

These apparently supernatural abilities were, Amiot asserted, thus explicable according to natural laws known to both Daoist masters and magnetic somnambulists. As he explained it, normal perception functions by means of "infinitely subtle agents," or "vital spirits," which pick up impressions of objects in the world and bring them back to the soul as representations. But one can also perceive things that are not present; memory, for example, works by willing the vital spirits "toward the place in your head" where past experiences have left their imprint. Daoists and somnambulists, however, can direct their vital spirits to access distant things that have not already left an imprint in the mind—but how? Amiot assumed the voice of a Daoist to resolve the challenge: "I explain easily and in a satisfying manner, for whoever does not want to close their eyes to a light that can show them the truth." The direction of vital spirits depends upon "the effluvious emanations or rays of the *taiji*"—for which one could substitute animal magnetism—"that my will puts into play." These rays are infinite in extension and duration, able to "penetrate everything, act on everything, give life or movement to all," borrowing a phrase from the pamphlet he had read on magnetic somnambulism. The Daoist or magnetizer can thus direct rays of *taiji* or animal magnetism from the third eye, through the continuum of time and space, and to any object of interest. These rays reflect off the target and return to the third eye, where they are processed and relayed to the imagination. The "adept" thus experiences them instantly and effortlessly, just like the recollection of a distant memory.[147]

Amiot investigated the practice as well as the theory. In a suburb a mile or two outside of the Beijing city walls, he had often noticed a peculiar kind of barber, or *titou*, making their shop in the roads and squares.[148] Their main employment seemed to be shaving the heads of Han Chinese men to maintain their state-mandated queues. For years, Amiot had ignored them, but after reading about magnetic somnambulism, he went back to determine, "by the light of the torch of this science," what they were up to. He paid one the customary fee and sat back to see what would happen. The barber—now revealed to be a "magnetizer"—gave Amiot a magnetic treatment, which he evocatively recounted as a performance in three "acts." Shaving his head was only the prelude. The magnetizer then began act one: planting himself half a foot in front of Amiot, he stared the missionary squarely between the eyes to "put himself in rapport" with his client, then gently rubbed his clean-shaven head. Act two was a full-body massage, in which the magnetizer pushed, squeezed, and caressed along his body, beginning with the shoulders and the arms and working his way down. To begin act three, the magnetizing barber waved a hand in front of Amiot's face as though swatting a fly, pinched the bridge of his nose, and executed an elaborate series of orchestrated touches, replete with "adagios and allegros," "duets and solos," culminating finally in a "grand choir" played out in taps on Amiot's face.[149]

The reported results were exceptional. Amiot experienced "the most agreeable sensations in the body and calm of the passions in the soul." He felt "nothing but thoughts of goodwill toward all the world, including

FIGURE 4.7 1. *Barber*, in Breton de la Martinière, *China: Its Costume, Arts, Manufactures, &c.*, volume 3 (1812). Courtesy of the University of California Library.

those detractors of magnetism of whom I disapprove." The only problem was that the effects were temporary, and he wished to avoid the fate of some other patients who had become addicted, requiring magnetic treatments no less than sleep or food. Chinese magnetizers had much to teach their European counterparts. After all, in China, they had been magnetizing for centuries, where in Europe "one only began, so to speak, yesterday." French magnetizers should not despair that they had not yet perfected the skills: "Everything takes time."[150]

Amiot himself attempted some magnetic feats, but he soon gave up, citing two reasons. The first was simple discouragement—try as he might, he could not produce “anything that one could call by the name of victory in any extension of the term.” The other was fear: “You exhort me ‘to magnetize’: I avow my weakness! I would not dare.” Perhaps his confreres at the North Church would have disapproved had they seen him engaging in exactly the kind of practice that ten generations of European missionaries had categorically maligned. But Amiot was still more worried about the opinion of his Chinese acquaintances. Animal magnetism, no less than other new sciences like gas balloons and electrical machines, seemed potentially dangerous and decidedly unorthodox: “Our literati, cast in the same mold as your European *beaux esprits,* assume a scornful affect,” he wrote. If he appeared to be “one of the adepts of a sect,” the consequences would be dire. It was not that he feared ridicule, he insisted; but in China, “propriety” was everything, and to breach it was to put the entire mission at risk.[151]

To prove the point, Amiot recounted the story of a Chinese magnetizer who had gone too far. A few years earlier, a Daoist hermit had come down from the mountains and presented himself at the palace gates with a message for the emperor: “In my solitude, dreaming upon the virtue over all of nature of the different combinations of *yin* with *yang* and of *yang* with *yin,* I have found a method by which I can, with a single gesture, a single regard, enchant the most ferocious beasts.” He would reveal this secret to the emperor alone, after giving a demonstration at the palace menagerie as proof of his sincerity. The message reached a certain Manchu prince, who granted the request: if the Daoist succeeded, it would be proof of his talent; and if he were eaten, it would be punishment for his audacity. The Daoist was duly lowered into a tiger pit. The beasts were unchained and proceeded to do absolutely nothing. For an hour, they were poked, prodded, and provoked, all to no effect. Finally, when a piece of meat was placed directly on the Daoist’s head, one of the tigers, enticed by the savory smell, jumped clear over him and swept it away, leaving only a tiny scratch. Satisfied with the demonstration, the prince relayed the petition to his father the emperor for further instructions. “My orders are already given,” the emperor replied: the Daoist was promptly beheaded.[152]

Whispers traveled through the court, where opinions on the outcome were split. “A small number of philosophers” discussed their feelings in private. They found nothing unseemly or unnatural about the accomplishment. Tigers are ferocious only because the currents of *yin* and *yang* flowing within them are rapid. The Daoist possessed the “power of accumulating within himself a quantity of *qi* (both *yin* and *yang*) strong enough to resist the *yin-yang* of the tigers, to repel them, and to send it back flowing against them,” rendering them sweet and timid. It was a rare skill, to be sure, but acquired like any other, through meditation and experimentation. The Daoist had been stripped nude, and no trace of any intoxicating potion or perfume had been found. It was wrong, those few philosophers believed, to execute "a man who had committed no other crime than that of being more enlightened than ordinary men." Their opinion, however, was a minority one; most of the court held that the Daoist had acted "against the laws, the sane doctrine of the ancients, and good customs." His powers must have been achieved through magic, which was rightly proscribed as harmful to society, and no punishment was too harsh for practicing it. Ever cautious, Amiot refused to "suggest a manner of reflection," but it was clear where his sympathies lay: "There are dreamers in China as there are elsewhere; Eh! Yes, sir, where does one not dream?" Notably, dreaming was also significant in both these esoteric sciences.[153]

In the end, Amiot came around to defending even the one thing that had long made Daoism most suspect of all: the search for immortality. He expressly conceded that "European authors, who copy each other, have said with a common voice: 'the sectarians of the Dao have the imbecility to believe that there is an elixir that can prevent them from dying.'"[154] This criticism, he believed, stemmed from a universal misunderstanding. The Daoists did not deny that the dissolution of the body is inevitable. The only kind of immortality that they sought is "purely mystical, relative to the dogmas they profess"—that is, the idea of metempsychosis, or the transmigration of souls.[155] After physical death, the body and soul disaggregate, and their component parts are reassigned to a new form. Every living thing occupies a position on the "ladder of beings"; for example, it is more dignified to be a spirit or an immortal than an insect or a quadruped.[156] For the Daoists, to be reconstituted after physical death as a being of lesser dignity would be a "metaphysical death," and this fate is what was avoidable.[157]

The so-called elixir of immortality, then, was merely something that helps to ensure that the Daoist's body and soul are recombined into a being of appropriate rank. The process is already determined by the action of *yin* and *yang* and the power of *tian*, which for the Daoist means "nature" or "chance." The elixir of immortality could only streamline the processes and prevent mistakes. Amiot compared it to a passport, signaling to the spirits presiding over the generation of new life that the deceased had already fulfilled the moral obligations incumbent upon him. It might indeed sometimes take the form of a beverage, but even then it was only a symbol that its drinker, "by his virtues and by his merits," had earned the right to be reborn as something dignified, like a human being, a spirit, or an immortal. Properly understood, the elixir of immortality could thus not prevent death—but its drinker, or at least the parts of which he was once composed, might indeed become an immortal.[158] No Jesuit in the two-hundred-year history of the China mission could have ever countenanced such an idea.

Amiot remained to some degree a skeptic, but a very generous one. He insisted that while it was true that the Daoists professed many incorrect views, they were judged for other views they did not actually hold. As he put it with respect to the elixir of immortality: "This absurdity, as you see, is quite different from that which is ascribed to them; it derives from their opinion on the nature of beings; and if this opinion were founded on the truth, the consequence they take from it would not be at all absurd."[159] Amiot came to believe that the Daoists had to be understood on their own terms. His conclusion was significant in two ways. First, it explains the context in which he wrote the "Treatise on the sect of the *Daoshi*," which soon became the longest published discussion of the topic in any European language and provided the basis for a brand-new interpretation of Chinese thought. Second, even though most of this material was never published, it demonstrates how a new link between European esotericism and non-Western philosophy was forged.

Indeed, Amiot tentatively embraced the discourse of occult science, despite some self-consciousness about it. He emphasized that the Daoists had been condemned by their enemies as "magicians, sorcerers, disciples of the spirit of shadows."[160] He expected that one French correspondent would therefore not wish "to be initiated into the mysteries of the Dao, much less become one of the adepts."[161] As for the mesmerists, he wrote Mellet, "I call them only *friends*, although I pass for being honored by them as *brother*, after the initiation to which you had the kindness to admit me."[162] Recounting to Bertin his exchanges with Mellet, he confessed that he harbored thoughts "concerning the philanthropic science into which he wanted to initiate me; but we are separated by too many lands and seas to dare confide to paper everything that I think."[163] Given how controversial his disclosures about the topic were, one can only wonder what he was still withholding. In his last letter to Mellet, he wrote: "Either you are a prophet, or you were in perfect somnambulism, when you wrote me."[164] In such a moment of characteristically detached humor, Amiot revealed his own mental scaffolding. Magic, sorcery, prophecy; shadows, adepts, initiations—this was the mode in which he had formulated his *yin-yang* theory of animal magnetism.

Amiot knew just how opposed his theory was to the spirit of the times, and in his final letter to Mellet, he lamented it. Although he never sent a Manchu version of the "Organs Section," he did send a Manchu version of the *Classic of Changes*—"the most ancient literary monument there is in the universe"—hoping that someone would "make better use of it than have the Chinese, our *Yijingist* missionaries, or M. Leibniz himself." Standing in the way, he wrote, were only "bad critics and disbelievers. You know, sir, that these two types of men are not rare in the century in which we live."[165] It was clear to both Amiot and Mellet that the Enlightenment had left no place for either animal magnetism or Chinese natural philosophy. The result of their conversation was to create one.

The French Fate of Chinese Magnetism

The correspondence between Amiot and Mellet took some time to percolate into broader European conversations; nevertheless, during the Revolutionary and Napoleonic period, it did become somewhat known within the small but growing world of French esotericism. China was no longer so much in vogue, but occult science very much was. It is not hard to imagine how a curious French reader might have encountered the *yin-yang* theory of animal magnetism. It had already been cast in the language of occult science and tied to new ideas like animal magnetism and magnetic somnambulism that the scientific establishment had rejected. This connection was not lost on other orphans of the Enlightenment, who embraced the theory for the very same reasons. Their engagement with Chinese natural philosophy constitutes the most enthusiastic European engagement with Chinese "physics" and "metaphysics" to date, and through it the foundations of a new understanding of "the sciences of China."[166]

Even if others might have found value in the new research on Chinese natural philosophy, they by and large did not have access to it. The French gatekeepers to Chinese studies found the discussions too controversial, or maybe just too weird, to print. Several pieces that had discussed mesmerism were published in the *Mémoires concernant les Chinois*, but mostly with those discussions specifically excised. For example, in one letter to a French doctor, Amiot snuck in a comment about Mesmer's "marvelous" discovery and its resonance with Chinese natural philosophy—"for among us, animal magnetism and everything that accompanies it are nothing but simple accessories of what the Chinese call *yin-yang*"—but this passage did not appear in the printed version.[167] The editor justified the decision to Amiot by claiming that animal magnetism was important "in its true principle," but only "if it were stripped of all charlatanism"; likewise, *yin-yang* medicine was "not one of the least interesting things, but it is not yet time to speak of it in our *Mémoires*."[168] Amiot replied that that was just as well, since it freed him from "overly rigorous circumspection."[169] In total, he mentioned mesmerism in over a dozen letters, addressed to Mellet, Bréquigny, Bertin, Bertin's brother, and probably others as well. Many more people demonstrably knew about those letters, including Bertin's secretary Desvoyes, the doctor Charles-Jacques Saillant, the young orientalist Louis Langlès, and the old orientalists de Guignes and Deshauterayes. None of them seems to have been eager to publish.

But the research was not as obscure in its time as it later became. For a while, it was known only from manuscripts. No doubt the major distributor was Mellet himself. The first two letters, on *yin-yang* medicine and natural philosophy, entered circulation almost immediately upon arrival in France in the mid-1780s. Multiple copies taken in different hands survive in various Paris archives. According to one mesmerist author of the time, it was unnecessary to recount their contents in detail since they were already well known to "persons of consideration"—that is, friends of their recipients.[170] Two decades later, one of the letters on magic and alchemy resurfaced in the possession of an unknown patient at the Hôpital Saint-Louis.[171] Many of the letters did eventually see publication in some form, but in this way, even those that did not still contributed to the development of esotericism in France.

The idea of a connection between animal magnetism and Chinese natural philosophy was already in the air, and it is not hard to see why. Mesmerists made much of animal magnetism's being ancient, and China was certainly that. For the philosophes, Chinese science was a relic of the past; for orphans of the Enlightenment, this was exactly why it was valuable. For example, a 1785 publication subtitled "Magnetic knowledge and procedure among various ancient peoples" speculated that Daoist doctors cured diseases by manipulating a "magnetic fluid" directly with their hands. The author seems to have formulated the idea based exclusively on knowledge of the earlier Jesuit accounts that had been published by Jean-Baptiste du Halde.[172] Likewise, the Celticist Jacques Cambry gave a mesmerist interpretation of Chinese natural philosophy in his *Traces of Magnetism* of 1784, instead based mainly on the studies of the early eighteenth-century scholar Nicolas Fréret. His main argument was that "what the ancients called spirit, we call magnetism."[173]

For Cambry, too, the idea of ancient wisdom facilitated the merging of occult and non-Western philosophy. He posited that there existed a common doctrine of the mutual influence between the elements, the planets, and humans, known to ancient Indian gymnosophists and Greek Pythagoreans, elaborated by medieval alchemists like George Ripley and Thomas Bungay, and still professed in modern times from Africa to Japan.[174] To demonstrate its manifestation in China, Cambry achieved a fairly sophisticated understanding of Neo-Confucian natural philosophy. As he explained it, the universe is composed of a single substance in which all entities participate. This substance, while ontologically necessary and prior to all entities, does not exist by itself, but only through the entities that it creates. Generation and corruption are, strictly speaking, illusory, being just the manifestation of changes in the relationships between things. Certain relationships are perceived by our senses; we call those matter. Others are perceived by thought; we call those spirit. But these differences inhere only at the level of perception, not at the level of being. Spirit and body are tied to a "same and single, infinite and inalterable existence." This "absolute existence" of the Chinese philosophers is nothing but "our universal fluid."[175] In conclusion, "it is the solar ray divined by Fuxi, the *li* of the Chinese, the *Lenge Cherire* of the Indians, the *Barhala-May-Capal* of the Philippine islands"; and it is also what Mesmer called animal magnetism.[176]

The similarities between Cambry's and Amiot's mesmerist interpretations of Chinese natural philosophy were striking, but they were probably convergent rather than genealogical. In both of them, Chinese metaphysics posits a primary substance that infuses all things through the action of its "rays," and explains microcosm-macrocosm harmony and enables extrasensory perception. On the other hand, there were also significant differences: Cambry was a materialist, or at least a monist, and identified animal magnetism as *li*, whereas Amiot identified it as *taiji* in part to avoid *li*'s materialist overtones. Cambry made no mention of *yin-yang*, which was essential in Amiot's interpretation, and which he might have been expected to do had he read about it, given his strong interest in polarity. Communication between the two would also be difficult to reconcile temporally. Amiot's first letter on animal magnetism reached Paris no earlier than the spring of 1784, and Cambry's book was published that same year. What their complementary interpretations thus suggest is that mesmerism and Chinese natural philosophy were already primed for a mutual embrace.

The first published mention of the correspondence between Amiot and Mellet appeared several years later in a gossipy miscellany about Paris high society published in 1787. Apparently, the author had obtained a manuscript copy of Amiot's 1784 letter, from which he excerpted the story of the Chinese doctor healing the sick wife of an official with a bamboo stick, though stripped of all details concerning *yin-yang* natural philosophy or Chinese medicine. The main point was only that superstitious practices went all the way back to ancient times, and that animal magnetism was not required to explain them. The similarity between the two theories, it was implied, was both irrelevant and inconsequential. The "mesmerists and company" had pillaged ancient doctrines and distorted them at will, contributing to true understanding of neither.[177] But if the story surfaced for the first time in the context of critique, it escaped its confines quickly enough. One anonymous reader of that first published mention drew the opposite conclusion from the story it reported, even with no additional information to go on. This fourth-hand reviewer concluded that magnetism had indeed been practiced in China for thousands of years, as proven by an "ancient Chinese book"—that is, the "Organs Section."[178]

The first substantive published discussion of the *yin-yang* theory of animal magnetism was composed around 1785. It appeared in a work entitled *Physical and Metaphysical Research on Celestial Influences, Universal Magnetism, and Animal Magnetism*, which bore the subtitle: "*The practice of which is found since time immemorial among the Chinese.*"[179] The author, one Robert de Lo-Looz, was a mysterious figure, in some ways quite similar to Mellet, whom he likely also knew through their membership in the exclusive Order of Saint-Louis. Born in Liège in 1730, he became a soldier of fortune, rising to the rank of colonel in the army of Sweden before taking up residence in France.[180] He then ventured into the arts and sciences, for example with a jaunty military history entitled *Soldiers Beyond the Ganges*.[181] His magnum opus, *Research on Solar and Lunar Influences to Prove Universal Magnetism*, was cut short by his death in 1786 and published posthumously in 1788. A complete draft of the second volume was already at the censor when Lo-Looz received manuscript copies of Amiot's first letters. While acknowledging this "gift to the supporters of animal magnetism," he immediately sought to distance himself from it; his own book was nearly complete, so there should be "no suspicion of connivance" between him and the missionary.[182] Nevertheless, he pulled the proofs and added a new subsection, entitled "Magnetism among the Chinese."[183]

Lo-Looz too associated Chinese natural philosophy with mesmerism, and through that, ancient wisdom with non-Western knowledge. In his version: "The doctrine of the Patriarchs, escaping the universal cataclysm, was transmitted to China; I found its debris among the brutish nomads of America; thus the dogma of creation, contested in Europe, is found widespread in the other parts of the globe."[184] That doctrine was an "Adamic tradition," perhaps borrowing a term from Mellet, that had been known to the peoples of the "*Monde primitif*," expressly echoing Court de Gébelin.[185] To explain it, Lo-Looz gave a fairly detailed summary and analysis of Amiot's letters on "the macrocosmic physics of the Chinese," including *yin-yang*, *taiji*, and *li*. It was not especially careful. For example, he consistently mislabeled "*li*" as "*I*," probably having taken the letter "l" in Amiot's romanization for a French definite article.[186] Such sloppiness, however, in no way spoke to a lack of enthusiasm. The main point for Lo-Looz was that ancient Chinese natural philosophy explained the system of "harmonic and magnetic correspondence" between the heavens, the earth, and all things in nature.[187] "*Voilà* the doctrine of Mesmer, contained in thousand-year-old manuscripts"—referring, though he did not know it, to the "Organs Section" of the *Complete Collection of Ancient and Modern Figures and Texts*.[188]

The book is still more remarkable for actively taking up the proposal to combine ancient Chinese natural philosophy and contemporary European science. Amiot had suggested that *li* could be adopted by Newtonians as a principle of attraction. For Lo-Looz, "this reflection, is very right, but its application is too limited"—*li* could furthermore point the way for "all philosophers who will want to have a certain compass to penetrate the vast sanctuary of nature." For example, Amiot's description of the mixture of *yin* and *yang* by *li* as a sort of "fuel" suggested to Lo-Looz that it must have been a means of storing energy; he therefore interpreted it as a kind of electric battery, where *yin* is alkaline and *yang* is acid.[189] Looking the other direction, modern science could also guide the interpretation of ancient natural philosophy. Lo-Looz maintained that the illustration of the "Universal Figure" that Amiot had sent was "contrary to the opinion even of the Chinese."[190] Understanding the *taiji* diagram instead as a kind of cosmography, he thought since *yang* is rare and *yin* is dense, the white area of *yang* should be bigger than the black area of *yin*, according to "the law of specific weights."[191] Chinese and European natural philosophy could not only explain, but even correct each other.

By 1788, then, two books had been published that dealt at length with Chinese natural philosophy and Mesmer's animal magnetism. Given that their authors were working with similar themes at around the same time, it would be surprising if Cambry and Lo-Looz did not cohabit in a shared scholarly community, and there is one further detail that suggests they did: they shared a publisher.[192] The catalogue of Pierre-Denis Couturier had a particular specialty in animal magnetism, including a pro-Mesmer *Report of the Report* of the royal committee, and a letter on the topic that had been sent to Court de Gébelin.[193] On their title pages, many of these works gave a fictitious place of publication, a common practice in the volatile world of late eighteenth-century printing. What is striking though, and quite uncommon for the time, is that the purported place of publication for several of them was given as Beijing.[194] The editor of a leading series on mesmerism seems to have decided that China deserved special attention.

It was not, however, until the early nineteenth century that the connection between *yin-yang* philosophy and animal magnetism was solidified, through the efforts of one of Mesmer's most important successors—the very marquis de Puységur whose theory of magnetic somnambulism had enchanted Amiot in the late 1780s. Unlike Cambry's and Lo-Looz's now completely forgotten books, Puységur's were widely read and broadly influential. His doctrine of magnetic somnambulism proved foundational for what later came to be called hypnotism—thus giving English the word "mesmerize"—and, by shifting the focus from physical treatments to psychological ones, has also been seen as an important step in its development into clinical psychology.[195] Puységur had undoubtedly learned about the *yin-yang* theory of animal magnetism from Mellet, probably as early as the 1780s. But he did not mention anything about it in print until 1807, after his friend was safely deceased.[196]

Puységur's most important theoretical work, *On Animal Magnetism*, contained two pieces and a brief commentary on the *yin-yang* theory. First was a short letter from Mellet, describing the beginning of his correspondence with Amiot and reproducing the illustration of the "universal figure" in print. Second was an extract of Amiot's second letter, in which he described the *yin-yang* theory and recounted the story of the Chinese doctor.[197] The editor's brief comments revealed that Amiot's enthusiasm for Puységur did not prove reciprocal: "The letter of Father Amiot only proves the state of stagnation in which the sciences of China have rested since time immemorial." Future exchange might yet be promising, but likely in the opposite direction to the one Amiot had proposed: "Once we and the Chinese will understand each other in *philosophy*, it will be easy for us, I believe, to get them to adopt our little magnetic catechism."[198] The ultimate result of this publication was nevertheless to connect Chinese natural philosophy with modern esotericism and to introduce the conversation to a global audience, not all of whom shared the editor's opinions. In 1879, another foundational figure of modern esotericism, Helena "Madame" Blavatsky, then living in Madras, founded the monthly journal the *Theosophist* in open embrace of Asian philosophy. The very first volume enthusiastically quoted from the letter written by Amiot, sent to Mellet, and published by Puységur, on the topic of "Magnetism in Ancient China."[199]

Meanwhile, the connection between Chinese natural philosophy and European esotericism was also beginning to make its way back from works on animal mesmerism into works on China. In 1808, a new edition of the Jesuit missionaries' *Lettres édifiantes et curieuses* appeared. The recently returned refractory priest who edited it seems to have known most or all of Amiot's late unpublished letters.[200] He affirmed that "between the

(385)

satisfaisante, et il me promit une plus grande instruction pour l'année d'après. Vous trouverez ci-joint un extrait de la seconde, et je vais, si vous voulez bien le permettre, vous tracer ici, de mémoire, l'essentiel de la première, qui servira d'explication à l'espèce d'exorde de celle que je viens de recevoir.

Les Chinois peignent la nature, qu'ils appellent *Tay-ki*, sous la forme d'un œuf, à peu près semblable à celui d'*Isis*. Cet agent général, ce *Tay-ki*, forme un tout mâle et femelle, c'est-à-dire, renfermant les principes opposés, qu'ils nomment *yang* et *yn*, et c'est par l'action de cet *yn-yang* que tout est produit, que tout naît pour être détruit, et se détruit pour renaître sous de nouvelles formes.

Figure du TAY-KI.

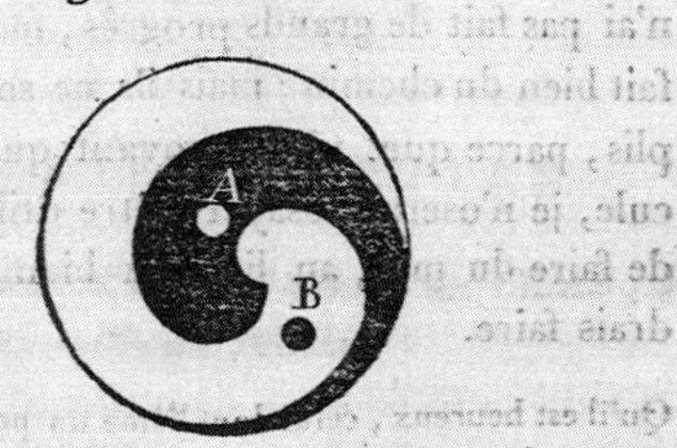

L'*yn* A a un foyer ou *ly-yang*; et l'*yang* B a un *ly* ou moteur *yn*.

Bb

FIGURE 4.8 *Figure of the* Taiji, in Chastenet de Puységur, *Du magnétisme animal* (1807), 2nd edition (1809). Courtesy of the Bibliothèque interuniversitaire de Santé.

doctrine of the bonzes and that of our *illuminés* of Europe, between the practices of kung-fu and those of European magnetism, the traits of resemblance are marked."[201] In order to help explain "the Chinese system," he promised supplemental notices on the esoteric philosophy of Cagliostro and Emmanuel Swedenborg. The Daoists, he thought, were superstitious through and through: "What they could not discover in the study of physics, they took to searching for in the study of astrology and magic."[202] These occult sciences were no substitute for natural science, yet China might have something to teach Europe after all: "The sect of horoscope readers, of Cagliostro, of our European *illuminés*, is still only at the alphabet of the science, in comparison to the progress boasted by the sect of Daoist bonzes."[203] In a way, this contention was exactly what Amiot and Mellet had proposed.

Orphans of the Enlightenment were able to believe that Europe could learn from China about natural science because they did not think that natural science was necessarily modern, nor European. For them, the ancient wisdom concealed in a Chinese book was not just compatible with new scientific discoveries but actually the same. All true knowledge, whether it was found in the past or the future, in the East or in the West, was one. Putting what they took to be ancient Chinese natural philosophy and modern European science together, they created a truly hybrid theory. But in so doing, they also drove Chinese and European knowledge further apart. Their *yin-yang* theory of animal magnetism was self-consciously at odds with established authorities, who accepted it as an account of both Chinese and esoteric natural philosophy but refused to recognize either as science.

The Yin-Yang *Theory of Animal Magnetism*

Following upon their early exchanges between Europe and China, orphans of the Enlightenment expanded from a historical account of ancient wisdom into a scientific one. Mellet reached out to the missionaries in Beijing, hoping to put European and Chinese theories into conversation. Amiot embraced that program with unprecedented excitement. The work that ensued was made possible by new intellectual, social, and material configurations that had taken shape only during the previous decade. In Paris, Chinese knowledge had already been cast as a source of ancient wisdom, and those who became interested in the former were usually drawn to it by the latter. In Beijing, Amiot had learned that ancient wisdom was something that China could provide, and post-Jesuit circumstances opened up previously inaccessible or impermissible ways of attempting to recover it.

Together over the course of a decade, Amiot and Mellet connected the ancient Chinese concepts of *yin* and *yang* with the new European theory of animal magnetism. They identified many similarities between European and Chinese ideas, microcosm-macrocosm correspondences and polar forces chief among them, and used each to explain the other. Just as Mellet relied on Amiot for new works on China, Amiot relied on Mellet for the latest research on mesmerism. Each of them was put in the vanguard of the other's field of expertise. Mellet looked to China for both facts and theories. He read Amiot's letters, responded with further inquiries, and sought translations of Chinese texts to develop his own scientific projects. Amiot responded by conducting original research, first on aspects of *yin-yang* cosmology and medicine, then expanding to encompass magic, alchemy, and religious Daoism. They argued that the *yin-yang* theory of animal magnetism was both ancient wisdom and modern science at the same time.

This account was quickly disavowed by the scientific establishment of the late Enlightenment, but it found new devotees among the occult and esoteric circles where the sprouts of Romanticism were coming into bloom. While historical theories of ancient wisdom had to some degree interested even philosophes like Voltaire, fleshed-out accounts of what that wisdom was gained no such traction among figures of establishment science such as Le Roy. It was telling that Bailly, who had promoted ancient wisdom as history during the 1770s, could not accept dowsing rods and magnetic somnambulism as science in the 1780s. The *yin-yang* theory of animal magnetism successfully connected ancient Chinese knowledge to occult and esoteric discourses from Europe's own past. For this reason, authorities at the Académie des sciences rejected it, while those outside the scientific establishment tightened their embrace.

The result of this effort to put European and Chinese natural philosophy together was thus ultimately to drive them further apart. By the end of the Enlightenment, Europeans had determined that China had no science. The question that followed was what it had instead. Amiot and his correspondents had introduced new primary and secondary sources on aspects of the Chinese tradition that had been almost entirely unknown in Europe until this time. His last letters arrived in France amid the throes of Revolution, and their immediate impact was somewhat muted. But when a new generation of scholars sought to reorder the structure of knowledge in the decades that followed, they built modern academic disciplines, including philosophy and orientalism, on the groundwork that had been laid by orphans of the Enlightenment.

⁕ 5 ⁕

The Invention of Eastern Wisdom

The End of Early Modern Chinese Studies

The Revolution brought Chinese studies in France to a generational pause. A few holdovers tried to keep the torch burning through tumultuous times. The last volume of the *Mémoires concernant les Chinois* came out in 1791, and new essays continued to appear in the publications of the Académie des inscriptions for a while after that. But by the end of the year 1800, the scholars who had been responsible for these projects, including Bréquigny, de Guignes, and Deshauterayes, were all dead.[1] Their aristocrat patrons, meanwhile, had already decided that China was far from their most pressing concern. Bertin emigrated in the summer of 1791. He died the next year in Belgium, bringing an abrupt end to the "literary correspondence" between Paris and Beijing.[2] The comte de Mellet fled around the same time. He proved to be an "old soldier" after all, rising to the rank of field marshal in the Army of Condé and fighting against the Revolution through to his seventies, despite a rifle shot to the leg and recurrent bouts of rheumatism.[3] He died in exile in Konstanz in 1804.[4]

Amiot survived to hear about the chaos engulfing France, and it rapidly accelerated his own decline. The septuagenarian missionary knew that the country he had left some forty years earlier was changing beyond recognition. He lamented the loss of "these ancient *mœurs*, this piety, these virtues" of his youth and condemned the "frivolity, levity, inconsistency" that seemed to have replaced them.[5] He spent his final days mostly alone and rarely left his quarters except to visit a park in what was then still the Beijing suburb of Haidian.[6] He called the place a "pleasure palace"; in fact, it was a cemetery.[7] Explaining himself to his brother in Toulon, he wrote, "I have no other company than that of the dead: you know that the dead are not importunate and, tranquil in their tombs, they only wait for us to go to them, without bothering to come to us." Old age had tempered his vivacity, but not his acerbity; the comment was apparently intended as a rebuke to his brother, who, very much alive, had asked him for a loan.[8]

When Amiot's death finally came, it was unexpected to no one. Writing to his sister in the spring of 1793, he predicted: "If I suffer even a slightly serious attack tomorrow, I sense that, my forces at the age of seventy-six being capable of no more than a certain resistance, I will certainly succumb. It will be as it pleases God."[9] The letter survives only because it arrived in Toulon too late for his sister to burn it along with the rest of her correspondence in fear of the advancing Republican troops who brutally occupied the city that winter.[10] On the morning of October 8, Amiot woke up, recited his breviary, and wept—this, too, seems to have become a habit. He spent the day as usual, walking alone among the headstones of Jesuits past. When he returned home to the North Church, he received the shocking news of the regicide. Thereupon, as his superior later recalled, he somberly recited a last mass for "his" king, Louis XVI. In the middle of the night, "sensibly affected by the events that took place in France," he succumbed to "phlegms."[11] In a sense, Amiot's death thus marked an end to both early modern studies of China and the early modern period in France.

By coincidence, it also marked a key moment in not only the creation of modern Sinology but also the birth of modern China. About two months earlier, a British ambassador, George Macartney, had arrived in Beijing to establish diplomatic relations with the Qing empire. He wanted to meet with the famous missionary—"a man of such probity and universal clarity, that his opinion is entitled to considerable respect"—but Amiot was already too sick to see him in person.[12] Over the course of the next month, Macartney met with the Qianlong emperor for a series of negotiations that has long been studied as a textbook case of cross-cultural miscommunication.[13] To the ambassador, China was an "immobile empire," unwilling or unable to acknowledge the value of modern science and industry. To the emperor, Britain was merely insignificant: "I set no value on objects strange or ingenious, and have no use for your country's manufactures," he declared in a letter to King George III.[14] With the delegation considering its next moves, Amiot made contact once more: "His judgment," Macartney summarized, "was that it would be most for our interests, at present, to signify my wishes to return home as soon as I could conveniently set out."[15] On October 7, 1793, Macartney left Beijing. Amiot died the next night.

The aim of the British embassy was to open China. Although it failed disastrously in the short run, it succeeded beyond imagination in the long. Macartney himself is the person most often credited with the apocryphal saying that the suns never sets on the British Empire.[16] The trade imbalance between Britain and the Qing was already reversing, thanks to the

FIGURE 5.1 The emperor receiving the embassy, William Alexander, 1793. By permission of the British Library.

growing demand for opium from India.[17] A reversal in geopolitical fortunes, too, soon became apparent, with British soldiers encroaching on the southwest Qing frontier.[18] The implications in the realm of intellectual exchange were felt almost immediately. In 1807, Robert Morrison, often considered the first Protestant missionary scholar of China, began preaching in Canton. At first, he tried to adopt Chinese customs as the Jesuits had, but he soon gave up to resume a more European lifestyle among the burgeoning expatriate community of the bustling port city.[19] The contrast with Matteo Ricci, entering the Forbidden City dressed in the robes of a Confucian scholar more than two hundred years before, is stark. The terms of the encounter had reversed. Orientalists left the service of one empire and entered into that of another.[20]

The rapid development of Chinese studies around this time is therefore not surprising in retrospect, but no one could have predicted it in advance. In Europe, interest in China had reached an all-time low. At the end of the year 1800, there was once again no scholar in Britain, France, or Germany who could even read Chinese.[21] It is debatable when exactly the transition from sinophilia to sinophobia took place.[22] In any case, that it had done so appeared evident to contemporary observers by the early nineteenth century. As one of them put it: "Unfavorable ideas of the Chinese are not new, but they are newly widespread and believed."[23] In China, things were more dire. The few remaining ex-Jesuits in Beijing had grown old and isolated, and the more recently arrived missionaries from other orders did not share their scholarly inclinations. The Qianlong emperor died in 1799, and the Jiaqing emperor who succeeded him did not share his scholarly inclinations, either. In 1811, the Qing expelled most of the European missionaries from the city. The four churches were abandoned one by one, until the last ex-Jesuit, remarkably still in office at the Astronomical Bureau, died in 1821.[24] In sum, conditions on both sides appeared to be less favorable to meaningful exchange than at any point arguably since the Renaissance.

And yet, professional Sinology as a branch of modern orientalism was indeed established at around this time. Edward Said, still its most insightful and influential critic, distinguished between two related meanings of "Orientalism": an approach, attitude, or ideology—the "ontological and epistemological distinction made between 'the Orient' and (most of the time) 'the occident'"—and a framework, profession, or discipline—"the corporate institution for dealing with the Orient."[25] Much of what Said wrote concerning the European study of the Islamic world could equally be said for that of China. Orientalism became a self-conscious discipline in the first decades of the nineteenth century, primarily in Britain and France. It adopted Christianizing imperatives and Romantic perspectives from its early modern predecessors. Most important of all, it cast both its subject and its object in brand-new terms: "The modern Orientalist was, in his view, a hero rescuing the Orient from the obscurity, alienation, and strangeness which he himself had properly distinguished."[26] This whole package got wrapped up so tightly and with such speed that it is easily taken as a given. The question then is, how did the orientalist come to see his mission this way in the first place?

And this question is actually much more interesting than it may at first appear. Historians have demonstrated that at the end of the eighteenth century, orientalism, at least as Said described it in its mature nineteenth-century form, did not really exist.[27] This finding in no way undermines his overarching point. It only drives home the question of how such a stark reversal took place, and in such a short amount of time. For Europeans at the end of the Enlightenment, China was already, in Said's terms, "old and distant"—but far-off places were more accessible than ever before, and the past seemed to be becoming ever more comprehensible.[28] China had not yet become obscure, alien, or strange; to the contrary, its oldness and distance were easily interpreted and discounted by progress theory. The key point is that the modern trope of the "mysterious Orient" required work to create.[29] And in the particular case of China, that work was exceptionally difficult, because it went against what nearly everyone in Europe had been saying for over two centuries—except, that is, for the orphans of the Enlightenment.

During the 1810s and 1820s, modern scholars drew a dividing line between a western present and an eastern past that made the knowledge of each incommensurable with that of the other. To do so, they drew from research that had been produced only recently, not in order to reject ancient Chinese knowledge, but to reclaim it. Orphans of the Enlightenment set the terms upon which new academic disciplines were founded. The first professional Sinologist, Jean-Pierre Abel-Rémusat, built on Amiot's studies of Daoism to create an account of "ancient eastern wisdom" that he believed to have been shared by all the peoples of ancient world. The philosopher Georg Wilhelm Friedrich Hegel, an acquaintance of Abel-Rémusat and a reader of Amiot, accepted that account, not as philosophy but as magic. Despite their disagreements, both concurred that the knowledge of the past and the East were the same as each other but different from that of the present and the West. Hegel's Daoist sage took the place of Voltaire's Confucian philosopher—not as a source of familiar models, but of exotic alternatives. In this way, non-Western knowledge was brought back into modern progress theories.

Jean-Pierre Abel-Rémusat and the Birth of Modern Sinology

The sad state of Chinese studies during the Revolutionary and Napoleonic period was perfectly clear to Abel-Rémusat, who is often regarded as the first modern Sinologist. "Recall, sir," he enjoined not long after, in 1822, "what was still thought about the Chinese in 1812; the disputes about them; the ignorance and the prejudice that even the missionaries' writings had not been able to completely dispel."[30] As he saw it, a new discipline of orientalism, replete with all the intellectual and institutional trappings, had emerged in the span of only about a decade. How did the study of China in France survive the Revolution and the Terror, the rise and fall of the First Empire, and the most devastating wars the continent had ever seen, only to emerge more coherent and productive than it had ever been before? In the foundation myth that he wrote for himself, Abel-Rémusat rises like a phoenix from the ashes to become the first to study China in a way that is professional, academic, scientific, or, even, modern.[31] But as with the other branches of modern orientalism, Sinology's break with the past was not nearly so decisive as the first Sinologists tried to make it appear.[32]

Abel-Rémusat was born in Paris 1788 to a well-known surgeon. He was on track to follow in his father's footsteps when, as an eighteen-year-old medical student, he stumbled across a Chinese herbal in the library of Charles Philippe Campion, abbé de Tersan. Probably it had been sent to France by the late ex-Jesuit Cibot. Fascinated by the illustrations, Abel-Rémusat resolved to read the text.[33] The difficulty of learning the language made a lasting impression. There was no Chinese dictionary in print in any European language, and the librarian in charge of Paris's principal Chinese collection refused him access. Only through great persistence did he gather together the translated extracts from European-language publications, late imperial Chinese reference works, and printed Manchu dictionary compiled by Amiot that he managed to make use of instead.[34] In 1813, he successfully defended his dissertation in medicine: a short comparative study of European and Chinese diagnostic techniques based on the condition of the tongue.[35] The very next year, he was hired not as a doctor, but as Europe's first professor of Chinese and Manchu.[36]

Abel-Rémusat wasted no time in establishing the institutional framework for the study of China in Europe, which is certainly the most compelling reason to call him the first modern Sinologist. In 1814, the new king, Louis XVIII, approved the creation of a chair at the Collège royal in "Chinese and Tartar-Manchu Language and Literature," and Abel-Rémusat was placed in it.[37] He thus became the first person in Western Europe paid to write and teach about China in a major academic institution with those duties reflected in the job title. On January 16, 1815, he commenced the first formal course on the Chinese language in France and over the next two decades taught an entire generation of students. He also played an instrumental role in the creation of a foundational periodical, the *Journal Asiatique*. In its inaugural issue of 1822, he published a "Letter to the Editor on the state and progress of Chinese literature in Europe," which celebrated the creation of a new field.[38] Here we can easily recognize all the trappings of a mature academic discipline: university courses, scholarly journals, postgraduate training, and prestigious professorships, some of which still exist even today.

The story is thus revealed as an archetypical foundation myth: a gifted teenager begins a quest to achieve the impossible (learn Chinese); finds a magic weapon (the methods of modern scholarship); prevails in brutal combat against all challengers (lesser contemporary scholars); assumes his well-earned throne (literally called a "chair"); and bequeaths his legacy to his descendants (professional Sinologists). Like all foundation myths, it survives through its retelling among a living community. After Abel-Rémusat's death, his position at the Collège de France passed down in an essentially unbroken chain of succession to: Stanislas Julien, Léon d'Hervey de Saint-Denys, Édouard Chavannes, Henri Maspero, Paul

FIGURE 5.2 *Jean-Pierre Abel-Rémusat*, Jules Léopold Boilly, 1821. By permission of the Trustees of the British Museum.

Demiéville, Jacques Gernet, Pierre-Étienne Will, and Anne Cheng. The two most recent chairs both gave talks at a recent bicentenary conference held there in his honor.[39] The Collège de France today claims that Sinology "may be considered a French invention, as Europe's very first chair dedicated to Chinese studies was established at the Collège de France as early as 1814."[40] Regardless of the extent to which the myth is true, it is Abel-Rémusat who both wrote and starred in it.

Abel-Rémusat's reputation rests less on what he said than on how he said it. He carefully cultivated the habitus of a modern academic. His style was wary of reductionism, oversimplification, and essentialization; more analytic than synthetic; more critical than constructive; more careful than courageous; learned and subtle but recondite and diffuse; and stylized tending to the obscure. Almost since his own time, historians have understood his work as reflecting the commitment to secular reason of a new positivist age. Early modern orientalism had been religiously motivated. Though he was a practicing Catholic, he did not betray it in his writings.[41] The greatest barrier to Chinese studies had always been the language. He overcame it and became, according to one early biographer, "the first who tackled Chinese as a true philologist."[42] And yet, Abel-Rémusat's beliefs about the Chinese language were exactly what led him into errors that many of his contemporaries avoided, while his interest in Chinese religion was just what led to the works that they generally considered his most important. It is notable that the distinguishing areas where he applied his supposedly modern and secular approach were also the ones where his debts to the past were the greatest.

Some historians have further thought that Sinology, as the word itself suggests, was supposed to be somehow more scientific than whatever preceded it. Yet to say that Sinology became a science would be extremely anachronistic. The first mention of *Sinologie* to appear in print did have some connection with Abel-Rémusat, but it was used only as an inflection of the older word *sinologue*, meaning simply "somebody who knows Chinese."[43] In the 1830s, the Académie française defined the term not as a discipline or a field of study, but rather as referring to a discrete work relating to the Chinese language.[44] And if Sinology did not yet carry its modern meaning in French, neither did "science." In 1798, the term could still refer generically to "certain and evident knowledge," and it was not defined by the Académie française as a "system of knowledge on some matter" until 1832.[45] In any case, no one seems to have described "Sinology" as "a science" until the late nineteenth century, reflecting the ideas and ambitions of a much later era, and Abel-Rémusat never invoked either term to refer to what he was doing.[46]

It is true nevertheless that Sinology took on some of the institutional, social, and intellectual trappings of the natural sciences. Abel-Rémusat embraced for the Collège de France and the Société Asiatique the same kind of policing and sanctioning functions that were already being performed by institutions like the Académie des sciences and the Faculté de médecine.[47] Professionalization was to be guided by the torch of empirical methods: Chinese sources were the only permissible evidence, and other Sinologists were the only reliable witnesses. His works showed an almost hypertrophied scholarly apparatus, replete with endnotes, footnotes, tables, appendixes, and indexes, often including long passages accurately printed in Chinese characters. The purpose in theory was to make the original sources accessible to other scholars, few as they were at the time.[48] As he put it: "An honorable change was at work in the spirit of those who cultivate Chinese literature. They feel the need to have collaborators and seek them with all their might."[49] The declaration was perhaps a bit hypocritical, given that he never acknowledged a collaborator in any of his own works. But it also shows how much he was thinking about his own legacy.

Another way to conceptualize the birth of Sinology, then, is to consider its family tree. We actually know a lot about whom Abel-Rémusat believed his predecessors to be, since he said a lot about them. As a professional academic, he liked to make his points by critiquing those of others. He thought that the Jesuits, despite their extraordinary talents, had distorted Chinese traditions in pursuit of their own religious programs.[50] His opinion of Amiot was fairly representative: the missionary was commendable for "immense Chinese erudition," and condemnable for "inconceivable unseriousness."[51] The sinologue scholars were held to be less biased, but basically incompetent: "The missionaries, who had made very great progress, had not given their secret to the savants of Europe." Fourmont, probably the best among them, "made the honorable mistake of trying to do too much," but his students, de Guignes and Deshauterayes, were too ignorant to even be properly ashamed. In any case, neither the missionaries nor the academicians had survived into the nineteenth century.[52] Abel-Rémusat thus believed from early on that it fell on him to rebuild Chinese studies mostly from scratch.

In fact, there were others like him: siblings in Sinology spread throughout the continent. At the time, they still formed a single community, and its international center was France.[53] In Russia and the Netherlands, Chinese studies had long been an economic and political priority, for trade and diplomacy in the first case, and for colonial administration in the second. Both governments had promoted technical language training already in the eighteenth century, but neither nation saw academic positions established until after 1850.[54] The German lands, too, had produced their share of missionaries and orientalists, from Johann Schreck and Johann Adam Schall von Bell to Athanasius Kircher and Theophilus Siegfried Bayer—but none of these famous figures had pursued their careers in Germany. This pattern was not yet to be broken. The biography of the second-most-preeminent continental Sinologist of Abel-Rémusat's generation, the Prussian Julius Klaproth, is illustrative. Around 1810, Klaproth was advised by a Polish friend whom he knew from St. Petersburg that any serious scholar should aim to work in Paris.[55] In 1814, he visited Napoleon in exile on Elba to market his talents, but his efforts seem to have gone nowhere. In 1816, he was appointed to the first Chinese professorship in Germany. After securing special permission to fulfill his duties remotely, he quickly moved to Paris and published exclusively in French until his death some three decades later.[56]

Another group of French Sinology's closest relatives could be found in Britain. From the perspective of Anglophone Chinese studies today, George Staunton, Robert Morrison, and James Legge are just as influential as Abel-Rémusat, if not more so. The importance of British Sinology for European intellectual culture at the time was, however, lesser than that the French branch, due to two related characteristics: first, it had a more practical orientation, and second, it developed comparatively late.[57] British efforts grew mostly in the field, under the auspices of empire, with the ambition to create knowledge that was useful for trade, diplomacy, and government. In 1810, Staunton, who had been Macartney's page as a teenager, published a landmark edition of the *Great Qing Code*, by some measures the first work ever translated directly from Chinese into English.[58] In 1818, Morrison founded the Anglo-Chinese College in Malacca, having finally published the first really usable Chinese–European language dictionary.[59] James Legge, the only one of the British group who can really be compared with Abel-Rémusat as a scholar of classical texts, spent thirty years in Hong Kong before finally becoming the first professor of Chinese at Oxford only in 1876—more than half a century after the creation of the chair at the Collège de France.[60]

Then, too, there were Abel-Rémusat's closest collaborators, cousins in related fields who together with him comprised the first generation of modern orientalists.[61] Like "Sinology," the word "orientalism" was not yet in common usage, but the term "orientalist" absolutely was. Like "sinologue," it simply meant "someone who is versed in the knowledge of oriental languages."[62] Other key figures involved in founding the seminal *Journal Asiatique* included the Arabist Antoine Isaac Sylvestre de Sacy, who had been instrumental in bringing Abel-Rémusat to the Collège de France, and the Indologist Antoine-Léonard de Chézy, who joined him there the following year. It is significant too that their new chairs in Chinese and Sanskrit had been added to the older ones in Syriac and Arabic, previously held by de Guignes and Deshauterayes. In the early modern period, orientalists had studied the Bible, while historians studied the Greek and Roman classics. Each discipline was responsible for a different part of the genealogy of the European present.[63] When modern orientalism first expanded to include places like India and China, early efforts were made to fit them, too, into more familiar stories.

A final way to consider the family of Sinology, then, is by reference to what it excluded. Just as there is a foundation myth of the early nineteenth-century creation of Sinology, so too is there one about philosophy that took place around the same time. Given the composition of university philosophy departments today, it is perhaps surprising that it was only around then that philosophers began to see their discipline as distinctively European. This development was both a cause and a result of academic professionalization.[64] While the international community of Sinology was uniting around Abel-Rémusat, that of philosophy was doing likewise around the figure of Hegel. And in the field of philosophy, German intellectual influences were gaining strength in France. In the same year that Abel-Rémusat took up his chair at the Collège de France, Hegel's French disciple, Victor Cousin, began teaching philosophy at the École Normale Supérieure. His question—"Has there been or has there not been philosophy in the Orient?"—would have been quite a strange thing to ask just a generation or two earlier.[65] Even for progress-oriented philosophes like Diderot, the answer to it, as indicated by the very title of his *Encyclopédie* article on "The Philosophy of the Chinese," had been obvious.

But as philosophers professionalized around the turn of the century, they too wrote themselves a new history for their discipline, and China was not a part of it. The major Enlightenment textbook on the subject had been the Lutheran minister Johann Jakob Brucker's *Critical History of Philosophy*, published in Latin in the 1740s and recycled many times over, notably throughout the *Encyclopédie*. This work had included a long discussion of the philosophy of China, covering everything from Confucian ethics to Neo-Confucian natural philosophy, along with references to both primary- and secondary-source texts.[66] Fifty years later, with the evolution of education systems throughout Europe, the history of philosophy was a growth industry, and the new options that became available, such as Wilhelm Gottlieb Tennemann's *History of Philosophy* and Joseph-Marie Degérando's *Comparative History of the Systems of Philosophy*, hardly mentioned China at all.[67] In the story upon which professionals were coming to agree, philosophy was a more technical term than it had once been before, with a narrower historical purview. It had developed linearly, or even organically, in time and space, from its birth in ancient Greece to its adulthood in modern Europe. Introductory courses are still often taught this way in European universities today. But in the early nineteenth century, it was a very recent, and indeed a somewhat controversial, development.[68]

Philosophers typically wanted to forbid orientalists from poaching on their lands, but since philosophers had little ability to interlope on the domains of orientalists, the reverse was not always the case. Abel-Rémusat noticed that the institutionalization of the two disciplines was pushing them further apart, and he lamented it.[69] Sinology was established in France, and philosophy beyond the Rhine; the split was "prejudicial to the interests of science."[70] Surely, he believed, with rapid developments in both fields and easy communication after the Congress of Vienna, a reconvergence was on the horizon. This was not to be. But that Abel-Rémusat hoped for it explains a great deal. If natural philosophers had already concluded that China had no natural science, now philosophers were arguing that China had no philosophy, either. A foundational question for academic Sinology then became what kind of knowledge it had instead.

In a state-of-the-field essay published in the first issue of *Journal Asiatique* in 1822, Abel-Rémusat laid out what he believed to have been the most noteworthy achievements of the previous decade. Some of these reflected the completion of the program that was begun by the sinologue scholars, like the mining of Chinese texts for new empirical facts about history and geography. But two more accomplishments in particular stood out to him. First, the "most ancient inscription of China has been decoded," granting insight into the language of "the remotest epochs of history." Second, a pan-Eurasian doctrine—"I would almost say the secret of the mysteries"—had been revealed in a Chinese book of the fifth century BCE.[71] The first comment referred to the Stele of King Yu; the second, to the philosophy of Daoism. Abel-Rémusat explicitly rejected the work of the orphans of the Enlightenment, but the sources and analyses they produced had been baked into his own thinking. If he dismissed Court de Gébelin and Bailly's stories of Egypt and Atlantis, he still maintained that "few subjects, in the domain of ancient History, properly give rise to more curiosity than the antique relations and now nearly forgotten connections that must have existed between these nations whose origin goes back to the first ages of the world."[72] And although he denied Amiot's account of ancient science in China, he insisted at the same time that "one must not necessarily accuse the ancients of telling us lies, when they recount events that no longer take place in our days."[73] The studies of the Stele of King Yu and the philosophy of Daoism that Abel-Rémusat touted as among the signal achievements of his nascent discipline thus reflected the accomplishment of an already-established research agenda.[74]

The History of Yu the Great

As the turmoil in France began to calm down a little, then burst outward with the rise of Napoleon, an international competition broke out over the mantle of Chinese studies that the missionaries and academicians of the Enlightenment had dropped. Almost immediately after it was decided by common consensus in favor of Abel-Rémusat, the conclusion seemed to have been practically preordained. Historians have generally disparaged the other contestants or, more often, simply ignored them.[75] Abel-Rémusat regarded the majority of his older contemporaries with nothing but contempt. It is true that these nearly forgotten figures did not win the acclaim of learned Europe in their time, much less in later ones. But their struggles were fruitful nevertheless, because they served to transmit materials, ideas, and questions from one generation to the next. In the process, the Stele of King Yu—indeed, the very rubbing that Amiot had sent to Paris—became for a time the most contested object in European Chinese studies.

During the 1790s, preparations for the Macartney embassy had put a heavy burden on Europe's few China experts, and its disappointing results only proved their inadequacy. The felt need for a decent Chinese dictionary was not by any means new.[76] Over the years, the missionaries had composed many lexicographical works, and many scholars had at least considered trying to get one published. The saga of the Chinese language in early modern Europe was long and eventful. Each generation had entertained its own vision for what the characters might provide.[77] For Kircher, they were a repository of lost Hermetic symbols. For Fourmont, they were the keys to a universal language. Now, in the early nineteenth century, it was their historical and philological evolution that were at stake. And it was for this reason that attention turned eventually to the tadpole-script writing of the Stele of King Yu.

In 1800, a certain Joseph Hager announced in the British *Monthly Magazine* his ambition to publish a Chinese dictionary. Born to Austrian parents in Milan in 1757 and arrived in Britain by way of Germany, his qualifications for the project were unclear from the start.[78] "Of Dr. Hager's learning and talents few will entertain a doubt," wrote an early critic, "but the propriety of his first deciding on the publication of a Chinese dictionary, and then commencing the study of the language, will be doubted by many."[79] Nor has posterity been kind to "Dr. Hager." According to one later historian, his corpus was "not even interesting today as a curiosity."[80] The only book he published in Britain, *An Explanation of the Elementary Characters of the Chinese,* was a disorganized paraphrase of various theories from orientalists including de Guignes, orphans of the Enlightenment such as Bailly, and most especially the works of Amiot.[81] That such a mediocrity could get as far as he did is itself significant—such was the state of Chinese studies at the time. Only his one idea of publishing a Chinese dictionary seems to have ever been considered a good one, and there he soon had competition.

Hager's first rival was similar to him in some ways: Antonio Montucci was both an international academic adventurer and an amateur when it came to Chinese studies. Born in Siena in 1762, he had moved to England in 1789 serve as Italian tutor for the family of the industrialist Josiah Wedgwood.[82] He then began to study with two Chinese Catholic seminarians who had also recently come from Naples in order to prepare for service as interpreters in the Macartney embassy.[83] According to his own dubious report, Montucci was tapped to produce the written Chinese version of King George III's message for the Qianlong emperor.[84] In 1804, apparently feeling some pressure from word of Hager's project, Montucci responded with his own prospectus for a Chinese dictionary, which also appeared in the *Monthly Magazine*. The rival proposal brought the two would-be lexicographers into direct competition, and soon they were waging an all-out pamphlet war in English, French, and German. Montucci's colorful nom de guerre: "Sinologus Berolinensis."[85]

Hager won the first battle. In 1802, one of Napoleon's agents invited him to the city on commission to compile the Chinese dictionary. Though Sinology was by this point a thoroughly international affair, Paris remained its center, if only because of the extraordinary resources available there. Upon arrival, Hager began to explore the collections of the Bibliothèque nationale. There, he found not only thousands of primary-source Chinese language documents, but also a vast repository of unpublished secondary work by earlier orientalists, Jesuit missionaries, and now ex-Jesuits as well. In 1795, the Committee for Public Safety had assigned the task of sorting through Bertin's confiscated Chinese collection to two of his former associates: François Desvoyes, his secretary who was later confused with both his brother and Mellet; and Louis Langlès, his protégé who later rebuffed the young Abel-Rémusat at the recently renamed Bibliothèque nationale. Together, they completed a haphazard catalog and turned over the choicest objects for a new national museum.[86] In 1815, a printed notice advertised the auction of over three hundred items from the "Chinese cabinet of the late M. Bertin."[87] Much of the physical collection—porcelains, vases, pottery, glasswork, calligraphies, carvings, jades, and much more—was thus dispersed.[88] But fortunately for scholars, much of the written material found its way into the national library, where it was when Hager arrived. Now, it hardly mattered whether he could read Chinese or not. Swimming in a sea of easier opportunities, he quickly lost sight of the Chinese dictionary and instead put out in rapid succession a few derivative studies dedicated to his new patron, the emperor of France.[89]

The first thing Hager published upon arrival in Paris was entitled *Monument of Yu, or the most ancient inscription of China*.[90] The centerpiece was a magnificent reproduction of the characters from the Stele of King Yu. Hager had been interested in the subject ever since he had read about it in an essay by Cibot and come across a different version in Japanese.[91] "I had hardly had the good fortune to be called to Paris," he explained, "and to gain access to the national library, as rich in literary treasures as in distinguished savants to whom their protection is entrusted, when I was pleasantly surprised to find an original manuscript of Father Amiot"—referring here to the letter Amiot had sent to Paris along with his package way back in 1777.[92] Hager published the missionary's entire French translation and commentary alongside the tadpole-script characters, each printed on its own folio page, with no trace or mention of the septenary layout. He concurred with Amiot and the Kaozheng scholars, contra certain "Chinese antiquarians," that the antiquity of the monument was dubious. The tadpole-script characters were quite unlike any other known scripts; any purported interpretation that Chinese scholars had given was "as vague and arbitrary as certain etymologies proposed by Vargas or Court de Gébelin."[93] There is no evidence that Hager knew that Court de Gébelin was the very person for whom Amiot's original manuscript had been written, even as he finally realized the savant's ambition to print a version of the rubbing large for the benefit of the learned public.

With Hager moving on from the dictionary to stake out new terrain, Montucci followed in hot pursuit. He responded to the Paris publications with a series of letters and pamphlets, "in order that his gross ignorance may be clearly demonstrated, and HIS MOST BASE AND ARTFUL PLAGIARISM fully exposed, and mathematically proved."[94] Montucci pointed out, quite correctly, that Hager's *Monument of Yu* as well as his *Explanation of the Elementary Characters of the Chinese* were both lifted directly from Amiot. The few points on which his analysis differed from that of the missionary were further evidence of "*artful* and *malicious*" manipulation, intended to conceal the true source.[95] Hager was a "contemptible abortion of the literary world."[96] He had fundamentally misunderstood the Chinese characters, incorrectly identifying the names of various scripts and confusing their inscribed and printed forms.[97] This line of attack was especially important for Montucci, because the point was still to show that Hager had no business working on a Chinese dictionary.

Montucci was in the end only half successful: he managed to discredit Hager but failed to replace him. By 1806, Paris authorities finally noticed that work on the promised dictionary was not going anywhere, and an inquest was begun. Hager put on a loud display of indignation, then quietly fled France. After a brief stint at Oxford, he returned to finish his career in Italy.[98] In the meantime, Montucci had somehow won a rival dictionary commission from Friedrich Wilhelm III of Prussia, but that project was also ill-fated. No sooner did he arrive in Berlin than imperial French troops, fresh off their victory at Jena, took the city.[99] At this point, it briefly seemed like Montucci might end up working on the Chinese dictionary in Paris after all. Antoine Isaac Silvestre de Sacy, a minor confidant of Napoleon, even recommended him for the job. But Napoleon passed him over again, this time apparently because he thought it might be a good idea to hire a Frenchman.[100]

The commission thus finally fell to a third man, Chrétien-Louis-Joseph de Guignes, who was in many ways a more obvious candidate than either of the other two. He had begun studying Chinese at a young age with his father, the academician Joseph, and in 1783 obtained, through the patronage of Bertin, a post in the French consulate in Canton. His duties there mostly involved supervising commerce, though his father had paraded a few lackluster studies of related topics like Chinese dyes and porcelains before the Académie des inscriptions.[101] When the Revolution broke out, the younger de Guignes stayed on precariously in Canton, again relying on his father's connections to get himself hired as interpreter for a Dutch embassy to Beijing that followed close on the heels of Macartney.[102] When he finally returned to Europe upon the death of his father in 1801, he thus immediately gained the surprisingly rare distinction among contemporary China experts of having actually spent time in the country and by some accounts even being able to speak the language. He further capitalized on his experiences by publishing a popular account of them in 1808, *Voyages to Beijing, Manilla, and Isle de France*, which described him in the subtitle: "*By M. de Guignes, Resident of France in China, Attaché to the Minister of Foreign Relations, First- and Third-Class Correspondent of the Institut.*" It was likely his newfound literary acclaim as much as anything else that won him the dictionary commission in the same year.[103]

Despite his many apparent qualifications, de Guignes proved to be a very poor scholar by practically everyone's account. Amiot had once held high hopes that de Guignes fils would someday follow in his famous father's footsteps. His only concern had been that serious study would prove difficult in uncouth and remote Macau, where the French delegation had been temporarily relocated.[104] But the aged missionary became frustrated as the wayward student began to ignore his well-intentioned but patronizing advice: "I do not know what fly has stung the young de Guignes," he complained; "I even declined to give opinions, restricting myself to flattery."[105] Eventually, his quiet resentment turned into open condemnation of the man's "indifference" to both Chinese studies and himself.[106] Amiot would not have been surprised that de Guignes failed to live up to whatever academic promise he might have once displayed. In 1813, the French imperial press released his *Chinese, French, and Latin Dictionary, Published by the Order of His Majesty the Emperor and King Napoleon the Great, by M. de Guignes*. This publication, however, proved to be little more than a sloppy edition of a manuscript that had been composed by a Franciscan missionary over a hundred years earlier. After that, de Guignes the younger contributed little to Chinese scholarship.[107]

Meanwhile, as the terrain disputed by Hager and Montucci was shifting, another contestant had entered the field. Klaproth's story began much like Abel-Rémusat's: born in 1783 to a prominent chemist, he started studying Chinese on his own as a teenager with an old Jesuit dictionary and some Chinese books that he found in the Royal Library of Berlin. But the restless Prussian's career soon diverged from that of his sedentary Parisian contemporary. In 1805, he signed up with a Russian diplomatic mission to travel thousands of miles through Siberia all the way to the Mongolian border of the Qing empire.[108] Returning to French-occupied Germany, he began pamphleteering. "In this species of warfare," wrote an early biographer, "he displayed an ardor and a skill which were invincible." His talent was matched only by his intemperance; the same biographer speculated that "he had resided so long amongst rude and unpolished people, that he had, insensibly, imbibed a tincture of their manners."[109] In 1810, Klaproth picked a fight with Montucci.[110] The next year, he published the unambiguously titled "Tomb Stele on the Grave of the Chinese Learning of Dr. Hager."[111] At one point, he even proposed the long-deceased Amiot for ceremonial execution after the Autumn Assizes.[112]

At stake in these battles was the evolution and interpretation of the Chinese language—and that came down to competing interpretations of the Stele of King Yu. In 1811, Klaproth published his own study, *Inscription of Yu*, featuring a Chinese title, printed large in characters across the front, to advertise the author's mastery of the language: "*The Correct Meaning of the Sacred Inscription of Yu*," it read.[113] His immediate target was Hager and, behind him, Amiot: he argued that the scholar's doubts about the monument's antiquity were baseless, but the missionary's French translation had been based upon an inaccurate modern Chinese one. By way of correction, Klaproth meticulously traced the development of each of the seventy-seven characters from tadpole script to modern forms and produced his own original German translation. He gathered together passages from late imperial Chinese sources, including the *Records of the Unification of the Great Ming* and the *Huguang Gazetteer*, and reprinted excerpts both in German translation and in Chinese original.[114] His conclusion was that flood of Yu the Great was a demonstrable historical event that had taken place around 2300 BCE, as confirmed by the records of both the Egyptians and Chaldeans, and thus, while it was true that the oldest surviving stone version of the Stele of King Yu dated only to the Song, all of them were cut according to an intelligible ancient original.[115]

Klaproth's *Inscription of Yu* put the matter to rest, at least for the time being. Abel-Rémusat wrote a short review hailing the study as "indisputably one of the most interesting that has appeared for a long time in Europe on Chinese literature."[116] For him, the main takeaway was that there was a traceable genealogical relationship between the tadpole script and modern Chinese characters, making the Stele of King Yu "one of the most ancient monuments of the art of writing among the Chinese, since it goes back nearly 41 centuries."[117] Today, Chinese epigraphers agree that first Stele of King Yu to bear the seventy-seven character tadpole-script inscription was probably inscribed during the Northern or Southern Song.[118] Ironically, it was Abel-Rémusat and Klaproth, not Hager and Amiot, who were most thoroughly taken in by the forgery. Modern Sinology was moving past some of the traps that had ensnared its early modern predecessors, but it was also falling into new ones all its own.

The Philosophy of Daoism

It was Abel-Rémusat's pioneering research on Daoism that most distinguished his work from that of his predecessors. He maintained that this was so, his contemporaries thought it too, and historians have more or less agreed ever since.[119] The first missionaries in China had presented the orthodox Confucianism of the literati elite as paradigmatic of Chinese philosophy. For the next two hundred years, admirers praised it as sensible and practical, while detractors criticized it as dull and prosaic. In focusing on Daoism, Abel-Rémusat thus departed from almost every European who had written about Chinese philosophy before, from Jesuits and Dominicans to philosophes and physiocrats. By the 1830s, the stereotype had completely reversed: Confucius was replaced by Laozi as China's spokesman in Europe.[120] The Confucian scholar-official and the Daoist mystic-sage were both figments of European fantasy, insofar as they represented anything essentially Chinese. That the former was swapped so swiftly for the latter only confirms this. Abel-Rémusat's deepest insights pertained not to China itself, but to its possibilities for post-Enlightenment Europe. Where Confucianism had failed as ancient philosophy, Daoism might succeed as what he came to call "ancient eastern wisdom."

Already by 1816, Abel-Rémusat had come to believe that "there is perhaps no work of philosophy in any country, whose antiquity and purity are as completely free of suspicion as the book of *Laozi*."[121] He probably became interested in Daoism first through his studies in medicine.[122] Amiot and Cibot had suggested the connection explicitly in their essays on kung-fu, deer-blood elixirs, and pulse diagnostics, among others, all of which had been published in the *Mémoires concernant les Chinois*, still the best source available to the young medical student in the 1800s. His initial impression was confirmed in the 1820s when he took over as librarian at the again-renamed Bibliothèque royale, where he discovered the archival manuscripts of the Figurist missionaries.[123] When it came to secondary European scholarship on Daoism, the works of the Jesuit and ex-Jesuit were pretty much all that there was.[124] As he pointed out, most others in Europe had dismissed its partisans as materialists, atheists, and nihilists on the one hand, or as astrologers, sorcerers, and magicians on the other.[125] Only recently were descriptions coming to seem contradictory, which posed new problems that the missionaries had never had to confront.[126] The upshot is that there were still very few studies of the subject, and essentially no complete translations by which questions might be resolved.

Abel-Rémusat aimed to redress the situation with his magnum opus, the *Memoir on the Life and Opinions of Laozi* of 1823. According to his star pupil, Stanislas Julien, the book produced "a lively sensation in Europe."[127] At its core was a partial translation and scholarly study of the *Daodejing*, or *Classic of the Way and Virtue*, traditionally attributed to the semi-mythical sixth-century-BCE philosopher Laozi, though probably composed mostly later, in the Warring States period.[128] The study presented both the first substantial European-language translation of a Daoist classic—replete with original Chinese passages and running commentary throughout—and by far the most positive evaluation of Daoism to be published in Europe to date. It was the beginning of a deliberate effort to rehabilitate Laozi's reputation: "Instead of the patriarch of a sect of charlatans, magicians, and astrologers, searching for the elixir of immortality and the method of flying through the air to heaven, I found in his book a veritable philosopher, judicious moralizer, eloquent theologian, and subtle metaphysician."[129] For Abel-Rémusat, Laozi was nothing less than a "Chinese Plato," speculating about the fundamental nature of the universe, and his doctrine of Daoism was essential in the global history of ancient philosophy.[130]

As philosophy, Daoism provided Abel-Rémusat with a response to the newly forceful critique that Chinese thought was only practical, not theoretical. The basic idea dated back to the earliest missionaries, who made it in order to pitch their own contributions in abstract sciences. Confucius's focus on the here and now had in fact been a major selling point for the

(41)

» le néant des êtres avant la création), pour apprécier
» ce qui existe à présent, ou l'univers, on peut dire qu'il
» tient la chaîne de la raison. »

sine imagine, imperspicuum. Obviàm is illi, non vides ejus caput; sequeris illum, non vides posteriorem. Tenens veterum rationem, ad regendum hujusce temporis esse, potest nosse vetus principium. Hoc est vocatum rationis catenatio.

視之不見名曰夷聽之不聞名
曰希搏之不得名曰微此三者
不可致詰故混而爲一其上不
皦其下不昧繩繩不可無復歸
於無物是謂無狀之狀無像之
像是爲惚怳迎之不見其首隨
之不見其後執古之道以御今
之有能知古始是謂道紀

F

FIGURE 5.3 *The Classic of Power and the Way*, in three languages. *Mémoire sur la vie et les opinions de Lao-Tseu*, Jean-Pierre Abel-Rémusat, 1823. Courtesy of Harvard College Library.

worldly philosophes of the high Enlightenment. But toward the end of the eighteenth century, especially after Immanuel Kant, European philosophy took its own turn, and for those who wished to defend Chinese philosophy, Confucius had to be either reinterpreted as a metaphysician, or abandoned in favor of one. Abel-Rémusat chose the latter approach.[131] Confucius, he claimed, "touted the virtue and the reason on which, according him, all human conduct should be regulated; but he is silent on the fundament of this purely human reason."[132] By contrast, Laozi, whose school was literally called the "Doctrine of Reason," addressed the "primordial reason" at the fundament of everything.[133] "In the end," he later wrote, Daoism was "a science, a rational ensemble of often ingenuous and sometimes sublime ideas on the constitution of the universe, the action of the first cause and second causes."[134] It was complete with a "refined metaphysic"—long held as distinctive to European philosophy, but only recently deemed definitional to philosophy writ large.[135] If Laozi had speculated about topics that Confucius ignored, that was now a point in his favor: "The opinions of the Chinese philosopher on the origin and the constitution of the universe present neither ridiculous fables nor shocking absurdities, but bear the impression of a noble and elevated mind."[136] This defense of a Chinese philosopher was very different from those that had come before. In a way, it was the opposite, since it praised Laozi for exactly the qualities that Confucius had always been thought to lack.

As history, too, Daoism represented a way for Abel-Rémusat to write China into what was fast becoming a uniquely European story. This aim was indicated in the very subtitle of his study of the book of Laozi: *Chinese Philosopher of the Sixth Century before Our Era; Who Professed Opinions Commonly Attributed to Pythagoras, Plato, and Their Disciplines.* As his student Julien again aptly observed, "to tell the truth, the main point of his memoir on Laozi is to prove this conjecture, and to establish from there that there had been communications between the West and China from the sixth century before Jesus Christ."[137] The opening line of the study was indeed quite up front about it: "Among the facts relative to the peoples of East Asia, those that attest to the existence, between these peoples and the nations of the West, of ancient communications and connections before those that arose in the middle ages, appear to us to merit a particular attention."[138] According to Abel-Rémusat, Laozi was not just a Chinese Plato; he also been in indirect contact with the Greek one. Legend held that Laozi traveled west to Persia or even Syria in order to learn "the explication and the development of principles that had been brought to him previously," just as Plato purportedly went to Egypt to study those that had been previously brought to him.[139] In ancient times, there was an open channel of learning, flowing in all directions and sustained over a long period of time.

The rehabilitation of Daoism as both philosophy and history moved naturally to a conception that was supposed to explain both: a single philosophical ur-tradition. Both comparison and connection were at play. According to Abel-Rémusat, the "opinions commonly attributed" to ancient Greeks alluded to in the subtitle had likely arisen in the Fertile Crescent, whence they spread west to Greece and east to China, brought there by a lost tribe of Israel, or maybe a restless predecessor of Pythagoras.[140] Here was the explanation for why the doctrines of China and Greece had so much in common: a chain of being built upon the monad; man as microcosm and copy of a divine archetype; individual souls as emanations of the ether to which they return after death; and a universal soul as harmony between matter and intelligence.[141] The word *dao* was identical with the *nous* of Hermes Trismegistus and the *logos* of Plato, meaning "sovereign being," "reason," and "language" as a noun, or "to speak," "to reason," and "to justify" as a verb.[142] Such details were "too surprising, too certain, too many, for one to see in them anything other than the effects of a communication."[143] As we have seen, Abel-Rémusat was not the first to see congruity between supposedly Chinese Daoist and Greek Platonic or Egyptian Hermetic doctrines, nor was he the first to propose a genealogical explanation for it—in fact, it is noteworthy that in the entire history of early modern European studies of China, no one seems to have completely denied either point.[144] His distinctive contribution was in the evidence he used and the significance he attached to it.

Although no extensive translation of the *Classic of Power and the Way* had yet been published in Europe, one brief passage from its chapter 42 was already somewhat well-known.[145] Abel-Rémusat's French translation of it, rendered literally into English, read: "Reason produced one; one produced two; two produced three; three produced all things. All things rest on matter and are enveloped by the ether; a vapor or a breath which unites them and maintains them in harmony."[146] While the early modern missionaries had focused on the first sentence for its resonance with the Christian trinity, the late modern Sinologist was more interested in the second for its supposed evocation of Greek matter theory. But if Abel-Rémusat's interpretation was perhaps more secular, it was certainly no less manipulative. His tendentious interpretation of *yin* and *yang*, which were referred to in the Chinese, as simply "matter" is one example. More revealing, his translation of the Chinese word *qi* as the Greek word "ether" may be accurate enough, but to translate one technical word with another because of an apparent similarity, and then to insist that the translation proves that similarity, might be called begging the question.[147] The accommodationist program that the Jesuit missionaries had brought to bear on the Daoist classic continued in full force, only with a different tradition that needed to be accommodated.

The most important evidence for Abel-Rémusat was a lesser-known passage of the *Classic of Power and the Way*, taken from its chapter 14. His translation read: "That which you look at and do not see is called *I* [*yi*];

that which you listen to and do not hear is called *Hi* [*xi*]; that which your hand reaches for but cannot grasp is called *Wei* [*wei*]. These are three beings that one cannot comprehend, and which taken together are but one."[148] Here, too, the Figurist missionaries had thought they found a triune God in the three ineffable entities *yi*, *xi*, and *wei*: that which cannot be seen, heard, or touched. Abel-Rémusat criticized their interpretation as "a little venturesome"; but here, too, his was hardly less so.[149] He claimed that all the previous commentators in Europe and China alike had fundamentally misunderstood the Chinese passage, and for a very simple reason: it was not actually Chinese. The three crucial characters had no meaning at all; they were merely a transliteration of sounds from a foreign tongue. Combining them into a single word, *I-Hi-Wei* (*Yixiwei*), revealed nothing other the biblical name of God, the tetragrammaton: the four Hebrew letters, הוהי, *yod*, *he*, *vav*, *he*, YHWH, *Yahweh* or *Jehovah*. Better still, the Chinese characters indicated the pronunciation of both vowels, lacking in Hebrew, and aspirations, lacking in Greek. It was therefore "remarkable that the most exact transcription of this famous name is found in a Chinese book."[150] God was there in the Chinese text after all.

Yet for Abel-Rémusat, this was an important passage in the *Classic of Power and the Way* not because it revealed ancient Chinese piety, but rather because it was the one that was, "more than any other, suitable for going back to the sources from which the author drew."[151] In his words, it testified to an "antique science, this science that was the object of the researches of all enlightened men in ancient times." [152] The first peoples of Eurasia had all referred to the divine the same way: a triad, representing its triune essence—being, intelligence, and life—and its tripartite extension—through the past, the present, and the future.[153] The tetragrammaton could be found not only in the *Classic of Power and the Way* and the Bible, but also in the religious texts of the Romans, the Phoenician Sanchuniathon, and the Greek works of the Pythagoreans. Daoism was the best surviving expression of "the mythology of the Greeks, the Egyptians, or the Indians, stripped of their allegorical veils, taken from their enigmatic language, purged of their incoherent myths and local legends, directly addressing intelligence and reason." Daoism not only demonstrated that China had proper philosophy—that is, metaphysical—and not only showed that China had participated in its agreed-upon history—that is, European—but revealed the ultimate origins of both. The writings of the Daoists Laozi, Liezi, and Zhuangzi all testified to "the height to which the philosophy of China was known to rise before Confucius, and, behold, of which remain the almost-effaced vestiges and scattered tatters." It was in these texts that "savant Europe should seek the precious memories and debris of the primitive traditions of high Asia."[154] This was exactly the point made by the orphans of the Enlightenment, and in the very the same language.

Toward the end of his career, Abel-Rémusat coined a new term for the primitive traditions at the origins of the history of philosophy: "*antique sagesse orientale*," or ancient eastern wisdom.[155] Underpinning his historical and philosophical account of ancient eastern wisdom lay a claim about progress. The ancient peoples of Asia were not as backward as they had been made to appear: "We are perhaps a little too disposed to blame on their ignorance that which is really an effect of our own."[156] At the end of a revised version of the essay originally published as the introduction *Memoire on the Life and Opinions of Laozi,* he cited a quotation that he attributed to one of the Daoist master's many disciples:

> A bright light illuminated high antiquity; but scarce few rays have reached all the way to us. It seems to us that the ancients were in shadows, because we see them through the thick clouds from which we have just emerged. Man is a child born at midnight; when he sees the sun rise, he believes that *yesterday* never existed.[157]

The study concluded on this quotation with no further reflection, for the words that Abel-Rémusat placed in the mouth of this unnamed Daoist might as well have been his own. The ancient wisdom buried in past and the orientalist's knowledge to be uncovered in the future were one and the same. Early modern *prisca scientia* had become late modern *sagesse orientale.*

But the transformation involved a difference. The orphans of the Enlightenment had believed that the knowledge of the ancient eastern past was the same as that which was yet to be discovered in an imminent European future. Whether it was the ancient tarot-card magic of the Egyptian priests, the astronomical calculations of the Atlantan sages, or the mesmerist hypnotherapy of the Chinese magnetizers, it was supposed to be discoverable by modern kabbalists, physicist, and doctors in their own time. Since knowledge for the Enlightenment was unitary, a final convergence was assured. Abel-Rémusat, on the other hand, was not so sure. Ancient eastern wisdom was not supernatural, mythological, or allegorical—it was after all a doctrine of reason, science, and enlightenment, consistent within its own cosmological framework.[158] But on the other hand, it was also mixed up with "occult sciences" and "mystic practices" that would never be revealed through the independent methods of European philosophy, still less the natural sciences.[159] The knowledge of different times and different places was, in some essential sense, different. This point was critical, and Abel-Rémusat most likely discussed it with a person who sought make his own idea of progress to account for it.

Georg Wilhelm Friedrich Hegel and Unhistorical History

While Europe's leading Sinologist was striving to bring philosophy and orientalism back together, its preeminent philosopher was working to drive them further apart. In his earliest works, Hegel, like most late Enlightenment progress theorists, had not been especially interested in non-European knowledge traditions. It was only in the 1820s that he began to study China seriously, and when he did, his two main sources were Amiot and Abel-Rémusat.[160] They convinced him that it was Daoism, not Confucianism, that mattered most for his own studies—but not as philosophy, or even as history. From Amiot, he learned to think of it as a form of magic; from Abel-Rémusat, as an expression of ancient eastern wisdom. Hegel brought them together to consider what made modern European thought unique. In this way, post-Enlightenment progress theory finally made sense of the account of non-Western knowledge that the orphans of the Enlightenment had created. The past and the East were now interesting again, but only insofar as they were incommensurable with the modern West.

Hegel and Abel-Rémusat likely met in Paris during the summer of 1827.[161] Their encounter was a key moment when foundational figures in orientalism and philosophy could have bridged the widening divide between them. The two men were roughly contemporaries—Abel-Rémusat was eighteen years younger than Hegel, but relatively precocious—and they shared many acquaintances and collaborators in each other's disciplines, including Klaproth, Georg Friedrich Creuzer, Wilhelm Schlegel, and Wilhelm von Humboldt. Nor is it difficult to imagine how a rapprochement might have taken place. On September 19, Hegel wrote to his wife in Berlin, mentioning that Abel-Rémusat had invited him to attend a session of the Académie des inscriptions scheduled for the upcoming Friday and that he planned to attend.[162] This brief letter is the only concrete evidence that we have of their meeting in person. It seems also that the topics that were up for discussion at the Académie on that day concerned Gallo-Roman archaeology and the Peloponnesian War.[163] But there is a great deal of circumstantial evidence that suggests the other conversations they may have had while Hegel was in Paris, and in any case explains what happened right after he left.

Chief among their mutual friends was the philosopher Victor Cousin.

During the 1820s, Cousin and Abel-Rémusat saw each other often at the biweekly sessions of the Académie des inscriptions. Their relationship was personal as well as professional. For example, Abel-Rémusat invited Cousin to join him after one meeting for what he called "*un petit diner Mystique*."[164] Cousin was particularly interested in the new research on Daoism; it was he who suggested that Abel-Rémusat's protégé Julien develop his own research in that direction in 1826.[165] By this time, Cousin was also France's most famous Hegelian. Twice in the previous decade, he had traveled to Germany to pay respects to the master, and he eventually returned the hospitality with the invitation that brought Hegel to Paris in the summer of 1827. Cousin played host for the duration of the month-long visit.[166] And it was in the academic year immediately following that he gave a notable lecture at the Sorbonne: "Has there or has there not been philosophy in the Orient?"[167] Presumably, he had taken advantage of a recent opportunity to discuss the question with the people at the forefront of each relevant field.

Hegel was not solely responsible for the European rejection of non-Western philosophy, but he has often been thought to mark its completion.[168] The result seems almost overdetermined by the rise of modern empires and the concomitant emergence of full-blown racism. On the one hand, the exclusion of Chinese thought from the history of philosophy was indeed the end of a process that had begun already in the time of Voltaire and was always especially pronounced in Germany, where Herder once likened China to an "embalmed mummy."[169] On the other hand, when Hegel received his first academic post in 1801, the status of Asian philosophy within the German academy was far from settled, as the prominence of the orientalist-philosopher brothers Schlegel attests.[170] Some historians have suggested that Hegel rejected Chinese thought in order to distance his own from a potential critique, while some philosophers have maintained that there really are underlying similarities between the two.[171] Neither of these arguments is widely accepted, but that either has been proposed points toward Hegel's true significance. Although his claim that China had no philosophy was not especially original, his proposal of what it had instead was.

Now, while French philosophes had not developed a conceptual tool-kit for dealing with radical alterity, German philosophers working in the aftermath of Kant were beginning to do just that.[172] Theirs was in fact the conceptual vocabulary that would be adopted by postcolonial theorists in their later critiques of the Enlightenment project.[173] Johann Gottfried Herder's theory of cultural individuality, Johann Gottlieb Fichte's distinction between the I and not-I, and Hegel's own dialectical method were all episodes in this development. All of them addressed the Romantic problem of subjectivity, which could also be phrased as a question of epistemology: If individual societies were truly and completely distinct, did that mean that their thought itself was distinct as well? This question had not been very much at issue for Enlightenment philosophes from Voltaire through Condorcet, for whom human nature was in some sense always and everywhere the same.[174] But as the French and German philosophical worlds drew closer to together after the Napoleonic era, it became increasingly important for everyone.

As heir to both Franco-British Enlightenment and German Romantic traditions, Hegel attempted to answer it through a novel theory of progress.[175] He broadly adopted the terminology of stadial theory, but his model was not really stagist in the same way that those of the philosophes had been. In a composite caricature covering both Scottish conjectural history and French philosophical history, societies move naturally and inevitably from one stage to the next; you may not be able to see where you are going, but you can always see where you came from. By contrast, in Hegel's thought, discrete cultures or societies could not be so many beads threaded together along a single string of progress, because reason itself changes as it evolves through space and time. As he gnomically proclaimed, "history in general is therefore the development of Spirit in *Time*"; and, vice versa, "philosophy, too, is *its own time comprehended in thoughts*."[176] History and philosophy were in principle inseparable—to lack either was to lack the other. On the one hand, this view of progress was Eurocentric in a way that even Condorcet's had not been, since both history and philosophy, properly understood, only inhered in certain times and places. But on the other hand, this view also turned attention back upon the rest of the world since places outside of progress could no longer be written off as merely absent. If they had no history or philosophy, then they must have had something else instead. Tellingly, Hegel explored non-Western thought most thoroughly not in his lectures on the history of philosophy, nor the philosophy of history, but on the philosophy of religion.

At the center of all of Hegel's discussions of China was an essay by Amiot: the "Treatise on the sect of the *Daoshi*," completed in Beijing in 1787 and published in the *Mémoires* in 1791.[177] This study was Amiot's longest touching on Daoism by far, written while his attitude toward the subject was in deep transition, and the only one that he ever intended for publication. Here, as in his private letters, he described a doctrine of *taiji* and *yin-yang*, material souls and metempsychosis. Notably absent, however, was any vindication of Chinese natural philosophy in terms of animal magnetism or magnetic somnambulism. Instead, he made mention of a different set of European traditions: "magic, cabala, judicial astrology, &.c"—"which we understand under the general denomination of *occult sciences*."[178] Amiot used this term three times in the essay, and nowhere else in any of his writings. He made no effort to defend Chinese "masters," "adepts," or "initiates." His only indication of any ambivalence was in his defensiveness about the choice to write about them in lieu of other topics. He stood by his decision on the grounds that Daoism was "the first to corrupt the ancient national doctrine, in joining to the noble simplicity of the primitive doctrines, the most absurd principles of the grossest errors."[179] That is to say, it preserved the vestiges of ancient wisdom.

To illustrate the supposed origins of Daoism, Amiot recounted a story set in the twelfth century BCE about the fall of the Shang and the establishment of the Zhou. In the climactic sequence, the general Jiang Ziya, working on behalf of the king of Zhou, receives a magical list of "the names and the offices of the *shen* who would be the new protectors of the empire."[180] Climbing upon an altar on Mount Qi, he uses it to invest the various spirits with their supernatural titles and powers, subjugating the forces of heaven itself to the earthly authority of the new dynasty and establishing a heavenly bureaucracy as its counterpart. A spirit was assigned to each of the twenty-four divisions of the calendar year. Some were given charge of the winds, the rains, and the clouds, and another was appointed to supervise all of those. Five were ordered to oversee infectious diseases, and five more were appointed "to head the fire department." [181] According to Amiot, this brief episode revealed something that would become crucial for later commentators: "The right to fire, demote, and punish the refractory or negligent *shen* does not belong solely to the *shen* superior to them; it belongs furthermore to the Emperor, as Son of Heaven and father of his people."[182] The supernatural authority of the emperor was the one thing that linked together all the religions of China, from antiquity all the way until present times.[183]

Fundamentally, the interpretation of Daoism that Amiot put forward in the "Treatise on the sect of the *Daoshi*" was not really coherent. Students of literature will immediately recognize the story as the plot of the popular Ming novel, *Investiture of the Gods*.[184] Amiot's retelling followed the last two chapters quite closely, sometimes word for word, but he never mentioned the fictional source. Narrating it instead as a historical account of the origins of Daoism, he paved the way for a great deal of later confusion.[185] He erased the distinction between the abstract philosophy and the popular religion, and, furthermore, got their chronology wrong. For example, Laozi appeared in the essay neither as the philosopher of the Warring States, nor as the apotheosized god of the Han, but as just one

"Master of the Dao" in an esoteric lineage extending back to the Shang.[186] In fact, the story of the *Investiture of the Gods*, though set five hundred years before the birth of Laozi, more properly reflects the beliefs of five hundred years after his death. Moreover, no Chinese reader would likely have taken it for a description of Daoist philosophy, religion, or history. Nevertheless, it became the centerpiece of the longest discussion of the subject published in Europe until Abel-Rémusat's book came out in 1823.

Hegel addressed the thought of China for the first time at length in his lectures on the philosophy of religion in 1824. At this point, his exclusive source was the essay by Amiot, and the conclusion that he drew from it was that the thought of ancient China was magic.[187] In his developmental typology, it fit along a spectrum between the primitive magic of Africa and North America and the true religion of pan-Asian Buddhism. The key evidence was the list that contained "the names and offices of the genii, known as *shen*, who were to be the new administrators of the empire in the natural world, the invisible officials over the natural world, as the mandarins are in the world of consciousness."[188] According to Hegel, the episode revealed the general feature of magic—direct human power over nature. It also revealed a more particular feature—the association of the individual with the divine. The emperor was revealed as a supreme magician: "This time it is the emperor of China, source of all laws in the present world, but also the lord of nature."[189] Driving the point home, he enumerated the natural phenomena entrusted to the various spirits on the emperor's authority: the mountains and the seas, the four seasons, the weather, the five elements, and so on.[190] The story of the *Investiture of the Gods*—which, again, he took for the ancient history of Daoism, rather than the plot of an early modern novel—formed the basis for his first analysis of Chinese thought.

Then, at some point in the following three years, Hegel read Abel-Rémusat's *Memoir on the Life and Opinions of Laozi*. The groundbreaking study seems to have thrown a wrench in his initial interpretation of Chinese thought. Now, he had to reconcile two very different sets of primary and secondary sources that were both nominally about the same subject. The first was a précis of the *Investiture of the Gods*, set at the beginning of the Zhou; the second was a partial translation of the *Book of Power and the Way*, set in the Spring and Autumn Period. The first, as interpreted by Amiot, described an occult science; the second, as interpreted by Abel-Rémusat, a rational philosophy.[191] Both, however, described Daoism as the most representative doctrine of ancient China, esoteric and mysterious, with origins going back to something still more ancient and pristine. In other words, Amiot and Abel-Rémusat, in the two most extensive published discussions of the topic to date, both presented it as ancient Eastern wisdom.

Just before departing for Paris in 1827, in a revised version of his lectures on the philosophy of religion, Hegel returned to the story recounted by Amiot, this time equipped with the interpretive lens of Abel-Rémusat.[192] "The Dao," he now believed, was both magic *and* reason: "an orientation to abstractly pure thinking, which orientation constitutes the transition to the second form of nature religion."[193] Hegel denied that ancient China could have known the true "laws of nature"—an idea that he credited specifically to Bailly.[194] Nevertheless, he granted Abel-Rémusat's argument that Daoism concealed the name of God and the trinity, even adding further evidence in support: the "symbol of the Dao is on the one hand a triangle, and on the other hand three horizontal lines one above the other."[195] This confused observation was in fact lifted from a different essay by Amiot, in which he had been referring not to the character *Dao* but rather to the character *Tian* from the Confucian Classics—which, it must be said, resembles a triangle only slightly more.[196] The conflation is significant. In 1824, Hegel had not explicitly named either Confucianism or Daoism; in 1827, he drew the distinction precisely in order to discount it. The emperor, as Son of Heaven, possessed both natural and moral characteristics; thus, in a stunning reversal, even Confucianism was revealed to be magical.[197] Amiot had insisted that the occult sciences were alien to the pristine doctrine of the ancients.[198] Hegel concluded instead that "the Chinese are the most superstitious people of the world."[199] His position, exactly antithetical to that of the high Enlightenment, was constructed with work that had been produced by its orphans.

After returning from Paris, in a revised version of his lectures on the philosophy of history, Hegel dealt directly with the question of ancient eastern wisdom.[200] He astutely pointed out that those who studied "oriental literature" were essentially interested in excavating the "advanced condition of the knowledge of God, and of other scientific, e.g., astronomical knowledge" in ancient times. Hegel did not believe that such knowledge had ever existed; nor did he believe in any of the hypothetical ur-civilizations to which it might have been attributed. As he put it: "The biblical account by no means justifies us in imagining *a people*, and a historical condition of such a people, existing in that primitive form; still less does it warrant us in attributing to them the possession of a perfectly developed knowledge of God and Nature." The problem with such claims, he wrote, was that they "seem to be awaiting the issue of an historical demonstration of that which is presupposed by it as historically established." What exactly Hegel meant by this is, frankly, a little opaque. It appears to have been related to his broader idea of progress: since "History" starts when "Rationality begins to manifest itself in the actual conduct of the World's affairs," there could be nothing properly called knowledge before it. Nevertheless, in this discussion, he specifically praised Abel-Rémusat for his "meritorious investigations of Chinese literature." Furthermore, he maintained, "we owe to the interest which has occasioned these investigations"—that is, the interest in ancient eastern wisdom—"very much that is valuable."[201] The important point was that it lay outside the realm of knowledge, science, or philosophy—that is, outside the purview of progress.

Hegel's late lectures on the history of philosophy reveal just how much European conversations about China had changed since the end of the Enlightenment. For Voltaire, the opposite of superstition had been reason; for Hegel, it was freedom. This reframing suggested new answers to many old questions. First, about China's political system. Rational religion and oriental despotism, once apparently at odds, could now each explain the other. In both the supposedly Daoist story of the *Investiture of the Gods* and the Confucian conception of the Son of Heaven, natural and supernatural authority were consolidated in the form of single person. For the missionaries, this unity had made the emperor a high priest; for Hegel, it made him an autocrat—the only independent agent, the only free subject, indeed, the only true "Substance."[202] Second, about China's natural science. If it had none, there was no longer any reason to expect otherwise—for without freedom, there could be no science of any kind whatsoever. Hegel was not overly concerned about the question that had so troubled Voltaire, partly because the place of the natural sciences in his own broader system was somewhat vexed.[203] At one point, he suggested that the language was a crucial inhibiting factor.[204] But more generally, China lacked what he called the "free ground of subjectivity, and that properly scientific interest, which make them a truly theoretical occupation of the mind."[205] In Greece, freedom and philosophy had been born attached at the hip; but in China, there was never any grounds for political or philosophical life.[206]

Imbricated in these discussions was a new conception of history itself as axiomatically European. As the development of spirit in time, history was in principle inseparable from philosophy—to lack either was to lack the other.[207] The idea of ahistorical China had always been potentially implicit in the trope of static China. Hegel certainly did not invent the latter; indeed, Abel-Rémusat viewed it already as a cliché. But his original conception of "History" led him to develop it in a new direction.[208] For Voltaire, China had been curiously missing progress, but possessed of history in abundance; for Hegel, having progress was a precondition for calling anything properly historical. Voltaire had called the East the "cradle of all the arts"; Hegel called it the "childhood of history."[209] The metaphor was the same, but the evaluation was reversed. When Hegel said that "Europe is absolutely the end of History, Asia the beginning," the term "absolutely" carried the real thrust of the point.[210] Accordingly, he elaborated: "China and India lie, as it were, still outside the World's History, as the mere presupposition of elements whose combination must be waited for to constitute their vital progress."[211] Hegel's creativity in finding different ways to phrase this same point was almost astonishing. China was an "immovable unity," belonging to "mere space" and not time; or, in other words, to "unhistorical History."[212] The East was no longer merely stuck in time; it was now truly timeless.

It is nevertheless remarkable that along the way to reaching that conclusion, Hegel ended up saying more about China than any of the philosophes had since the high Enlightenment. Historians have often focused on the negative results of his project: once prized for its history and philosophy, China was stripped of both. But the positive element—its past construed as timeless and opaque, its doctrines as magical and metaphysical—was also significant; and this positive element was the part that required the work of Amiot and Abel-Rémusat. Hegel challenged the motivations of their research but drew his own implications from it. He excoriated Confucius as "only a man who has a certain amount of practical and worldly wisdom—one with whom there is no speculative philosophy." It would have been better for his reputation, Hegel quipped, if he had never been translated at all.[213] Instead, it was Laozi who mattered for European thought going forward. Daoism now presented a mode of thinking not in the stadial progress of society, but in the transformative development of reason in time. The Enlightenment had placed hyper-historical China outside progress theory. Now that it was placed outside history itself, it could be brought back in again.

Science and Civilization in China

At the outset of the first volume of *Science and Civilisation in China* in 1954, Joseph Needham set out a "Plan of the Work" that would ultimately extend over twenty-five volumes and fifty years. "Modern Sinology did not really begin until the nineteenth century," he wrote, "and here the inaugural discourse of J.P. Abel Rémusat delivered in the Collège de France in 1815 is still worth reading. I cannot refrain from quoting here what he then said." In Needham's translation from the French, Abel-Rémusat said:

> Many Westerners have been brought to believe that the Chinese have remained at the first stages of civilization. If I may run the risk of being reproached for partiality towards a people to whose literature I have given many years, I would like to try to bring them back to a less unfavorable opinion. Most Europeans smile at hearing of the geometry, astronomy, or natural history of the Chinese. But if it is true that the recent progress of these sciences among us has dispensed us from the necessity of having recourse to the knowledge of these far-off peoples, ought we on that account to refuse to examine the present state of their knowledge, and especially what it anciently was in a nation which has never failed to cultivate and honor it?[214]

Needham offered only one comment after reciting the rest of the page-long quotation: "How far Rémusat's facts and judgements were correct will perhaps appear in the sequel."[215] Thus began the series that would almost by itself establish a subfield in Europe and the Americas. For, as Needham himself realized, it was with Abel-Rémusat that the foundations for his own historical subject—*Science and Civilisation in China*—were established.

It must be noted at the outset that the "Needham question"—why did modern science fail to develop in China?—was neither original to Needham, nor essential to what might be called the "Needham project." In his own telling, his interest in the subject of the history of science in China did begin from the question, occasioned by the arrival of his future wife, the biochemist Lu Gwei-Djen, in Cambridge in 1938. Granted, too, he raised the question at the very beginning of *Science and Civilisation in China,* and returned to it near the final conclusion.[216] But it is not insignificant that Needham's most commonly cited formulation of the question was published elsewhere, in a journal tellingly titled *Science and Society*.[217] And in fact, the answers that he gave to it were remarkably consistent over the years, never veering far from his lifelong commitments to Marxism.[218] As he wrote in the 1990s: "In sum, I believe that the analysable differences in social and economic patterns between China and Western Europe will in the end illuminate, as far as anything can ever throw light on it, both the earlier predominance of Chinese science and technology and also the later rise of modern science in Europe alone."[219] Few of the volumes of *Science and Civilisation in China* had much to say about those matters. Therefore, the project did not address the question in any way that Needham himself seemed to think would have been satisfactory. The Needham question was "the grand question which has partly inspired the whole work," but it was not the question that it ultimately answered.[220]

A direct statement of that question, however, lies also in plain sight, even before any mention of the more famous one, on the first page of the first volume of the entire series: "What exactly did the Chinese contribute, in the various historical periods, ancient and medieval, to the development of Science, Scientific Thought and Technology?"[221] Already implicit in this initial phrasing were the philosophical underpinnings of the whole project. First, the spelling of 'Science' with a capital 'S,' indicating a scientific realism about which Needham was never shy. Second, the temporal restriction to ancient and medieval periods, reflecting the axiom that modern science was different from everything that had come before. Third, the invocation not of achievements but of contributions, revealing a broader story of genealogical connections. All three points were summed up in Needham's evocative metaphor of modern science as a universal ocean into which all the cultural rivers of premodern science flowed.[222] For all the focus on the question of why China never developed modern science, it is easy to overlook how much Needham saw China as playing a role in its formation elsewhere. It was thus very true when he plainly declared at the outset that "the scientific contribution of Asia, and in particular of the 'Central Country,' China, is the theme of this work."[223] Again, the content of the volumes themselves is revealing, bearing titles like *Physics and Physical Technology: Civil Engineering* and *Nautics* and *Spagyrical Discovery and Invention: Magisteries of Gold and Immortality*. If the Needham question was about what had not happened in China, the question that *Science and Civilisation in China* actually answered was about what had.

It was fitting that this effort to recover the history of science in China should begin with Abel-Rémusat. Just as few Europeans before had asked if China had a history of philosophy, few had asked if it had a history of science—but due to an opposite starting consensus. In his inaugural lecture at the Collège de France in 1815 from which Needham quoted, Abel-Rémusat invoked the history of "the natural and exact sciences" specifically to argue against the received opinion that China had remained at "the first steps of human reason." To the contrary, he claimed, its astronomers had kept careful and accurate records, while its botanists had been "as much above the Latin or Medieval naturalists as they were inferior to Linnaeus, Jussieu, or Desfontaine." The properties of the right triangle had been discovered there long before Pythagoras, and the tides had been explained there by "the love of the moon for the earth" long before Newton.[224] Abel-Rémusat developed the theme ten years later in a lecture entitled "Observations on the state of the natural sciences among the peoples of East Asia." Apparently given as the beginning of a longer project that he had planned, its particular subject was natural history.[225]

In this field, he maintained that the Chinese language had been a positive advantage for science, with two-character labeling conventions that were "absolutely analogous to the principle of Linnaean nomenclature."[226] His conclusion was that "the advance and progress of the sciences of observation" in China were so significant that Europeans might look there to learn about botany and zoology even in their own day.[227]

Abel-Rémusat openly posed the question: "Does there exist in the Orient knowledge that we can honor with the name of science?"[228] But he never gave a direct answer. Even in his analysis, the study of nature in China had never made the crucial leap from observation to theory. Chinese biology was characterized by "an almost complete ignorance of the internal structure of beings and the laws of organization"; for example, whales were classified with fish, and rats were thought to transform into quails in the spring.[229] Chinese physics was not just wrong, but actively "mendacious," apparently on account of its materialism: "Everything is explained by the action of causes that are considered to be natural, even though they are entirely imaginary." He rejected several answers to Voltaire's question about what had stalled Chinese science, including the nature of the language itself—which was also the subject of a high-profile conversation he was carrying on with Wilhelm von Humboldt at around the same time.[230] Instead, he laid primary blame on the Song philosophers' "scientific jargon that one might think was borrowed from our scholasticism of the Middle Ages," favoring words over ideas and arrangement over analysis.[231] Interestingly, Needham, too, described Zhu Xi as a Chinese Aquinas – "because he was, after all, a man of the medieval age"—but he also asserted that "the Neo-Confucians deserve a much greater place in the history of science than that accorded them."[232] For Abel-Rémusat, the rehabilitation of premodern science was not an option: "Very few, among those of antiquity and the Orient, maintained, in the study of nature, the disinterest that constitutes its charm and dignity."[233] Perhaps he gave up his planned history of Chinese science when he concluded that, in this specific domain, the identification between the past and the East brought honor to neither.

The trope of ancient China had not changed, but the idea of ancientness itself was changing. As Abel-Rémusat pointed out, "If there is one notion accredited, one fact recognized, one point unwaveringly fixed in the mind of Europeans, it is the enslavement of the people of Asia to ancient doctrines, to primitive usages, to antique customs."[234] The ancient, primitive, and antique had already been associated with the East; but now, as Hegel was then explaining, these antonyms of progress perforce meant enslavement. As a result, the Enlightenment's admiration for China despite its purported lack of progress was no longer viable. Science had already been linked to progress, but it was now linked to something more. When Abel-Rémusat referred to "the stationary state in which some allege that civilization and the sciences have remained among the Chinese," he explicitly invoked another new term with a late Enlightenment origin story of its own: *civilization*.[235] It is striking, too, that in *Science and Civilisation in China*—both in the "Plan of the Work" at the beginning, and in the "General Conclusions and Reflections" at the end—Needham explicitly invoked J. B. Bury's landmark 1920 history of the idea of progress, not just in the footnotes but in the text itself.[236] After the Enlightenment, the idea of progress was what linked the stories of science and civilization inextricably together.

Thus it was that Abel-Rémusat's commitment to the knowledge of the past and the East led him to an open and sustained critique of progress. The course of history, he claimed, had no absolute direction, either in China or in Europe, either in the past or in the future; everywhere had seen "centuries of splendor and centuries of decadence."[237] With all the pieces of timeless China now in place, he became the first to describe it not as a historical fact but as a historiographical trope: "There is yet one point on which the enthusiasts and the detractors of the Chinese are more or less in accord: I want to speak of the state of the sciences and the arts which, brought by them since highest antiquity to a certain degree of mediocrity, have, following our authors, remained since that epoch in a perpetual infancy and absolutely stationary."[238] This was a point on which Abel-Rémusat strongly disagreed, but he could never quite figure out the best way to deny it. In some places, he suggested that there really was change in the East. In others, he questioned whether change was desirable in the first place. Neither argument was entirely convincing. For, as he himself pointed out, the absence of progress outside of Europe was in his own time already taken for granted, and the value of progress in Europe was even more so.

The first argument, essentially empirical, was that China did have change after all. This was an explicitly revisionist part of Abel-Rémusat's agenda: "The immutability of the Orient has, so to speak, become a proverb," he observed; "dare I, first braving general conviction, come to disturb the security we enjoy in this regard?"[239] Indeed, he did, and he was clear about the implications of doing so: "The Chinese themselves, this people thought to be so obstinately attached to their practices and habits, were no more strangers to the variations of fashion than to political revolutions, which, no doubt, infinitely honors them for a good number of our compatriots."[240] In another familiar academic move, he argued his point by critiquing the coherence of the subject: "the Chinese empire, vast aggregation of states and nations of all races."[241] This basic fact of temporal and spatial diversity seems so obvious that one might wonder why it had not been noticed before. The answer is that it likely had been—but for opposite purposes, since uniformity had long been ranked among China's virtues. The idea of progress had changed the conversation: "Europeans, who have taken a prodigious taste for change in all respects, believe that I praise Asians in describing their variations, and I fear passing for an excessive panegyrist of Orientals in making myself guarantor of their inconstancy."[242] This comment was classic Abel-Rémusat. At once, he impugned those who disagreed with him, feigned acknowledgment of a criticism he did not accept, and explicitly denied his intention while implicitly avowing it.

The second argument then, more properly philosophical, was to challenge the desirability of change itself. In a truly remarkable phrase, Abel-Rémusat openly impugned the progress paradigm: "Who does not know what one can produce with these magic words: *perfection, amelioration, progress of knowledge, intellectual development* . . . [*sic*] sonorous terms, harmonious phrases for nineteenth-century ears, and which, like the enchantment formulas of ignorant times, change the face of objects and fascinate the most perceptive eyes." Two points stood out to such a perceptive observer. First, progress was both markedly and recently ascendant. Its enchantment worked in a distinctive way over those of his own time. Second, progress had come to resemble the very superstition that it was supposed to supersede. Lest there be any confusion as to his opinion of "magic words," he concluded: "I would not want to rely too much on their mysterious effect."[243] By pointing out the magic power of progress specifically in his two speeches on "Oriental literature," he suggested also an element of self-critique: "These Europeans, so disdainful, so prideful of the progress that they have made in the arts and in the sciences for the last three hundred years, are constantly informing themselves on how people they regard as being very inferior to them in all regards think, reason, and feel."[244] Despite the recent progress of orientalism for which he claimed no small part of the credit, orientalism was also positioning itself as at odds with the recent ascension of progress.

This conviction led all the way to a limited indictment of European imperialism.[245] By the early nineteenth century, the diffusion of progress had become a major justification for the expansion of empires.[246] Abel-Rémusat noted that the authorities in London and Moscow regarded the encroachment of their traders and diplomats upon Chinese territory as "infinitely advantageous to the progress of civilization."[247] Their view, he believed, was both disingenuous and wrong. The Qing emperors and Tokugawa shoguns should keep their borders closed to Europeans indefinitely—"if these governments are well-advised in the interest of their peace and their independence."[248] Not coincidentally, their taking that advice would also be good for the orientalists. Abel-Rémusat's anti-imperialism did not stem from an unwavering commitment to self-determination; to the contrary, it was plainly self-serving. For, during this brief moment, empire appeared to be a hindrance rather than a help to orientalism. As he put it: "It will be too late to study men, when there are none on earth but Europeans."[249] The discipline itself would cease to be if imperialism ever grew too successful in enforcing Europe's new claim to a monopoly on progress. Orientalism was already the handmaid of empire—as indeed he pointed out—but it could also be a surprisingly disgruntled one.[250]

In light of his views of modern European progress, the real meaning of Abel-Rémusat's idea of ancient eastern wisdom finally becomes clear. Progress presented problems that the orientalist was uniquely positioned to solve. If China could not provide progressive knowledge, ancient eastern wisdom was what it offered instead. For this reason, Abel-Rémusat made no effort whatsoever to revive Voltaire's dead Confucius. Indeed, he threw more stones upon his grave: "His metaphysics is vague and incoherent, and what little there is of theology or psychology in his writings has the fault of leading to the most contradictory conclusions." Confucius, once praised as the "interpreter of right reason," was now condemned for having abused it.[251] His successors, the Neo-Confucians, had completed his rejection of that which was most valuable in their own tradition, exchanging "the recent for the ancient, the later for the primeval, the opinions of a writer of the middle ages for the beliefs of high antiquity."[252] The ancient, the primeval, the antique—the alternatives to progress were precisely what had to be recovered. When Abel-Rémusat wrote with tongue in cheek that "Asia is the domain of fables, of reveries without object, of fantastic imaginations," it was ambiguous whether he was referring to the stories told *in* Asia or the stories told *about* it.[253] Either way, they were the stock-in-trade of the orientalist. What Said called the "the nineteenth-century academic and imaginative demonology of 'the mysterious orient'" was intended rather as an angelology.[254] Ancient eastern wisdom was conceived not in order to condemn Asia, but to valorize it.

But intentions do not determine outcomes, and it was not long before Sinology fell in lockstep, as Indology and Arabic studies had already, with new narratives of progress and the rising ambitions of empire. Abel-Rémusat was never more wrong than when he predicted that "despite the wishes and threats of certain disgruntled diplomats and some ambitious geographers for whom conquests cost nothing, there is no appearance that China could be soon dismembered by the tea merchants of Canton."[255] By the time his prognostication was published, the dismemberment had already begun. In one still-influential historiographical tradition, the Treaty of Nanjing that ended the First Opium War in 1842 is taken to mark China's entry into the modern world.[256] The intellectual relationship between China and Europe was radically reconfigured by the dramatic increase of cross-cultural communication in both directions. The next generation of European Sinologists became the first to actually spend time in China, while the first generation of Chinese Europeanists took the opportunity to study abroad in Britain and France at around the same time. But the terms of the exchange had fundamentally reversed: for it was now the Chinese scholars, not the European ones, who looked afar in search of models for social and political reform.[257]

And yet, the idea of ancient eastern wisdom began a new life in the modern world. The nineteenth century saw what one historian called the "Oriental Renaissance": Europe was again transformed by an influx of classical texts, this time not Greco-Roman, but Asian.[258] Hegel was perhaps the biggest name in philosophy, but he was by no means the only one, not even in Germany.[259] Thus his rival Friedrich Schlegel—who was also a critic of the great apostle of progress, Condorcet—enjoined that "we must seek the supreme romanticism in the Orient."[260] Not much later, Arthur Schopenhauer, another critic of the Enlightenment idea of progress, became the first philosopher in Europe to call himself a Buddhist.[261] Meanwhile, in Britain and France, the Victorian era and the July Monarchy saw the full flowering of modern esotericism. The future of tarot-card fortune-telling, techno-utopian Atlantis legends, and mesmerist hypnosis would be bright indeed. At the very point of its apogee, progress theory finally united its own opposition. The East became especially important for new critics of the West—just as that term, too, was coming into use.[262]

Hegel and Abel-Rémusat died within a year of each other during the cholera pandemic of 1831–1832. Surveying the ideas of their successors, it might seem that the philosopher and the orientalist had each established a rival camp: one praising progress, the other condemning it. But more importantly, they had forged an epistemological consensus. Both agreed that the knowledge of the past and the East were not only essentially the same as each other, but also fundamentally different from their own. The East was no longer merely, in Said's apt phrase, "old and distant."[263] It was now also truly "Other." For the philosopher, this was a problem; for the orientalist, it was the point. At the time of Voltaire's death in 1778, the idea of incommensurable knowledge did not exist. By the time of Hegel's in 1832, it was pervasive. In the early modern period, only God had been inscrutable; in modern times, people could be, too.

The Invention of Eastern Wisdom

The professionalization of academic disciplines led to a new way of explaining global progress at a time when Europe itself was undergoing momentous historical change. As late as the Enlightenment, orientalism and philosophy had worked closely together. But as the two fields began to drift apart following the Revolutionary and Napoleonic period, their new leaders asked whether their objects of study were really the same. During the 1820s and 1830s, Hegel and Abel-Rémusat, still in conversation with each other, answered that question by drawing from the primary sources and secondary interpretations left to them by the orphans of the Enlightenment. Theirs were the terms upon which Chinese ideas entered the new order of knowledge.

What took place at the end of the early modern period was not only the birth of modern orientalism. That story could be told easily enough with similar texts and the same cast of characters. It might run chronologically, from the ideological writings of missionaries like Amiot, through the desultory studies of proto-Sinologists like de Guignes, to the fastidious scholarship of academics like Abel-Rémusat, tracking their most rigorous works—on music, politics, and grammar, say—in search of their positivist contributions to the accumulation of true knowledge of China. But in fact, these figures all held views of China that were equally different from those of today. The establishment of modern orientalism did not mean that the European understanding of China grew more accurate, correct, or comprehensive. It was rather a new way of thinking about other matters—history, philosophy, and science in particular—that would have the greater impact on the broad trajectory of European intellectual history.

In post-Enlightenment progress theory, whatever the ancient East had possessed, it was not the kind of knowledge, either scientific or philosophical, that was considered as distinctive to the modern West alone. That the rest of the world should have been excluded from the ambit of progress was by no means inevitable. To the contrary, China had once provided Europe with a vision not of a past that no longer was, but of a future that might yet be. The very ideas of science and philosophy to which China contributed during the Enlightenment were denied from it only by those who came after. The global history of science and philosophy became a story of progress. For modern scholars from Abel-Rémusat through to

Joseph Needham, the question was not what had made science in China the same as in the West, but, rather, why it appeared to have been so different. The essential element of orientalism was generated not by the idea of progress, but by the tensions within it.

And yet, modern progress theory ultimately reincorporated the work of the orphans of the Enlightenment. *Prisca scientia* did not work anymore; but *sagesse orientale* did. As one historian noticed almost a hundred years ago: "The masters of Christian illuminism disappear with the eighteenth century; their successors will establish the cult of the Orient in France."[264] No longer available as an alternative source of knowledge, China became instead a repository of alternatives to knowledge. This was the conclusion of half a century spent trying to figure out what to do with the thinking of distant times and places, providing new answers to old questions as Enlightenment universalism broke down. The transition from the early modern to the late modern period was rich with possibilities. Ideas of progress transformed not to exclude those possibilities but to accommodate them.

Conclusion

The idea of progress has defined the Enlightenment ever since it took place. In 1784, Immanuel Kant began his prizewinning essay, "An Answer to the Question: What is Enlightenment?," as follows: "Enlightenment is the human being's emergence from his self-incurred minority."[1] Now, consider again the phrase that Abel-Rémusat attributed to a disciple of Laozi: "Man is a child born at midnight; when he sees the sun rise, he believes that *yesterday* never existed." In thinking about progress, both philosophes and orphans of the Enlightenment cast history and epistemology in the same terms. In the words of Michel Foucault, they were "looking for a difference: what difference does today introduce with respect to yesterday?"[2] Enlightenment, or illumination, is about how we understand the knowledge of the past, in the present. This is why the progress narratives told by Amiot, Court de Gébelin, Bailly, Mellet, and Abel-Rémusat must be placed alongside those of Voltaire, Diderot, Condorcet, and Hegel. I have focused on where their views were different, rather than where they were the same. But I have done so in order to reveal their shared assumptions and to show that they ultimately converged.

From the Enlightenment to the present day, people have struggled to find a place for alternative forms of knowledge in their own stories of progress. The orphans of the Enlightenment were among the first to grapple with this problem. As experts on ancient wisdom, they were quick to notice what progress appeared to be leaving behind. In response, they mounted a defense of ancient and Eastern knowledge. Their successes made lasting contributions to esotericism and orientalism, while their failures informed the redefinition of science. The outcome was neither inevitable nor easy. The nonscientific, or nonprogressive, knowledge of ancient and foreign places had not always seemed ineffable. It was made so, in the space of just a few generations. The past became a foreign country, and a foreign country became the past. To understand either, you had to immerse yourself in a different worldview. The difference that today introduces with respect to yesterday, it seemed, might inhere in the knowledge itself; it might even make yesterday impossible to know. This was the problem that the orphans of the Enlightenment left to be explored in global progress thinking. It was eventually taken up not only by orientalists, esotericists, historians, and philosophers, but even by scientists themselves.

In the nineteenth century, the rift exposed by the orphans of the Enlightenment grew more apparent. The engines of progress gained steam alongside new epistemologies that were supposedly outside it: science and empire on the one hand, and esotericism and orientalism on the other. The Victorian age saw the establishment of national research organizations, doctoral programs, experimental laboratories, and the very term *scientist* joining together to assert science's epistemic imperialism.[3] The period also saw a new kind of geopolitical imperialism, with the Berlin conference of 1884–1885 marking the apogee of European global ambitions.[4] Yet alternative ways of knowing, too, were in the ascendant. This was the era of spiritualism, séances, mediums, and mystical experiences, when the very term "occultism" came into common use.[5] It was also the heyday of the celebrity orientalist, *The Book of the Thousand Nights and a Night* and the *Sacred Books of the East*, and widespread European interest in Asian art, religion, and philosophy.[6] That popular beliefs in progress and nonprogressive knowledge flourished together was a temporal coincidence precisely because it was not a conceptual one.

The link between Eastern and esoteric knowledge tightened, and for many of the same reasons that it had first been forged. As has been well documented, the association neatly suited the purpose of the modern orientalist. If Eastern texts were arcane and obscure, there was all the more need for men like Richard Francis Burton and Ernest Renan to decipher them. And if the East had no progress, there was all the more need for European empires to bring it there by force.[7] As is also well known, Orientalists provided new resources for esotericists to draw upon. Helena Blavatsky, who founded the Theosophical Society between travels in Egypt and India, argued that ancient occultism and Asian philosophy had both anticipated modern science.[8] Aleister Crowley, who traveled in China and made his own creative translation of the *Classic of the Way and Virtue* for his magnum opus *Magick*, opined: "What is magic today is science tomorrow."[9] Orientalists and esotericists did not deny the reality of progress on some level; to the contrary, they often believed that their own fields were undergoing it.

In the twentieth century, the problem of knowledge outside progress was finally confronted from within. Early impetus came from the fledgling social sciences.[10] The sociologist Max Weber placed Western science and reason in opposition to Eastern magic and enchantment in order to challenge "progress" in light of other human values.[11] The anthropologist Franz Boas combined scientific and historical theories to reject progressive models of cultural evolution.[12] The psychologist Sigmund Freud deployed the methods of science to undermine its own foundations, severing human flourishing from social advancement.[13] To understand progress, social science turned to non-Western ideas. Sociology, the science of society, was comparative and global by definition. Anthropology pioneered the techniques of fieldwork to make non-Western knowledge cognizable to scientific theory. Psychology drew from China in particular: the first study of lucid dreams was published anonymously by the chair in Sinology at the Collège de France, and Carl Jung claimed it was a Daoist classic that set him "on the right track" in his studies of alchemy and the collective unconscious.[14] In their engagement with non-Western and nonscientific knowledge, social scientists advanced the project begun by the orphans of the Enlightenment: not to reject progress narratives, but to reclaim them.

Around the same time, progress became a central concern in China as well. Toward the end of the Qing, reformers argued that progress was essential, and that modern China did not have it. The initial challenge for Confucian scholars like Kang Youwei was to make Westernization and natural science compatible with Chinese history and philosophy.[15] In the years surrounding the collapse of the dynasty in 1911, progress thinking became extremely diverse. Some scholars believed that traditional Chinese thought could be preserved. Yan Fu, the influential translator of Charles Darwin, proposed a universal law of change based on biological and social evolution while pointing to supposed parallels in the *Book of Changes*.[16] Others thought the whole edifice had to go. Chen Duxiu, who later founded the Chinese Communist party, conceived of the unfolding of history in Marxist terms, with Confucianism as its greatest impediment.[17] No matter their position on progress, all major Chinese figures from this time looked to European authors. Yan Fu was particularly interested in Hegel's philosophy of mind and its implications for the development of knowledge in time.[18] Chen Duxiu wrote that French thinkers around the time of the Revolution had invented modern civilization by changing the past.[19] In an interesting reversal, these supposedly Western conceptions had been built upon the very progress theories that Chinese ideas had earlier informed.

The May Fourth movement—called by some historians the "Chinese Enlightenment"—further catalyzed cross-cultural engagement and scientific exchange.[20] Cai Yuanpei, an early president of Peking University, studied in Leipzig with the founder of experimental psychology, Wilhelm Wundt, before returning to China in 1911. Hu Shi completed a PhD in philosophy with John Dewey at Columbia in 1917 before his own return. These were not the first Chinese figures to acquire a global education, but in a time of radical reform and experimentation, they were in a unique position to exploit it. On their invitations, John Dewey and Bertrand Russell gave lecture tours in China that influenced a generation. New approaches to progress never seemed more promising. The intellectual development of Liang Qichao, probably the most famous Chinese intellectual of the early twentieth century, is indicative. After participating in the late Qing reforms, he went into exile in the United States and Japan and promoted progress through the natural sciences. Touring Europe after the First World War, his faith in progress was shaken, and he embraced Buddhist metaphysics. The goal, he came to believe, was "to supplement Western civilization with ours so as to synthesize and transform them to make a new civilization."[21] On the eve of foreign invasion and civil war, that ambition was to be delayed.

But not long after, a new place for Chinese knowledge within the academy was created in the United States. In 1964, the Association of Asian Studies held a symposium on "Chinese Studies and the Disciplines." A generation of young scholars voiced their ambition to join up with other fields like economics and sociology and help contribute to a "universal social science."[22] On the defensive were the traditional Sinologists, who were caricatured as stuffy, pedantic, old-fashioned, and, essentially, European.[23] The upstart social scientists won: indeed, the term "Sinology" is so uncommon in the United States today that several excellent historians have thought that this book pertained not to China, but to signs. Even so, one and a half centuries of Sinological studies of the kind pioneered by Abel-Rémusat had made a lasting impression. As one participant put it: "Pointing up China's uniqueness is to call attention to the extraordinary longevity of the society or, put otherwise, to the extraordinary degree of continuity in the sociopolitical and cultural system." For this reason, he maintained, "the Chinese case is absolutely essential for an understanding of the development of civilization itself."[24] The point, at the end as at the beginning, was to account for the distinctive knowledge of China—still seen as ancient and unchanging—in a universal story of progress.

The problem of progress that emerged at the end of the Enlightenment—how to fit a mystical, exotic, inscrutable, and ancient way of knowing into a global and universal story of the development of knowledge—took on new, modern forms. In Europe and the United States, it became a signature issue of nineteenth- and early twentieth-century thought. It is not only true that esotericism reflected a development of Enlightenment ideas; scientific ideas, too, were informed by the development of esotericism. Likewise, orientalism exiled non-Western knowledge from the Enlightenment; but it also made it possible for other disciplines to later bring that knowledge back into the fold. In the twentieth century, Chinese thinkers, too, became interested in ideas of progress. They theorized with great sophistication, putting modern science and ancient philosophy into conversation again, and for a host of new social and political purposes. The direction of cross-cultural exchange reversed. It may yet do so again.

✷

I have given this brief historical sketch only to suggest the continued salience of the orphans of the Enlightenment; but the same point could also be made, perhaps more convincingly, with a gesture to the present. The question of progress has taken on new urgency in the world today, driving conversations both inside and outside the academy. So have arguments over what the Enlightenment was and what it should stand for. The problem of cross-cultural engagement and exchange in the natural sciences has never been more pressing. In the twenty-first century, these issues are debated not only in Europe and the United States, but in China as well. In the introduction to this book, I raised some of the problems that progress presents for present-day historians. Here in the conclusion, I propose a few ways we might consider the orphans of the Enlightenment when we think about them.

Our solutions to the problems of progress will not be the same as theirs were. Some people today do still believe in the theories that the orphans of the Enlightenment invented, but we can venture to concede that professional historians are generally not among them. I do not propose that tarot cards conceal Egyptian magic, that human civilization was built on the ruins of Atlantis, that animal magnetism flows throughout the universe, or that Chinese philosophy foreshadowed modern science. What I do propose is that these invented notions reveal something about how modern ideas of progress operate: who creates them and why, what they include and exclude, where they find their force, and how they themselves are subject to historical change. In contrast to the prophetic nature of progress thinking, many of the answers to those questions have been surprising and unexpected. The lesson of the orphans of the Enlightenment for historians today is about what we should be willing to take seriously:

both faith in progress and curiosity about what lies beyond it. If we remain too entrenched in our own professional commitments, a rapidly changing world might make orphans of us, too.

Our academic culture is in one way quite opposite to that of the late Enlightenment: contemporary scholars are less likely to be orphaned for believing in ancient wisdom than for retaining a commitment to progress. The halcyon days when social scientists thought they might be living through the "end of history"—a term borrowed directly from Hegel—are decidedly over.[25] In my ten years spent teaching college students, I have been amazed at the steady erosion of their belief in progress, accompanied by growing confusion about what progress even means. In 2009, when I began graduate school, Barack Obama proclaimed that "while the future is unknowable, the winds always blow in the direction of human progress."[26] In 2017, the year I completed my PhD, Donald Trump gave his inaugural address on the theme of "American carnage."[27] Given the recent history of political dysfunction, environmental disasters, and social inequality, it is hard to imagine Joe Biden declaring certainty in progress in the same way that his Democratic predecessor did. The rise within that party of a political movement self-identifying as "progressive" itself suggests that progress is no longer taken for granted.

Yet beliefs in progress have not simply disappeared from our intellectual scene, either. Their chief prophets are now to be found in Big Tech, where the historical reality of positive development in the recent past is virtually impossible to deny. Elon Musk's space program, Mark Zuckerberg's metaverse, and Peter Thiel's immortality quest—projects supported by a new clerical caste of "technology evangelists" and in-house philosophers—are only the most obvious examples of progress thinking. But although the seat of belief has shifted, its theorization remains substantially the same. In 2019, Steven Pinker published *Enlightenment Now: The Case for Reason, Science, Humanism, and Progress*.[28] Hailed by Bill Gates as "my new favorite book of all time," it became an immediate bestseller among the TED-talk tech set whose worldview it confirmed.[29] The book argued that the human condition—especially in quantitatively measurable terms of life expectancy, literacy rates, extreme poverty, and so on—has improved since the Enlightenment, and indeed because of it. As the subtitle explains, reason, science, and humanism are supposed to be the reasons for which progress has taken place. They are also allegedly the legacy of the European Enlightenment.

Enlightenment Now met with universal opprobrium from professional historians of the Enlightenment—I am aware of literally no exceptions.[30] Their objections were diverse. One line of critique had to do with its claims about progress. The data on social and economic improvement was selectively chosen from sources that were far from objective. Moreover, to assume the metrics of progress that such data supported was to beg the question of whether it exists. One might instead point to rising sea levels and rampant inequality, which are equally well documented by empirical evidence in the natural and social sciences, to argue that things have gotten worse. Another line of critique addressed its interpretation of the Enlightenment. The thought of the eighteenth century was not characterized by blind faith in science and humanism, but by the exploration of their limits through skepticism and religion. For all the discussion of proponents of progress like Voltaire, there was no mention of his equally influential opponents like Rousseau. In sum, the story of progress told in *Enlightenment Now* was not the one told during the Enlightenment; or, at least, not the only one.

That the book struck so many different nerves inside academia suggests that there was a deeper concern than merely lazy scholarship. Historians took particular umbrage at the needling claim that "intellectuals hate progress."[31] But it is interesting to note that none of them responded with a straightforward denial: "No, intellectuals love progress!" would have been a perfectly cogent, but not exactly accurate, reply. Nor, on the other hand, did they confess to the charge. It was instead a conservative journalist who quarreled with *Enlightenment Now* on the grounds that it was oblivious to forms of nonscientific and nonprogressive knowledge lurking at "The Edges of Reason."[32] It is true that historians today are skeptical of the idea of progress—rightly, in my view—but that is not the same thing as hating it. Their objections to the book reflected an overarching belief not that progress is not real, but that history is too complex to say anything so simplistic about it. They did not deny their own commitments to "reason, science, humanism, and progress;" instead, they argued that these values have always been contested. It was an argument that would have resonated not only with the philosophes, but also with those who first challenged their views of progress.

The revival in recent years of the very ideas that were created by the orphans of the Enlightenment is hard to ignore but easy to discount. In the late 2010s, the *New York Times* remarked upon an upsurge of interest in astrology, especially as it entered into the realm of self-care, therapy, pop psychology and the wellness industry.[33] The core conceit of mesmerism, that human bodies can be healed by the power of the earth and the stars, is alive and well. Tarot cards, too, seem to be having a moment, primarily for use in fortune-telling as first proposed by Mellet and Court de Gébelin. In 2021, the *Washington Post* ran an article, "Tarot Cards are

Having a Moment with Help from the Pandemic," while the library of the Massachusetts Institute of Technology recently established a dedicated tarot-card collection.[34] Also worth mentioning are the perennially popular updates to the Atlantis myth: for example, the television show *Ancient Aliens*, which explores the premise that a lost progenitor civilization gave its science and technology to all the peoples of the ancient world. Substitute Hyperboreans for extraterrestrials, and the rest would hardly have surprised Bailly. Mesmerism, tarot cards, Atlantis—China was present in the modern invention of all these supposedly ancient traditions. At the limits of progress, people continue to explore the lands pioneered two hundred years ago.

Inside its own borders, natural science is still defined by what it excludes—and what it excludes has remained largely the same since the end of the Enlightenment. Many scientists today subscribe to the theory of falsifiability: a theory is scientific if it can be disproved. Yet modern philosophers have been deeply dissatisfied with attempts to define science by its content alone.[35] It is often unclear what experimental results would constitute disconfirmation, and many scientific conclusions in disciplines ranging from evolutionary biology to theoretical cosmology are hardly susceptible to it.[36] The field of science and technology studies sets off from the observation that what counts as science is determined not by the raw facts of nature alone but rather by the social practices and intellectual commitments that produce consensus within the scientific community.[37] That is to say, the lines that divide pseudosciences like cryptozoology and cold fusion from mainstream biology and physics are still drawn by the experts themselves. And among the pseudosciences, "vestigial sciences"—theories that were once widely accepted, but no longer are—take pride of place. Ancient wisdom remains the type specimen for the kind of knowledge that does not count as science.[38]

Today, it is ideas from beyond the West, not theories from the European past, that seem to hold more promise for the science of the future. This is particularly true in the medical and environmental sciences, which deal with problems where recent progress is not always so clear. American doctors no longer endorse mesmerism yet endorse alternative practices ranging from Chinese acupuncture to Buddhist meditation.[39] This openness within the medical establishment reflects a common concern that modern medicine has come at some cost to human well-being. The environment presents an even more spectacular example of progress run amok. Here, industrial and economic development have brought about its opposite—degradation. Many theorists believe that in the age of the

Anthropocene, traditional ecological knowledge can be applied in diverse areas from farming techniques to resource management.[40] Within the medical and environmental sciences, some thinkers seek to reject modernity and embrace tradition, while others aim to integrate them. Yet the fields share something in common. They are fundamentally global in nature, as their most daunting present problems—global warming and the global pandemic—make abundantly clear. Cross-cultural exchange will inevitably be part of any effective solution.

Among the most ambitious efforts taking place anywhere in the world today toward reconciling ideas of progress with non-Western knowledge are those under way in the People's Republic of China. One need look no further than the leadership of the Chinese Communist Party for proof that beliefs in progress need not be accompanied by a commitment to the full collection of "Enlightenment values." At the 18th Party Congress in 2012, the nation's constitution was amended to include a new raft of principles, cumulatively called "Scientific Outlook on Development"—a title that almost sounds like a late eighteenth-century progress treatise.[41] At the 19th Party Congress in 2017, Xi Jinping invoked "human progress" and the "pursuit of progress" in the official statement of Xi Jinping Thought. But he added a new theme as well: "China's fine traditional culture, which was born of the Chinese civilization and nurtured over more than 5,000 years." Accordingly, the party now embraces traditional Chinese medicine and the *Classic of Changes* after decades of official ambivalence, while promoting cutting-edge genomics research and a new manned-flight space program.[42] The rising power of the twenty-first century aims to "promote cultural advancement along with the progress of history."[43] Progress may still be driven by Chinese science, but it is thought to be Western no longer.

I have said already that this book tells the story of how a certain way of thinking was excluded from the Enlightenment—but in conclusion, I wish to point out that it also suggests how that same way of thinking, in a remarkable turn, reclaimed the Enlightenment for itself. For whatever else ancient Eastern wisdom might connote today, it surely includes enlightenment. Outside the small realm of intellectual history, the term refers far more often to the traditional philosophies of Asia than to the cultural and intellectual movement of eighteenth-century Europe. Look again at the frontispiece of this book: a painting that Amiot obtained from a Tibetan lama in Beijing and sent to the comte de Mellet in Paris in 1789—as good a year as any to mark the beginning of the modern period. Amiot confessed to his own benightedness in an accompanying letter: "I filed this cosmographical map among the useless papers in my folder, only to take it out when the opportunity to learn about it should present itself."[44] In fact, it depicts the three realms of karmic rebirth—desire, form, and formlessness—alluding to a cyclical, nonlinear process of knowledge gained and lost, with transcendence lying just beyond. For many around the world, the illustration suggests what a global Enlightenment might look like now.

Acknowledgments

For this book, I wish first to thank those who taught me how to be a historian. In particular: Paula Findlen, Keith Baker, Dan Edelstein, Catherine Jami, Matthew Jones, Jessica Riskin, Matthew Sommer, and Kären Wigen.

I am also grateful to the many friends and scholars who made my time at Stanford such a delight. Among them: Brendan Ballou, Halley Barnet, Brad Bouley, Brian Brege, Joshua Brett, Wesley Chaney, Catherine Chou, Jinyi Chu, Jason Cieply, Frederic Clark, Mackenzie Cooley, Emily Davis, Max Friedman, Jonathan Gienapp, Blaine Greteman, Christopher Hutchinson, Eun So Jo, Sienna Kang, D. Brian Kim, Koji Hirata, Adrian LeCesne, Crystal Lee, Tiffany Lee, Martin Lewis, Hannah Marcus, Anton Matytsin, Thomas Mullaney, George Qiao, Peter Reill, Paul Robinson, John Rick, Stephan Risi, Gina Tam, Londa Schiebinger, Andrew Schupanitz, Yan Slobodkin, Anna Toledano, Corey Tazzara, Caroline Winterer, Yvon Wang, and Yiqun Zhou.

I am further fortunate to have enjoyed the companionship and assistance of others all around the world. Among them, in Paris: Bruno Belhoste, Karine Chemla, John Finlay, Isabelle Landry-Deron, Nathalie Monnet, Maike Song, Alexandra Steinleight, and Stéphane Van Damme. In Beijing: Han Qi, Pan Shuyuan, Sun Chengsheng, Wu Boya, Yang Danxin, and Zhan Duo. In Berlin: Sebastian Conrad, Andreas Eckert, Lisa Hellman, Hansun Hsiung, Mårten Söderblom Saarela, and the inaugural doctoral fellows of the Berlin Graduate School Global Intellectual History. In Los Angeles: Manuel Covo, Katherine Cox, Mordechai Feingold, Dena Goodman, Steve Hindle, Katherine Kadue, Penn Lawrence, Alexandra Romanoff, J. B. Shank, and the Huntington Library fellows of 2018–2019. In Wisconsin: Thomas Broman, Hanna Golab, Florence Hsia, Charles Kim, Galen Poor, and Katharina Steiner. Elsewhere: Lydia Barnett, Karen Darling, Christoffer Basse Eriksen, Henrietta Harrison,

Ben Heller, Nicholas Kelley, Henry Klementowicz, Suzanne Marchand, Darrin McMahon, Eric Rosenblum, Haun Saussy, Simon Schaffer, Jessica Wilson, and Glover Wright.

I owe a special debt to four historians who helped me figure out what this book was really about: Tristan Brown, David Cohen, Abram Kaplan, and Huiyi Wu.

My research was made possible by the generous support of institutions and the administrators who ran them: the American Council of Learned Societies, the Andrew W. Mellon Foundation, Sebastien Gottschalk and Julia von der Wense at the Berlin Graduate School Global Intellectual History, the École Normale Supérieure, the Embassy of France in the United States, the Georges Lurcy Trust, Juan Gomez at the Huntington Library, the Stanford Center at Peking University, John Groschwitz at the Stanford Center for East Asian Studies, Andrea Davies at the Stanford Humanities Center, Art Palmon and Rosemary Rogers at the Stanford University Department of History, the United States Department of State, Megan Massino at the University of Wisconsin Center for the Humanities, the University of Wisconsin–Madison Department of History, and the Holtz Center for Science and Technology Studies.

For everything, I thank my parents, Jay and Diana, and my sister, Rosie.

Notes

Introduction

1. Max Weber, *From Max Weber: Essays in Sociology*, eds. Hans Heinrich. Gerth and C. Wright Mills (New York: Oxford University Press, 1946): 350, 139.

2. Hans Heinrich Gerth and Charles Wright Mills, "Introduction," in Weber, *From Max Weber*, 51.

3. David Lowenthal, *The Past Is a Foreign Country—Revisited* (Cambridge: Cambridge University Press, 2015): 3–5.

4. Dipesh Chakrabarty, *Provincializing Europe: Postcolonial Thought and Historical Difference* (Princeton, NJ: Princeton University Press, 2000): 1–11.

5. Antoine Lilti, "La civilisation est-elle européenne? Écrire l'histoire de l'Europe au XVIIIe siècle," in *Penser l'Europe au XVIIIe siècle: Commerce, civilisation, empire*, eds. Antoine Lilti and Céline Spector (Oxford: Voltaire Foundation, 2014): 139–66.

6. Jennifer Pitts, "The Global in Enlightenment Historical Thought," in *A Companion to Global Historical Thought*, edited by Prasenjit Duara, Viren Murthy, and Andrew Sartori (Chichester, UK: Wiley Blackwell, 2010): 184–96; Charles W. J. Withers and David N. Livingstone, eds., *Geography and Enlightenment* (Chicago: University of Chicago Press, 2007); Larry Wolff and Marco Cipolloni, eds., *The Anthropology of the Enlightenment* (Stanford, CA: Stanford University Press, 2007).

7. Dorinda Outram, *The Enlightenment*, 4th ed. (New York: Cambridge University Press, 2019): 8–9.

8. Sebastian Conrad, "Enlightenment in Global History," *American Historical Review* 117, no. 4 (2012): 999–1027, at 1005–17.

9. Laura Hostetler, "Qing Connections to the Early Modern World: Ethnography and Cartography in Eighteenth-Century China," *Modern Asian Studies* 34, no. 3 (2000): 623–62, at 645–49; Vera Schwarcz, *The Chinese Enlightenment: Intellectuals and the Legacy of the May Fourth Movement of 1919* (Berkeley: University of California Press, 1986): 1–11.

10. Conrad, "Enlightenment in Global History," 999.

11. Eric du Plessis, "L'influence de la Chine sur la pensée française au dix-huitième siècle: état présent des travaux," *Dalhousie French Studies* 43 (1998): 145–60.

12. Stefan Gaarsmand Jacobsen, "Chinese Influences or Images? Fluctuating Histories of How Enlightenment Europe Read China," *Journal of World History* 24, no. 3 (September 2013): 623–60.

13. Nicolas Standaert, *The Intercultural Weaving of Chinese Texts: Chinese and European Stories about Emperor Ku and His Concubines* (Leiden: Brill, 2016); Wu Huiyi, *Traduire la Chine au XVIIIe siècle: Les jésuites traducteurs de textes chinois et le renouvellement des connaissances européennes sur la Chine (1687–ca. 1740)* (Paris: Honoré Champion, 2017).

14. Dan Edelstein, *The Enlightenment: A Genealogy* (Chicago: University of Chicago Press, 2010), 2; John Greville Agard Pocock, *Barbarism and Religion, Volume 1: The Enlightenments of Edward Gibbon, 1737–1764* (New York: Cambridge University Press, 1999): 9.

15. Jorge Cañizares-Esguerra, *How to Write the History of the New World: Histories, Epistemologies, and Identities in the Eighteenth-Century Atlantic World* (Stanford, CA: Stanford University Press, 2001); Alexander Bevilacqua, *The Republic of Arabic Letters* (Cambridge, MA: Harvard University Press, 2018); Blake Evan Smith, "Myths of Stasis: South Asia, Global Commerce and Economic Orientalism in Late Eighteenth-Century France" (PhD diss., Northwestern University, 2017), https://arch.library.northwestern.edu/concern/generic_works/m039k4989; John Gascoigne, *Encountering the Pacific in the Age of Enlightenment* (Cambridge: Cambridge University Press, 2014).

16. John Bagnell Bury, *The Idea of Progress: An Inquiry into Its Origin and Growth* (1932), reprint ed. (New York: Dover Publications, 1955): 2, 153–215.

17. David Spadafora, *The Idea of Progress in Eighteenth-Century Britain* (New Haven, CT: Yale University Press, 1990): 1–10.

18. Outram, *The Enlightenment*, 3–7.

19. Spadafora, *The Idea of Progress in Eighteenth-Century Britain*, 2.

20. Carl L. Becker, *The Heavenly City of the Eighteenth-Century Philosophers* (1932), reprint ed. (New Haven, CT: Yale University Press, 2003); Herbert Butterfield, *The Whig Interpretation of History* (London: G. Bell and Sons, 1931).

21. Max Horkheimer and Theodor Adorno, *Dialectic of Enlightenment: Philosophical Fragments* (1944) (Stanford, CA: Stanford University Press, 2007).

22. Peter Gay, *The Enlightenment: An Interpretation*, vol. 2, reprint ed. (New York: W. W. Norton, 1995): 84–215; *The Party of Humanity: Essays in the French Enlightenment* (New York: Alfred A. Knopf, 1964): 270–71.

23. Spadafora, *The Idea of Progress in Eighteenth-Century Britain*, xii.

24. Robert Nisbet, "Idea of Progress: A Bibliographical Essay," *Liberty Fund*, 1979, https://oll4.libertyfund.org/page/idea-of-progress-a-bibliographical-essay-by-robert-nisbet; Steven Pinker, *Enlightenment Now: The Case for Reason, Science, Humanism, and Progress* (New York: Viking, 2018).

25. Jürgen Osterhammel, *Unfabling the East: The Enlightenment's Encounter with Asia* (Princeton, NJ: Princeton University Press, 2018).

26. John Locke, *Second Treatise of Government* (1690), ed. C. B. Macpherson (Indianapolis, IN: Hackett, 1980): 29.

27. Edward Said, *Orientalism* (New York: Pantheon Books, 1978): 52.

28. Georgios Varouxakis, "The Godfather of 'Occidentality': Auguste Comte and the Idea of 'The West,'" *Modern Intellectual History* (October 11, 2017): 1–31, https://doi.org/10.1017/S1479244317000415.

29. Larry F. Norman, *The Shock of the Ancient* (Chicago: University of Chicago Press, 2011).

30. Isaiah Berlin, *The Roots of Romanticism*, 2nd ed., ed. Henry Hardy (Princeton, NJ: Princeton University Press, 2013).

31. John Bennett Shank, *The Newton Wars and the Beginning of the French Enlightenment* (Chicago: University of Chicago Press, 2008).

32. Peter K. J. Park, *Africa, Asia, and the History of Philosophy: Racism in the Formation of the Philosophical Canon, 1780–1830* (Albany: State University of New York Press, 2013).

33. Varouxakis, "The Godfather of 'Occidentality.'"

34. Darrin McMahon, *Enemies of the Enlightenment: The French Counter-Enlightenment and the Making of Modernity* (Oxford: Oxford University Press, 2001); Isaiah Berlin, "The Counter-Enlightenment," in *Dictionary of the History of Ideas*, vol. II, ed. Philip Wiener (New York: Scribner, 1973): 100–12.

35. Edelstein, *The Enlightenment: A Genealogy*, 16–39.

36. Berlin, *The Roots of Romanticism*.

37. Dan Edelstein, ed., *The Super-Enlightenment: Daring to Know Too Much* (Oxford: Voltaire Foundation, 2010).

38. McMahon, *Enemies of the Enlightenment*.

39. Thomas Fröhlich and Axel Schneider, *Chinese Visions of Progress, 1895 to 1949* (Leiden: Brill, 2020).

40. Amy Allen, *The End of Progress: Decolonizing the Normative Foundations of Critical Theory* (New York: Columbia University Press, 2017): 1–34.

41. Daniel Carey and Lynn Festa, *The Postcolonial Enlightenment: Eighteenth-Century Colonialism and Postcolonial Theory* (Oxford: Oxford University Press, 2013): 1–8.

42. Chakrabarty, *Provincializing Europe*, 16, 275.

43. Urs App, *The Birth of Orientalism: Encounters with Asia* (Philadelphia: University of Pennsylvania Press, 2010); Suzanne L. Marchand, *German Orientalism in the Age of Empire: Religion, Race, and Scholarship* (Cambridge: Cambridge University Press, 2010).

44. Said, *Orientalism*, 205–06.

45. *Gewu* 格物, *zhizhi* 致知, *daoxue* 道學.

46. *Lisuan* 歷算, *bencao* 本草, *xixue* 西學; Benjamin A. Elman, *On Their Own Terms: Science in China, 1550–1900* (Cambridge, MA: Harvard University Press, 2005): xxii–xxx, 5.

47. Joel Mokyr, *A Culture of Growth: The Origins of the Modern Economy* (Princeton, NJ: Princeton University Press, 2016): 287–320.

48. Nathan Sivin, "Why the Scientific Revolution Did Not Take Place in China—Or Didn't It?," *Chinese Science* 5 (1982): 45–66, at 51.

49. Needham and Robinson, *Science and Civilisation in China: Volume VII, Part II*, 208–10.

50. I. Bernard Cohen, "The Eighteenth-Century Origins of the Concept of Scientific Revolution," *Journal of the History of Ideas* 37, no. 2 (1976): 257–88.

51. Donald F. Lach and Edwin van Kley, *Asia in the Making of Europe*, 3 vols. (Chicago: University of Chicago Press, 1965–1993); André Gunder Frank, *ReORIENT: Global Economy in the Asian Age* (Berkeley: University of California Press, 1998); Matthew Mosca, *From Frontier Policy to Foreign Policy: The Question of India and the Transformation of Geopolitics in Qing China* (Stanford, CA: Stanford University

Press, 2013); Gregory Afinogenov, *Spies and Scholars: Chinese Secrets and Imperial Russia's Quest for World Power* (Cambridge, MA: Harvard University Press, 2020).

52. Elman, *On Their Own Terms*; Catherine Jami, *The Emperor's New Mathematics: Western Learning and Imperial Authority During the Kangxi Reign (1662–1722)* (Oxford: Oxford University Press, 2012).

53. Joseph Needham, *The Grand Titration: Science and Society in East and West* (1969), reprint ed. (Oxford: Routledge, 2005); Sivin, "Why the Scientific Revolution Did Not Take Place in China—Or Didn't It?"

54. George Sarton, *Sarton on the History of Science*, ed. Dorothy Stimson (Cambridge, MA: Harvard University Press, 1962).

55. Kapil Raj, *Relocating Modern Science: Circulation and the Construction of Knowledge in South Asia and Europe, 1650–1900* (New York: Palgrave Macmillan, 2007): 1–5.

56. Dagmar Schäfer, *The Crafting of the 10,000 Things: Knowledge and Technology in Seventeenth-Century China* (Chicago: University of Chicago Press, 2011); Carla Nappi, *The Monkey and the Inkpot: Natural History and its Transformations in Early Modern China* (Cambridge, MA: Harvard University Press, 2009); Mario Cams, *Companions in Geography: East-West Mapping of Qing China* (Leiden: Brill, 2017); Qiong Zhang, *Making the New World Their Own: Chinese Encounters with Jesuit Science in the Age of Discovery* (Leiden: Brill, 2015).

57. Fa-ti Fan, "The Global Turn in the History of Science," *East Asian Science, Technology and Society* 6 (2012): 249–58.

58. Benjamin Breen, *The Age of Intoxication: Origins of the Global Drug Trade* (Philadelphia: University of Pennsylvania Press, 2019); Lydia Barnett, *After the Flood: Imagining the Global Environment in Early Modern Europe* (Baltimore, MD: Johns Hopkins University Press, 2019).

59. Paula Findlen, ed., *Early Modern Things* (New York: Routledge, 2012); Pamela Smith, *Entangled Itineraries: Materials, Practices, and Knowledges Across Eurasia* (Pittsburgh, PA: University of Pittsburgh Press, 2010).

60. Simon Schaffer, Lissa Roberts, Kapil Raj, and James Delbourgo, eds., *The Brokered World: Go-Betweens and Global Intelligence, 1770–1820* (Sagamore Beach, MA: Science History Publications, 2009); Raj, *Relocating Modern Science*.

61. Angus Charles Graham, *Yin-Yang and the Nature of Correlative Thinking* (Singapore: Institute of East Asian Philosophies, 1986); Robin Wang, *Yinyang: The Way of Heaven and Earth in Chinese Thought and Culture* (New York: Cambridge University Press, 2012); Tristan G. Brown, "The Veins of the Earth: Property, Environment, and Cosmology in Nanbu County, 1865–1942" (PhD diss., Columbia University, 2017), https://clio.columbia.edu/catalog/12848655; Yuan Chen, "Legitimation Discourse and the Theory of the Five Elements in Imperial China," *Journal of Song-Yuan Studies* 44, no. 1 (2014): 325–64.

62. Marwa Elshakry, "When Science Became Western: Historiographical Reflections," *Isis* 101, no. 1 (2010): 98–109.

63. Frances Amelia Yates, *Giordano Bruno and the Hermetic Tradition* (Chicago: University of Chicago Press, 1964); Keith Thomas, *Religion and the Decline of Magic: Studies in Popular Beliefs in Sixteenth and Seventeenth Century England* (1971), reprint ed. (New York: Oxford University Press, 1997); David Bates, "Super-Epistemology," in *The Super-Enlightenment*, ed. Edelstein, 53–74; Peter Reill, "The Hermetic Imagi-

nation in the High and Late Enlightenment," in *The Super-Enlightenment*, ed. Edelstein, 37–51.

64. Robert Westman, *The Copernican Question: Prognostication, Skepticism, and the Celestial Order* (Berkeley: University of California Press, 2011); Lawrence M. Principe, *The Aspiring Adept: Robert Boyle and His Alchemical Quest* (Princeton, NJ: Princeton University Press, 1998); Rob Iliffe, *Priest of Nature: The Religious Worlds of Isaac Newton* (New York: Oxford University Press, 2017).

65. Jay L. Garfield and Bryan W. Van Norden, "If Philosophy Won't Diversify, Let's Call It What It Really Is," *New York Times*, May 11, 2016, https://www.nytimes.com/2016/05/11/opinion/if-philosophy-wont-diversify-lets-call-it-what-it-really-is.html.

66. Jason Ananda Josephson-Storm, *The Myth of Disenchantment: Magic, Modernity, and the Birth of the Human Sciences* (Chicago: University of Chicago Press, 2017).

Chapter 1

1. "De la seule raison salutaire interprète, / Sans éblouir le monde éclairant les esprits, / Il ne parla qu'en sage, et jamais en prophète; / Cependant on le crut, et même en son pays." Voltaire, "De la Chine," in *Questions sur l'Encyclopédie (C–E)* (1772), *Œuvres complètes* (Oxford: Voltaire Foundation, 1968), ARTFL Encyclopédie Projet, ed. Robert Morrissey, http://encyclopedie.uchicago.edu/.

2. Antoine Lilti, *The Invention of Celebrity*, trans. Lynn Jeffress (Cambridge: Polity Press, 2017): 16–19.

3. Voltaire, "Sur l'insertion de la petite vérole," in *Lettres philosophiques* (1730), *Œuvres complètes* (Oxford: Voltaire Foundation, 1968), ARTFL.

4. Shun-Ching Song, *Voltaire et la Chine* (Aix-en-Provence: Université de Provence, 1989): 3; David Porter, *Ideographia: The Chinese Cipher in Early Modern Europe* (Stanford, CA: Stanford University Press, 2001): 2.

5. Arnold H. Rowbotham, "Voltaire, Sinophile," *PMLA* 47, no. 4 (1932): 1050–65; René Étiemble, *L'Europe Chinoise II: De la Sinophilie à la Sinophobie* (Paris: Gallimard, 1989); Song, *Voltaire et la Chine*.

6. The topic is very well studied, but no overview has superseded the one written in 1932 (Virgile Pinot, *La Chine et La Formation de L'esprit philosophique* [1932] [Geneva: Slatkine Reprints, 1971]).

7. Stefan Gaarsmand Jacobsen, "Chinese Influences or Images? Fluctuating Histories of How Enlightenment Europe Read China," *Journal of World History* 24, no. 3 (September 2013): 623–60; Eric du Plessis, "L'influence de la Chine sur la pensée française: État présent des travaux," *Dalhousie French Studies* 43 (1998): 145–60.

8. David E. Mungello, *Curious Land: Jesuit Accommodation and the Origins of Sinology* (Honolulu: University of Hawaii Press, 1989): 134–73; Jonathan I. Israel, *Enlightenment Contested: Philosophy, Modernity, and the Emancipation of Man, 1670–1752* (Oxford: Oxford University Press, 2006): 640–62.

9. Anne Gerritsen and Stephen MacDowell, "Material Culture and the Other: European Encounters with Chinese Porcelain, ca. 1650–1800," *Journal of World History* 23, no. 1 (2012): 87–113; Peter J. Kitson, *Forging Romantic China: Sino-British Cultural Exchange 1760–1840* (Cambridge: Cambridge University Press, 2013).

10. David Porter, *The Chinese Taste in Eighteenth-Century England* (Cambridge:

Cambridge University Press, 2010); Bianca Maria Rinaldi, *Ideas of Chinese Gardens: Western Accounts, 1300–1860* (Philadelphia: University of Pennsylvania Press, 2016).

11. Larry Wolff, "Discovering Cultural Perspective: The Intellectual History of Anthropological Thought in the Age of Enlightenment," in *The Anthropology of the Enlightenment*, eds. Wolff and Cipolloni (Stanford, CA: Stanford University Press, 2007): 3–32.

12. Ashley Eva Millar, "Revisiting the Sinophilia/Sinophobia Dichotomy in the European Enlightenment through Adam Smith's 'Duties of Government,'" *Asian Journal of Social Science* 38 (2010): 716–37; Zhang Chunjie, "From Sinophilia to Sinophobia: China, History, and Recognition," *Colloquia Germanica* 41, no. 2 (2008): 97–110.

13. Jürgen Osterhammel, *Unfabling the East: The Enlightenment's Encounter with Asia* (Princeton, NJ: Princeton University Press, 2018); Alexander Bevilacqua, *The Republic of Arabic Letters* (Cambridge, MA: Harvard University Press, 2018).

14. Étiemble, *L'Europe Chinoise II*, 384; Basil Guy, *The French Image of China before and after Voltaire* (Geneva: Institut et Musée Voltaire, 1963): 337.

15. Du Plessis, "L'influence de la Chine sur la pensée française," 147; Millar, "Revisiting the Sinophilia/Sinophobia Dichotomy," 716–18.

16. Jean-Pierre Abel-Rémusat, *Programme du cours de langue et de littérature chinoises et de tartare-mandchou; précédé du discours prononcé à la première séance de ce cours, dans l'une des salles du Collège royal de France, le 16 janvier 1815* (Paris: Charles, 1815): 15.

17. To take only the three works that a recent historiographer considered most important (du Plessis, "L'influence de la Chine sur la pensée française," 145–60): the first concluded before the period (Pinot, *La Chine et La Formation de L'esprit philosophique*); the second discussed it in only a short chapter (Guy, *The French Image of China*) and the third gave it a five-page conclusion (Étiemble, *L'Europe Chinoise II*).

18. Michel Cartier, "Introduction," in *La Chine entre amour et haine: Actes du VIIIe Colloque de sinologie de Chantilly*, ed. Michel Cartier (Paris: Desclée de Brouwer, 1998): 7–13.

19. Guy, *The French Image of China*, 433.

20. Anne Gerritsen, *The City of Blue and White: Chinese Porcelain and the Early Modern World* (Cambridge: Cambridge University Press, 2020): 223–26.

21. Rule, "The Tarnishing of the Image: From Sinophilia to Sinophobia," in *La Chine entre amour et haine*, ed. Cartier (Paris: Desclée de Brouwer, 1998): 89–109, at 91.

22. Kenneth Pomeranz, *The Great Divergence: Europe, China, and the Making of the Modern World Economy* (Princeton, NJ: Princeton University Press, 2000).

23. Henrietta Harrison, *The Perils of Interpreting: China and the Rise of the British Empire* (Princeton, NJ: Princeton University Press, forthcoming).

24. Joanna Waley-Cohen, "China and Western Technology in the Late Eighteenth Century," *American Historical Review* 98, no. 5 (1993): 1525–44.

25. Millar, "Revisiting the Sinophilia/Sinophobia Dichotomy," 717–20.

26. Dan Edelstein, *The Enlightenment: A Genealogy* (Chicago: University of Chicago Press, 2010): 1–3; Michael Adas, *Machines as the Measure of Men: Science, Technology, and Ideologies of Western Dominance* (1989), reprint ed. (Ithaca, NY: Cornell University Press, 2015): 81–95.

27. Étiemble, *L'Europe Chinoise II*, 373; Song, *Voltaire et la Chine*, 225; Row-

botham, "Voltaire, Sinophile," 1057; Ernst Rose, "China as a Symbol of Reaction in Germany, 1830–1880," *Comparative Literature* 3, no. 1 (1951): 57–76, at 57–58.

28. Alain Peyrefitte, *The Immobile Empire*, trans. Jon Rothschild (New York: Knopf, 1992).

29. Peter Gay, *The Enlightenment: An Interpretation*, vol. 2, reprint ed. (New York: W. W. Norton, 1995): 83.

30. Antonella Romano, *Impressions de Chine: L'Europe et l'englobement du monde (XVIe–XVIIe siècle)* (Paris: Fayard, 2016): 53–79.

31. Jeffrey D. Burson, "Chinese Novices, Jesuit Missionaries and the Accidental Construction of Sinophobia in Enlightenment France," *French History* 27, no. 1 (2013): 1–4.

32. R. Po-chia Hsia, *A Jesuit in the Forbidden City: Matteo Ricci, 1552–1610* (Oxford: Oxford University Press, 2010): 1–25.

33. Katherine Park and Lorraine Daston, "Introduction," in *The Cambridge History of Science, Volume III: Early Modern Science*, eds. Katherine Park and Lorraine Daston (New York: Cambridge University Press, 2006): 1–17, at 1–6.

34. Li Zhizao 李之藻; Xu Guangqi 徐光啓.

35. Benjamin A. Elman, *On Their Own Terms: Science in China, 1550–1900* (Cambridge, MA: Harvard University Press, 2005): 90–98; R. Hsia, *A Jesuit in the Forbidden City*, 202–223.

36. Catherine Jami, *The Emperor's New Mathematics: Western Learning and Imperial Authority During the Kangxi Reign (1662–1722)* (Oxford: Oxford University Press, 2012); Kristina Kleutghen, *Imperial Illusions: Crossing Pictorial Boundaries in the Qing Palaces* (Seattle: University of Washington Press, 2015).

37. Elman, *On Their Own Terms*, 145–47; Jami, *The Emperor's New Mathematics*, 36–41.

38. Cornelius de Pauw, *Recherches philosophiques sur les égyptiens et les chinois* (Berlin: C. J. Decker, 1773–1774): vol. 1, 6.

39. Florence Hsia, *Sojourners in a Strange Land: Jesuits and Their Scientific Missions in Late Imperial China* (Chicago: University of Chicago Press, 2009): 30–50.

40. Han Qi, "The Jesuits and their Study of Chinese Astronomy and Chronology in the Seventeenth and Eighteenth Centuries," in *Europe and China: Science and Arts in the 17th and 18th Centuries*, ed. Luís Saraiva (Singapore: World Scientific Publishing, 2012): 71–79.

41. Joseph Needham with Wang Ling, *Science and Civilisation in China: Volume III: Mathematics and the Sciences of the Heavens and the Earth* (Cambridge: Cambridge University Press, 1959): 171–77.

42. Florence Hsia, "Chinese Astronomy for the Early Modern Reader," *Early Science and Medicine* 13, no. 5 (2008): 417–50, at 418–19, 438–40.

43. Minghui Hu, "Provenance in Contest: Searching for the Origins of Jesuit Astronomy in Early Qing China, 1664–1705," *International History* 24, no. 1 (2002): 1–36, at 3–8.

44. Elman, *On Their Own Terms*, xxv.

45. Song learning, *Song xue* 宋學; principles learning, *li xue* 理學; Song-Ming principles learning, *Song-Ming lixue* 宋明理學. The term "Neo-Confucianism," *xin ruxue* 新儒学, is contentious; in Chinese, it has only been used to refer to a twentieth-century philosophical movement.

46. Zhu Xi 朱熹; Four Books, *Sishu* 四書; Peter K. Bol, *Neo-Confucianism in History* (Cambridge, MA: Harvard University Press, 1997): 1–3, 78–90.

47. *Taiji* 太極; *li* 理.

48. *Yin-yang* 陰陽.

49. *Qi* 氣.

50. William Theodore de Bary and Irene Bloom, eds., *Sources of Chinese Tradition, Volume I: From Earliest Times to 1600*, 2nd ed. (New York: Columbia University Press, 1999): 697–99 (this section edited by Wing-tsit Chan); Joseph Needham with Wang Ling, *Science and Civilisation in China: Volume II: History of Scientific Thought* (Cambridge: Cambridge University Press, 1956): 455–73.

51. Bol, *Neo-Confucianism in History*, 90–91.

52. *Xingli daquan* 性理大全 [Great compendium on nature and principle]; Knud Lundbaek, "The Image of Neo-Confucianism in Confucius Sinarum Philosophus," *Journal of the History of Ideas* 44, no. 1 (1983): 19–30, at 21–22.

53. Quoted in Qian Mu, *Zhongguo sixiang shi* [An intellectual history of China] (1937), reprint ed. (Beijing: Jiuzhou chubanshe, 2012): 158–62; for an English translation of many relevant passages, see de Bary and Bloom, *Sources of Chinese Tradition, Volume 1*, 699–701.

54. Lundbaek, "The Image of Neo-Confucianism"; David E. Mungello, "The Reconciliation of Neo-Confucianism with Christianity in the Writings of Joseph de Prémare, S. J.," *Philosophy East and West* 26, no. 4 (1976): 389–410; René Étiemble, "Les concepts de Li et de K'i dans la pensée européenne au XVIIIe siècle," *Mélanges Alexandre Koyré II* (1964): 144–59.

55. Four Books, *Sishu* 四書; Classics, *Wujing* 五經. See de Bary and Bloom, *Sources of Chinese Tradition, Volume 1*, 24–29 (this section also edited by B. Watson and David S. Nivison) and 720–21.

56. Gianamar Giovannetti-Singh, "Writers of the Lost Ark: Reconstructing the Fight for Primacy in the Jesuit China Mission from the *Acta Pekinensia*, 1658–1707," *Modern Intellectual History*, forthcoming.

57. Mungello, *Curious Land*, 13–22.

58. Israel, *Enlightenment Contested*, 542.

59. Thierry Meynard, *The Jesuit Reading of Confucius: The First Complete Translation of the Lunyu (1687) Published in the West* (Leiden: Brill, 2015): 1–88.

60. *Tian* 天; *Shangdi* 上帝.

61. Philippe Couplet, ed., *Confucius sinarum philosophus, sive scientia sinensis: latine exposita* (Paris: Daniel Horthemels, 1687): xxxiv.

62. Couplet, *Confucius sinarum philosophus*, lxxxix.

63. Couplet, *Confucius sinarum philosophus*, xxxviii.

64. Knud Lundbaek, "Notes sur l'image du Néo-Confucianisme dans la littérature européenne du XVIIe siècle," in *Actes du IIIe Colloque International de Sinologie: Appréciation par l'Europe de la tradition chinoise à partir du XVIIe siècle* (Paris: Les Belles Lettres, Cathasia, 1983): 130–76, at 135.

65. Niccolò Longobardo, *Traité sur quelques points de la religion des Chinois* (Paris: Louis Guérin, 1701): 1.

66. Longobardo, *Traité sur quelques points de la religion des Chinois*, 13.

67. Paul Rule, "Moses or China?," in *Images de La Chine: Le Contexte Occidental de La Sinologie Naissante*, eds. Edward Malatesta and Yves Raguin (San Francisco: Ricci Institute for Chinese-Western Cultural History, 1995): 303–32, at 303–07.

68. John W. Witek, *Controversial Ideas in China and Europe: A Biography of Jean-François Foucquet* (Rome: Institutum Historicum S. I., 1982): 332.

69. Wu Huiyi, *Traduire la Chine au XVIIIe siècle: Les jésuites traducteurs de textes chinois et le renouvellement des connaissances européennes sur la Chine (1687–ca. 1740)* (Paris: Honoré Champion, 2017): 156–96.

70. Arnold H. Rowbotham, "The Jesuit Figurists and Eighteenth-Century Religious Thought," in *Discovering China: European Interpretations in the Enlightenment*, eds. Julia Ching and Willard G. Oxtoby (Rochester, NY: University of Rochester Press, 1992): 43–45.

71. Isabelle Landry-Deron, *La Preuve par la Chine: La "Description" de J.-B. Du Halde, jésuite, 1735* (Paris: Éditions de l'École des Hautes Études en Sciences Sociales, 2002): 283; Mungello, *Curious Land*, 300–15.

72. Michael Lackner, "Jesuit Figurism," in *China and Europe: Images and Influences in Sixteenth to Eighteenth Centuries*, ed. Thomas H. C. Lee (Hong Kong: Chinese University Press, 1991): 129–50.

73. David E. Mungello, "An Introduction to the Chinese Rites Controversy," in *The Chinese Rites Controversy: Its History and Meaning*, ed. David E. Mungello (Sankt Augustin: Institut Monumenta Serica and the Ricci Institute for Chinese-Western Cultural History, 1994): 3–14.

74. Wu, *Traduire la Chine*, 373–83.

75. Jean-Baptiste du Halde, *Description géographique, historique, chronologique, politique, et physique de l'empire de la Chine et de la Tartarie chinoise*, vol. 1 (Paris: P. G. Le Mercier, 1735): xx.

76. Landry-Deron, *La Preuve par la Chine*, 11–18.

77. Wu Huiyi, "Alien Voices under the Bean Arbour: How an Eighteenth-Century French Jesuit Translated the Doupeng xianhua 豆棚閒話 as the 'Dialogue of a Modern Atheist Chinese Philosopher,'" *T'oung-Pao* 103, no. 1–3 (2017): 155–205.

78. Florence Hsia, *Sojourners in a Strange Land*, 142–56.

79. Jean-Baptiste Du Halde, *Description géographique, historique, chronologique, politique, et physique de l'empire de la Chine et de la Tartarie chinoise*, vol. 3 (Paris: P. G. Le Mercier, 1735): 268.

80. Du Halde, *Description de la Chine*, vol. 3, 264.

81. Dominique Parrenin, "Lettre du Père Parennin à M. Dortous de Mairan . . . à Pékin, ce 11 août 1730," in *Lettres édifiantes et curieuses, concernant l'Asie, l'Afrique, et l'Amérique*, vol. 3, ed. Louis Aimé-Martin (Paris: Société du Panthéon Littéraire, 1843): 645–62, at 648.

82. Du Halde, *Description de la Chine*, vol. 3, 265.

83. Jean-Robert Armogathe, "Voltaire et la Chine: Une mise au point," in *La mission française de Pékin aux XVIIe et XVIIIe siècles* (Paris: Belles Lettres, 1974): 26–39, at 27–30; Rowbotham, "Voltaire, Sinophile," 1050–53.

84. Étiemble, *L'Europe Chinoise II*, 14; du Plessis, "L'influence de la Chine sur la pensée française," 154.

85. Song, *Voltaire et la Chine*, 170–74; Armogathe, "Voltaire et la Chine," 32–33; Rowbotham, "Voltaire, Sinophile," 1053.

86. Voltaire, *La Philosophie de l'histoire* (1764), *Œuvres complètes*, 155.

87. Voltaire, *La Philosophie de l'histoire*, 153.

88. Voltaire, *La Philosophie de l'histoire*, 156.

89. Voltaire, *La Philosophie de l'histoire*, 108.

90. Voltaire, "Lettre 5," in *Lettres chinoises, indiennes et tartares* (1775), *Œuvres complètes*; Étiemble, *L'Europe Chinoise II*, 292–93; Rowbotham, "Voltaire, Sinophile," 1065.

91. Song, *Voltaire et la Chine*, 189.

92. Voltaire, "Lettre 5."

93. Voltaire, "Chapitre 1," in *Essai sur les mœurs et l'esprit des nations* (1756), *Œuvres complètes*.

94. Voltaire, "Lettre 5."

95. Donald F. Lach, "Leibniz and China," *Journal of the History of Ideas* 6, no. 4 (1945): 436–55; David E. Mungello, *Leibniz and Confucianism: The Search for Accord* (Honolulu: University of Hawaii Press, 1977); Daniel J. Cook and Henry Rosemont Jr., "The Pre-Established Harmony between Leibniz and Chinese Thought," *Journal of the History of Ideas* 42, no. 2 (1981): 253–67; Franklin Perkins, *Leibniz and China: A Commerce of Light* (Cambridge: Cambridge University Press, 2004).

96. *Classic of Changes, Yijing* 易經; Gottfried Wilhelm Leibniz to Joachim Bouvet, May 18, 1703 [Letter J], in "Leibniz-Bouvet Correspondence," trans. and annot. Alan Berkowitz and Daniel J. Cook, 6–14, http://leibniz-bouvet.swarthmore.edu/.

97. Gottfried Wilhelm Leibniz, *Writings on China*, eds. Daniel J. Cook and Henry Rosemont (Chicago: Open Court, 1994): 75–138; Albert Ribas, "Leibniz' 'Discourse on the Natural Theology of the Chinese' and the Leibniz-Clarke Controversy," *Philosophy East and West* 53, no. 1 (2003): 64–86.

98. Leibniz to Bouvet, December 2, 1697 [Letter B], "Leibniz-Bouvet Correspondence," 15–20.

99. Quoted in Perkins, *Leibniz and China*, 42.

100. Israel, *Enlightenment Contested*, 640–62.

101. Edwin J. van Kley, "Europe's 'Discovery' of China and the Writing of World History," *American Historical Review* 76, no. 2 (1971): 358–85; Étiemble, "Les concepts de Li et de K'i."

102. Jonathan I. Israel, "Admiration of China and Classical Chinese Thought in the Radical Enlightenment (1685–1740)," *Taiwan Journal of East Asian Studies* 4, no. 1 (2007): 1–25, at 7–12.

103. Donald F. Lach, "The Sinophilism of Christian Wolff (1679–1754)," in *Discovering China*, eds. Ching and Oxtoby, 119–26.

104. Danielle Elisseeff, *Nicolas Fréret (1688–1749): Réflexions d'un humaniste du XVIIIe siècle sur la Chine* (Paris: Collège de France, Presses Universitaires de France, 1978): 52–63.

105. John Greville Agard Pocock, *Barbarism and Religion, Volume 2: Narratives of Civil Government* (New York: Cambridge University Press, 2001): 97–159.

106. Rowbotham, "Voltaire, Sinophile," 1057–65.

107. John Bagnell Bury, *The Idea of Progress: An Inquiry into Its Origin and Growth* (1932), reprint ed. (New York: Dover Publications, 1955): 148–53.

108. Voltaire, "Chapitre 39," in *Le Siècle de Louis XIV* (1751), *Œuvres complètes*.

109. Voltaire, *Essai sur les mœurs*, 195.

110. Peter Gay, *The Party of Humanity: Essays in the French Enlightenment* (New York: Alfred A. Knopf, 1964): 271.

111. John Anthony George Roberts, "L'image de La Chine Dans *l'Encyclopédie*," *Recherches Sur Diderot et Sur l'Encyclopédie* 22, no. 1 (1997): 87–108, at 105.

112. Denis Diderot, "Chine, la," and "Chinois, Philosophie des," in *Encyclopédie*,

ou dictionnaire raisonné des sciences, des arts et des métiers, eds. Denis Diderot and Jean le Rond d'Alembert, 1753, ARTFL.

113. Lundbaek, "Notes sur l'image du Néo-Confucianisme," 148.

114. Diderot, "Chinois, Philosophie des."

115. Peter Gay, *The Enlightenment: An Interpretation*, vol. 1, reprint ed. (New York: W. W. Norton, 1995): 132–41.

116. Jacques Barzun, "Diderot as Philosopher," *Diderot Studies* 22 (1986): 17–25.

117. Diderot, "Chinois, Philosophie des."

118. De Pauw, *Recherches philosophiques sur les égyptiens et les chinois*, vol. 1, 9; vol. 2, 234, 334.

119. De Pauw, *Recherches philosophiques sur les égyptiens et les chinois*, vol. 2, 221–22.

120. De Pauw, *Recherches philosophiques sur les égyptiens et les chinois*, vol. 2, 250.

121. De Pauw, *Recherches philosophiques sur les égyptiens et les chinois*, vol. 2, 198.

122. De Pauw, *Recherches philosophiques sur les égyptiens et les chinois*, vol. 2, 337.

123. De Pauw, *Recherches philosophiques sur les égyptiens et les chinois*, vol. 1, 5.

124. Joseph-Marie Amiot, "Observations sur le livre de M. de Pauw, intitulé *Recherches philosophiques sur les Égyptiens et les Chinois*" (1777), in Joseph-Marie Amiot, François Bourgeois, Pierre-Martial Cibot, Aloys Ko, Aloys de Poirot, Charles Batteux, M. de Bréquigny, A. I. Silvestre de Sacy, and Antoine Gaubil, *Mémoires concernant l'histoire, les sciences, les arts, les moeurs, les usages, &c. des Chinois* [*MCC*], vols. 1–15 (Paris: Nyon l'aîné, 1776–1791): vol. 6 (1780), 275–345; Pierre-Martial Cibot and Aloys Ko, "Remarques sur un écrit de M. de Pauw intitulé, Recherches sur les Égyptiens et les Chinois," in Amiot et al., *MCC*, vol. 2 (1776), 365–574.

125. Song, *Voltaire et la Chine*, 10; Armogathe, "Voltaire et la Chine," 28; Voltaire, "Article 2," in *Fragment sur l'histoire générale* (1773), *Œuvres complètes*.

126. Voltaire, "Lettre 2," in *Lettres chinoises, indiennes et tartares*.

127. Rowbotham, "Voltaire, Sinophile," 1050.

128. Voltaire, "Avant-propos," in *Essai sur les mœurs*; de Pauw, *Recherches philosophiques sur les égyptiens et les chinois*, vol. 1, 2.

129. Fortunato de Felice, "Chinois (de la Littérature des)," in *Supplément à l'Encyclopédie. Encyclopédie, ou dictionnaire raisonné des sciences, des arts et des métiers*, vol. 2 (Amsterdam: M. Rey, 1776): 401–02.

130. Jean-Antoine-Nicolas de Caritat, marquis de Condorcet, *Esquisse d'un tableau historique des progrès de l'esprit humain* (1794), in *Œuvres complètes de Condorcet*, eds. F. Condorcet O'Connor and F. Arago (Paris: Firmin Didot, 1847–1849): vol. 6: 1–513, at 57.

131. Hsia, *Sojourners in a Strange Land*, 52–58.

132. Catherine Jami, "Pékin au début de la dynastie Qing: Capital des savoirs impériaux et relais de l'Académie royale des sciences de Paris," *Revue d'Histoire Moderne et Contemporaine* 55, no. 2 (2008): 43–69.

133. Jami, *The Emperor's New Mathematics*, 385–89; Mario Cams, *Companions in Geography: East-West Mapping of Qing China* (Leiden: Brill, 2017): 156–76.

134. Jami, *The Emperor's New Mathematics*, 64–74, 104–76.

135. Witek, "Jean-François Foucquet and the Chinese Books in the French Royal Library: A Preliminary Survey," in *Actes du IIe Colloque International de Sinologie, Les rapports entre la Chine et l'Europe au temps des lumières* (Paris: Les Belles Lettres, Cathasia, 1980): 145–83, at 148.

136. Cécile Leung, *Étienne Fourmont, 1683–1745: Oriental and Chinese Languages in Eighteenth-Century France* (Leuven: Leuven University Press, 2002): 134–38.

137. Witek, "Jean-François Foucquet," 162.

138. Xu Minglong, *Huang Jialüe yu Faguo zaoqi hanxue* (Beijing: Shangwu yinshuguan, 2014).

139. Elisseeff, *Nicolas Fréret.*

140. Leung, *Étienne Fourmont.*

141. Jean-Pierre Abel-Rémusat, "Lettre au Rédacteur sur l'état et les progrès de la littérature chinoise en Europe," *Journal Asiatique* 1 (1822): 279–91, at 281.

142. Henri Bertin to Aloys Ko and Étienne Yang, January 18, 1774, in MS 1522, "36 Lettres de Mr. Bertin aux Missionnaires en Chine, 1773–1778," Institut de France (IF), Paris, France; Joseph-Marie Amiot to Henri Bertin, August 17, 1781, in MS 1516, "80 lettres du P. Amiot à Bertin," IF.

143. John Greville Agard Pocock, *Barbarism and Religion, Volume 4: Barbarians, Savages and Empires* (New York: Cambridge University Press, 2005): 99–153; App, *Birth of Orientalism*; 188–253; Nathaniel Wolloch, "Joseph de Guignes and Enlightenment Notions of Material Progress," *Intellectual History Review*, 21, no. 4 (2011): 435–48.

144. "Joseph de Guignes," in MS 15 CDF 142, "Joseph de Guignes," Collège de France (CF), Paris, France; "André Leroux des Hauterayes," in MS 15 CDF 187, "André Leroux des Hauterayes," CF.

145. Janine Hartman, "Ideograms and Hieroglyphs: The Egypto-Chinese Origins Controversy in the Enlightenment," *Dalhousie French Studies* 43 (1998): 101–18.

146. Voltaire, "Celtes," in *Questions sur l'Encyclopédie (C–E).*

147. Anonymous, "Notice Historique sur l'Académie des inscriptions et Belles-Lettres," *Comptes rendus des séances de l'Académie des inscriptions et Belles-Lettres* (1857): 1–43, at 22.

148. Joseph de Guignes, "Examen critique des annales chinoises, ou Mémoire sur l'incertitude des douze premiers siècles de ces annales, et de la chronologie chinoise," in *Mémoires de l'Académie des inscriptions*, vol. 36 (1769): 164–89, at 164–65; Joseph-Marie Amiot to Jérôme-Frédéric Bignon, December 20, 1769, in MS Bréquigny 7, "Mélanges sur la Chine et les Chinois," Bibliothèque nationale de France (BnF), Paris, France.

149. *Journal des sçavans* (October 1783), 1989.

150. Isabelle Landry-Deron, "De Guignes, Joseph," in *Dictionnaire Des Orientalistes de Langue Française*, ed. François Pouillon (Paris: Karthala, 2008): 468.

151. App, *Birth of Orientalism*, 196–97, 235–37; Alexander Statman, "The First Global Turn: Chinese Contributions to Enlightenment World History," *Journal of World History* 30, no. 3 (2019): 363–92, at 387–88.

152. Pocock, *Barbarism and Religion*, vol. 4, 111; App, *Birth of Orientalism*; Wolloch, "Joseph de Guignes and Enlightenment Notions of Material Progress," 435–48.

153. Joseph de Guignes, "Essai historique sur l'étude de la philosophie chez les anciens Chinois," in *Mémoires de l'Académie des inscriptions*, vol. 38 (1777): 269–311, at 290, 310; Joseph de Guignes, "Préface," in *Le Chou-king: Un des livres sacrés des Chinois, qui renferme les fondements de leur ancienne histoire, les principes de leur gouvernement & de leur morale*, trans. Antoine Gaubil, ed. Joseph de Guignes (Paris: N. M. Tilliard, 1770): i–xliii, at iii.

154. Voltaire, D9483 and D10246, in *Œuvres complètes, Correspondence and Related*

Documents, ed. Thomas Besterman (Geneva: Institut et Musée Voltaire, 1968–1977); de Pauw, *Recherches philosophiques sur les égyptiens et les chinois*, vol.1, 1–2.

155. App, *Birth of Orientalism*, 207–08.

156. Jeffrey D. Burson, "Unlikely Tales of Fo and Ignatius: Rethinking the Radical Enlightenment through French Appropriation of Chinese Buddhism," *French Historical Studies* 38, no. 3 (2015): 391–420, at 398–400.

157. Joseph Dehergne, "Les historiens Jésuites du Taoïsm," in *La Mission française de Pékin aux XVIIe et XVIIIe siècles: La Chine au temps des Lumières* 2 (Paris: Les Belles Lettres, Cathasia, 1976): 59–67, at 59–65.

158. Voltaire, "Chapitre 2," *Essai sur les mœurs*.

159. De Guignes, "Essai historique," 269.

160. De Guignes, "Essai historique"; App, *Birth of Orientalism*, 188–253.

161. De Guignes, "Essai historique," 273.

162. De Guignes, "Essai historique," 271.

163. De Guignes, "Préface," *Le Chou-king*, iv; Joseph de Guignes, "Lettre de M. De Guignes, interprète du roy," *Journal des sçavans* (January 1752): 812–14, at 812; Joseph de Guignes, "Idée de la littérature Chinoise en général, et particulièrement des historiens et de l'étude de l'histoire à la Chine," in *Mémoires de l'Académie des inscriptions*, vol. 36 (1769): 190–238, at 237.

164. Alexander Statman, "Fusang: The Enlightenment Story of the Chinese Discovery of America," *Isis* 107, no. 1 (2016): 1–25; Statman, "The First Global Turn."

165. The standard reference remains Lewis A. Maverick, "Chinese Influences upon the Physiocrats," *Economic History* (1938): 54–67.

166. Alexis de Tocqueville, *The Old Regime and the Revolution*, trans. John Bonner (New York: Harper and Brothers, 1856): 198.

167. Nicolas Baudeau, "Le Chou-King," in *Éphémérides du citoyen, ou Bibliothèque raisonnée des sciences morales et politiques*, vol. 7 (Paris: Lacombe, 1770): 139–69, at 140.

168. Virgile Pinot, "Les physiocrates et la Chine au XVIIIe siècle," *Revue d'histoire moderne et contemporaine* 8, no. 3 (1906–1907): 200–14; Étiemble, *L'Europe Chinoise II*, 322–33.

169. Tocqueville, *The Old Regime and the Revolution*, 162.

170. Stefan Gaarsmand Jacobsen, "Physiocracy and the Chinese Model," in *Thoughts on Economic Development in China*, edited by Ma Ying and Hans-Michael Trautwein (London: Routledge, 2013): 12–34.

171. Étiemble, *L'Europe Chinoise II*, 331.

172. Walter Watson, "Interpretations of China in the Enlightenment: Montesquieu and Voltaire," in *Actes du IIe Colloque International de Sinologie*, 16–37; Hermann Harder, "La question du 'gouvernement' de la Chine au XVIIIe siècle," in *Actes du IIIe Colloque International de Sinologie*, 80–91; Julia Ching and Willard G. Oxtoby, "Introduction," in *Discovering China*, eds. Ching and Oxtoby, xiv–xv.

173. François Quesnay, *Œuvres Économiques et Philosophiques*, ed. Auguste Oncken (Frankfurt: Joseph Baer and Jules Peelman, 1888): 564.

174. Quesnay, *Œuvres Économiques et Philosophiques*, 636.

175. Quesnay, *Œuvres Économiques et Philosophiques*, 656.

176. Quesnay, *Œuvres Économiques et Philosophiques*, 592.

177. Robert Nisbet, "Idea of Progress: A Bibliographical Essay," *Liberty Fund*, 1979, https://oll4.libertyfund.org/page/idea-of-progress-a-bibliographical-essay-by-robert

-nisbet; "Turgot and the Contexts of Progress," *Proceedings of the American Philosophical Society* 119, no. 3 (1975): 214–22, at 218.

178. Anne Robert Jacques Turgot, *Œuvres de Turgot*, vol. 2 (Paris: Guillaumin, 1844): 599.

179. Turgot, *Œuvres de Turgot*, vol. 1, 310–21.

180. Cited in Gerald J. Cavanaugh, "Turgot and the 'Encyclopédie,'" *Diderot Studies* 10 (1968): 23–33, at 28.

181. Georges Bussières, *Henri Bertin et sa famille: la production nobiliaire du Ministre, ses ancêtres, son Intendance à Lyon, ses ministères* (Périgueux: Imprimerie de la Dordogne, 1906), and Gwynne Lewis, *Madame de Pompadour's Protégé: Henri Bertin and the Collapse of Bourbon Absolutism c. 1750–1792* (Gloucester: Emlyn Publishing, 2011); Jacques Silvestre de Sacy and Michel Antoine, *Henri Bertin dans le sillage de la Chine: 1720–1792* (Paris: Éditions Cathasia, les Belles Lettres, 1970); John Finlay, *Henri Bertin and the Representation of China in Eighteenth-Century France* (New York: Routledge, 2020).

182. Gwynne Lewis, "Henri-Léonard Bertin and the Fate of the Bourbon Monarchy: The 'Chinese Connection,'" in *Enlightenment and Revolution: Essays in Honour of Norman Hampson*, eds. Malcolm Crook, William Doyle, and Alan Forrest (Aldershot: Ashgate, 2004): 69–90; Étiemble, *L'Europe Chinoise II*, 373–75.

183. Isabelle Landry-Deron, "Bertin, Henri-Léonard," in *Dictionnaire Des Orientalistes de Langue Française*, ed. Pouillon, 110–11.

184. Bruno Belhoste, *Paris savant: parcours et rencontres au temps des Lumières* (Paris: A. Colin, 2011): 117.

185. Henri Bertin to Joseph-Marie Amiot, December 20, 1787, in MS 1524, "24 Lettres de Mr. Bertin aux Missionnaires en Chine, 1783–1788," IF; Bertin to Amiot, December 21, 1785, in MS 1524, IF; Henri Bertin to Pierre-Martial Cibot, December 31, 1780, in MS 1523, "28 Lettres de Mr. Bertin aux Missionnaires en Chine, 1779–1782," IF.

186. Bertin to Ko and Yang, December 7, 1764, in MS 1521, "24 Lettres de M. Bertin aux Missionnaires en Chine, 1764–1772," IF.

187. Lewis, *Madame de Pompadour's Protégé*, 221–55; Georges Pédro, "Henri-Léonard Bertin et le développement de l'agriculture au siècle des Lumières," *Comptes Rendus Biologies* 335, no. 5 (2012): 325–33.

188. Bertin to Ko and Yang, December 31, 1766, in MS 1521, IF; Lewis, *Madame de Pompadour's Protégé*, 233.

189. Lewis, "Henri-Léonard Bertin and the Fate of the Bourbon Monarchy," 82–83.

190. Keith Michael Baker, *Inventing the French Revolution: Essays on French Political Culture in the Eighteenth Century* (New York: Cambridge University Press, 1990): 69–75.

191. Bertin to Ko and Yang, December 2, 1764, in MS 1521, IF.

192. Bertin to Ko and Yang, December 31, 1766, MS 1521, IF.

193. These were the names by which their European contemporaries always referred to them. Their apparent Chinese names, Gao Leisi 高類思 and Yang Dewang 楊德望, were likely calques from the French.

194. Anonymous, "Préface," in Amiot et al., *MCC*, vol. 1 (1776): i–xv; Henri Cordier, "Les Chinois de Turgot," in *Mélanges d'histoire et de géographie orientales*, vol. 2 (Paris: Maisonneuve, 1920): 31–39; Finlay, *Henri Bertin and the Representation of China*, 8–39.

195. Bussières, *Henri Bertin*, 71.

196. Cordier, "Les Chinois de Turgot."

197. Henri Bertin, "Memoire sur ce que les Chinois doivent voir en France avant de retourner à la Chine," in MS 1520, "Lettres des PP. Kô et Yang et de M. Brisson," IF; Henri Bertin, "Instructions pour M. Ko et pour M. Yang," January 16, 1765, in MS 1521, IF.

198. MS 1520–1522, IF.

199. Joseph-Marie Amiot to Henri Bertin, 1766, in MS 1515, "80 lettres du P. Amiot à Bertin," IF.

200. Jeffrey D. Burson, "Between Power and Enlightenment: The Cultural and Intellectual Context for the Jesuit Suppression in France," in *The Jesuit Suppression in Global Context*, eds. Jeffrey D. Burson and Jonathan Wright (Cambridge: Cambridge University Press, 2015): 40–64; D. Gillian Thompson, "French Jesuits 1756–1814," in *The Jesuit Suppression in Global Context*, eds. Burson and Wright, 189–98.

201. Jeffrey D. Burson and Jonathan Wright, "Towards a New History of the Eighteenth-Century Jesuit Suppression," in *The Jesuit Suppression in Global Context*, eds. Burson and Wright,1–10.

202. Henri Bertin to the missionaries, 1776, in MS 1522, IF; "État des Sommes envoyées aux missionnaires Chinois depuis 1773, et pièces justifications," in MS 1526, "30 mémoires par divers savants adressés aux missionnaires," IF; Camille de Rochemonteix, *Joseph Amiot et les derniers survivants de la mission française à Pékin (1750–1795)* (Paris: Alphonse Picard et fils, 1915): 209–50.

203. "Compte général des dépenses faites pour Mrs. Ko et Yang Chinois, 1764," in MS 1526, IF.

204. Henri Bertin to François Bourgeois, November 27, 1776, in MS 1522, IF.

205. "Extrait pour le Roy de lettres de Pékin, 1786–1787," MS 5790, BnF Site Arsenal.

206. Bertin to Bourgeois, November 27, 1776, MS 1522, IF.

207. Henri Bertin to Étienne Yang, November 27, 1776, in MS 1522, IF.

208. Henri Bertin to Montigny, February 24, 1776, in MS 1525, "60 lettres relatives aux missionnaires en Chine," IF.

209. Bertin to Cibot, December 15, 1779, in MS 1523, IF; Henri Bertin to de la Tour, October 8, 1785, in MS 1524, IF.

210. Lewis, *Madame de Pompadour's Protégé*, 249.

211. Bertin to Bourgeois, December 31, 1780, in MS 1523, IF.

212. Bertin to Bourgeois, November 30, 1777, in MS 1522, IF; Bernard-Maître, "Catalogue des objets envoyés de Chine par les missionnaires de 1765 à 1786," *Bulletin de l'Université l'Aurore* 9 (1948): 33–34.

213. Bertin to Amiot, November 30, 1777, in MS 1522, IF.

214. Bertin to Bourgeois, November 30, 1777, in MS 1522, IF.

215. Bertin to Amiot, December 15, 1779, in MS 1523, IF.

216. Joseph-Marie Amiot to Henri Bertin, October 16, 1790, in MS 1517, "80 lettres du P. Amiot à Bertin," IF; Bertin to Bourgeois, December 15, 1779, in MS 1523, IF; Bertin to Bourgeois, October 25, 1786, in MS 1524, IF.

217. Bertin to Bourgeois, October 25, 1786, in MS 1524, IF.

218. Kee Il Choi Jr., "Father Amiot's Cup: A Qing Imperial Porcelain Sent to the Court of Louis XV," in *Writing Material Culture History*, eds. Anne Gerritsen and Giorgio Riello (New York: Bloomsbury Publishing, 2014); John Finlay, "Henri

Bertin and the Commerce in Images between France and China in the Late Eighteenth Century," in *Qing Encounters: Artistic Exchanges between China and the West*, eds. Petra ten-Doesschate Chu and Ning Ding (Los Angeles: Getty Research Institute, 2015): 79–94.

219. Jean Baptiste Joseph Breton de la Martinière, *China: Its Costume, Arts, Manufactures &c. Edited Principally from the Originals in the Cabinet of the Late M. Bertin*, 2nd ed., trans. (London: J. J. Stockdale, 1812); "Commission Temporaire des Arts," MS F^17, Archives nationales (AN), Pierrefitte-sur-Seine, France; "Bertin et la Mission de Chine, 1762–1792," MS 131, Institut national d'histoire de l'art, Paris, France.

220. Luc-Vincent Thiéry, *Guide des amateurs et des étrangers voyageurs à Paris*, vol. 1 (Paris: Hardouin and Gattey, 1787): 134–36.

221. Finlay, *Henri Bertin and the Representation of China*, 107–29.

222. Bertin to Amiot, December 21, 1785, in MS 1524, IF; Bertin to Bourgeois, December 25, 1787, in MS 1524, IF.

223. *junzi bu qi* 君子不器; Amiot to Bertin, October 16, 1790, in MS 1517, IF.

224. Henri Bertin to Jean-Paul Collas, December 31, 1780, in MS 1523, IF.

225. Bertin to Cibot, December 15, 1779, in MS 1523, IF.

226. Bertin to Amiot, December 31, 1784, in MS 1524, IF; Bertin to Amiot, October 19, 1785 [?], in MS 1524, IF.

227. Bertin to Amiot, December 20, 1787, in MS 1524, IF.

228. Bertin to Amiot, December 21, 1785, in MS 1524, IF.

229. Bertin to Amiot, December 15, 1779, in MS 1523, IF.

230. François Desvoyes [?] to Joseph-Marie Amiot, December 14, 1792, in MS 1524, IF.

231. Bertin to Ko and Yang, January 18, 1774, in MS 1522, IF; Bertin to de la Tour, October 8, 1785, in MS 1524, IF.

232. Bertin to Amiot, November 22, 1787, in MS 1524, IF; emphasis original.

233. Abel-Rémusat, *Programme du cours*, 6–12; Henri Cordier, "La Suppression de la Compagnie de Jésus et la mission de Péking," *T'oung Pao* 2nd series, 17, no. 3 (1916): 271–347, at 283; Joseph Dehergne, "Une grande collection: Mémoires concernant les Chinois (1776–1814)," *Bulletin de l'École française d'Extrême-Orient* 72, no. 1 (1983): 267–98, at 268.

234. Amiot et al., *MCC*.

235. Dehergne, "Une grande collection," 267–72.

236. Jean-Louis Coster, ed., *L'Ésprit des journaux, françois et étrangers*, vol. 4 (Paris: L'Imprimerie du Journal, April 1777): 36.

237. In 1814, two more volumes appeared under the same title sharing little in common with the first fifteen.

238. *Journal encyclopédique*, vol. 4, part 1 (Paris: Bouillon, 1788): 27–28; *Journal des sçavans* (October 1783), 1989.

239. The best overview remains Dehergne, "Une grande collection."

240. Joseph-Marie Amiot, "Lettre sur les caractères chinois" (1764), in Amiot et al., *MCC*, vol. 1 (1776): 275–324, at 318; Amiot to Bertin, September 28, 1777, in MS 1515, IF; Joseph-Marie Amiot, "Extrait d'une lettre écrite de Péking le 20 mai 1786," in Amiot et al., *MCC*, vol. 13 (1788): 417–58, at 417.

241. Pierre-Martial Cibot, "Requête à l'Empereur pour la cérémonie du Labourage" (1767), in Amiot et al., *MCC*, vol. 3 (1778): 499–504, at 504.

242. Bertin to Amiot, December 31, 1780, in MS 1523, IF; Société archéologique de Touraine, *Bulletin trimestriel*, vol. 12 (Tours: Péricat, 1900): 297.

243. Baker, *Inventing the French Revolution*, 78–81.

244. Edelstein, *The Enlightenment*, 2, 81.

245. For an annotated table of contents, see Dehergne, "Une grande collection."

246. Dehergne divided the articles into two categories, "sinology" and "science," based on the extent to which they engaged with Chinese primary sources (Dehergne, "Une grande collection," 291).

247. Amiot to Bertin, September 1, 1788, in MS 1517, IF.

248. Louis Pfister, *Notices biographiques et bibliographiques sur les jésuites de l'ancienne mission de Chine*, vol. 2 (Shanghai: Imprimerie de la Mission Catholique, 1932): 890, 924, 926.

249. Amiot to Bertin, November 22, 1783, in MS 1516, IF.

250. Joseph-Marie Amiot, "Antiquité des Chinois, prouvée par les monuments," in Amiot et al., *MCC*, vol. 2 (1776): 1–364, at 8, 60.

251. Pierre-Martial Cibot, "Essai sur la langue et les caractères des Chinois II," in Amiot et al., *MCC*, vol. 9 (1783): 282–430, at 285.

252. Pierre-Martial Cibot, Aloys Ko, and Étienne Yang, "Essai sur l'antiquité des Chinois," in Amiot et al., *MCC*, vol. 1 (1776): 1–271, at 72; Amiot, "Antiquité des Chinois," 27.

253. Nicolas Standaert, "Jesuit Accounts of Chinese History and Chronology and Their Chinese Sources," *East Asian Science, Technology, and Medicine* 35 (2013): 11–88, at 11–12; Wu Huiyi, *Traduire la Chine*, 32–46.

254. Amiot, "Antiquité des Chinois," 194–45.

255. Benjamin A. Elman, *From Philosophy to Philology: Intellectual and Social Aspects of Change in Late Imperial China* (Cambridge, MA: Council on East Asian Studies, Harvard University, 1984): 17–33.

256. Amiot, "Antiquité des Chinois," 151–55.

257. Dehergne, "Une grande collection," 276–78.

258. Pierre-Martial Cibot, "Observations de Physique et d'Histoire naturelle de l'empereur K'ang Hi" (1771), in Amiot et al., *MCC*, vol. 4 (1779): 452–83; Catherine Jami, "Portrait of the Emperor as an Enlightened Monarch: The French Translation of Kangxi's Collection of the Investigation of Things in Leisure Time (1779)," presentation at the biennial conference of the European Society for the History of Science, September 14–17, 2018, London.

259. Quoted in Dehergne, "Une grande collection," 284.

260. Hanlin Academy, *Hanlin yuan* 翰林院.

261. Cibot et Al., "Essai sur l'antiquité des Chinois," 17; Joseph-Marie Amiot, "Extrait d'une lettre écrite de Péking le 13 juillet, 1778," in Amiot et al., *MCC*, vol. 15 (1791): 281–91, at 289.

262. *Siku quanshu* 四庫全書; R. Kent Guy, *The Emperor's Four Treasuries: Scholars and the State in the Late Ch'ien-Lung Era* (Cambridge, MA: Harvard University Asia Center, 1987); Amiot, "Extrait d'une lettre écrite de Péking le 13 juillet, 1778," 289.

263. Amiot to Bertin, November 15, 1784, in MS 1516, IF; Bertin to Amiot, November 22, 1787, in MS 1524, IF.

264. Elman, *From Philosophy to Philology*, 7–13.

265. Yu Minzhong 于敏中; Joseph-Marie Amiot, "Extrait d'une lettre écrite de Péking le 15 oct. 1773," in Amiot et al., *MCC*, vol. 1 (1776): 419–27, at 419; Amiot, "Lettre écrite de Péking le 26 sep. 1780," in Amiot et al., *MCC*, vol. 9 (1783): 45–59, at 47.

266. Arthur W. Hummel, ed., *Eminent Chinese of the Ch'ing Period (1644–1912)*, vol. 2 (Washington, DC: United States Government Printing Office, 1943): 942–44.

267. Dai Zhen 戴震; Minghui Hu, *China's Transition to Modernity: The New Classical Vision of Dai Zhen* (Seattle: University of Washington Press, 2015); Elman, *From Philosophy to Philology*, 20–21, 66–67; Qian Mu, *Zhongguo jin sanbai nian xueshushi* [History of Chinese scholarship of the last three centuries] (1937), vol. 1, reprint ed. (Beijing: Jiuzhou chubanshe, 2011): 396–414.

268. Amiot to Bertin, September 1, 1788, in MS 1517, IF.

269. Amiot, "Extrait d'une lettre écrite de Péking le 13 août 1780," in Amiot et al., *MCC*, vol. 9 (1783): 6–24, at 18; Amiot, "Lettre écrite de Péking le 26 sep. 1780," 56.

270. Pierre-Martial Cibot to Henri Bertin, October 26–28, 1778, in MS 1519, "Lettres de plusieurs missionnaires," IF.

271. De Bary and Bloom, *Sources of Chinese Tradition, Volume 1*, 236–38 (this section also edited by Harold Roth, Sarah Queen, and Nathan Sivin).

272. Cibot et Al., "Essai sur l'antiquité des Chinois," 258–61; Amiot, "Antiquité des Chinois," 14–15, 38–39.

273. Amiot, "Extrait d'une lettre écrite de Péking le 19 nov. 1787," in Amiot et al., *MCC*, vol. 14 (1789): 536–81, at 536; Cibot, "Requête à l'Empereur," 499; Amiot, "Extrait d'une lettre écrite de Péking le 20 mai 1786," 418.

274. Cibot et Al., "Essai sur l'antiquité des Chinois," 72, 96; Émile Littré, *Dictionnaire de la langue française*, 2nd ed. (Paris: Hachette, 1872–1877); Darrin McMahon, *Enemies of the Enlightenment: The French Counter-Enlightenment and the Making of Modernity* (Oxford: Oxford University Press, 2001): 32.

275. Cibot, "Tchong-yong, ou Juste Milieu," in Amiot et al., *MCC*, vol. 1, 459–89, at 474; Cibot and Ko, "Remarques sur un écrit de M. de Pauw," 385.

276. Cibot, "Essai sur la langue et les caractères des Chinois I," 231.

277. Cibot and Ko, "Remarques sur un écrit de M. de Pauw," 365; Cibot et Al., "Essai sur l'antiquité des Chinois," 124.

278. Cibot to Bertin, October 26–28, 1778, in MS 1519, IF.

279. Amiot to Bertin, November 25, 1784, in MS 1516, IF.

280. Joseph-Marie Amiot to François Desvoyes, November 2, 1792, in MS 1517, IF.

Chapter 2

1. Louis Pfister, *Notices biographiques et bibliographiques sur les jésuites de l'ancienne mission de Chine*, vol. 2 (Shanghai: Imprimerie de la Mission Catholique, 1932): 837.

2. Michel Hermans, "Biographie de Joseph-Marie Amiot," in *Les Danses rituelles chinoises d'après Joseph-Marie Amiot: Aux sources de l'ethnochorégraphie*, eds. Yves Lenoir and Nicolas Standaert (Namur: Presses Universitaires de Namur, 2005): 11–77, at 12–22.

3. Qian Deming 錢德明.

4. *Quanzong* 05-0109-024, Qianlong year 15, month 11, day 18, *Zongguan Neiwufu* [Records of the Imperial Household Department], First Historical Archives of China (FHA), Beijing.

5. Joseph-Marie Amiot, "Lettre du Père Amiot au Père Allart . . . le 20 octobre

1752," in *Lettres édifiantes et curieuses*, vol. 3, ed. Louis Aimé-Martin (Paris: Société du Panthéon Littéraire, 1843): 832–39, at 834.

6. Mårten Söderblom Saarela, *The Early Modern Travels of Manchu: A Script and its Study in East Asia and Europe* (Philadelphia: University of Pennsylvania Press, 2022): 214–16.

7. Grand Secretariat, *Neige* 內閣 (Michel Benoist, "Lettre du Père Benoist" [1772], in *Lettres édifiantes et curieuses*, vol. 4, ed. Louis Aimé-Martin (Paris: Société du Panthéon Littéraire, 1843): 209–17, at 211).

8. Jacques-François d'Ollières, "Extrait d'une lettre du Père d'Ollières, à son frère, cure de Lexie, près Longwi, 1780," in *Lettres édifiantes et curieuses*, vol. 4, ed. Louis Aimé-Martin (Paris: Société du Panthéon Littéraire, 1843): 275–82, at 282. The only surviving evidence of Amiot's employment in the Chinese record comes from an obscure document that must have been composed in 1783 or 1784, a "Name List of the Westerners at the Inner Xi'anmen Catholic Church," which states: "Qian Deming is knowledgeable in music and translates Russian Latin for the Mongolian Hall of the Grand Secretariat" (*Archives Concerning Western Catholic Missions* [undated], no. 557, FHA).

9. *Journal des sçavans* (March 1780), 436.

10. Jim Levy, "Joseph Amiot and Enlightenment Speculation on the Origin of Pythagorean Tuning in China," *Theoria* 4 (1989): 63–88.

11. Joseph-Marie Amiot, *Art militaire des Chinois, ou, Recueil d'anciens traits sur la guerre, composés avant l'ère chrétienne, par différents généraux Chinois*, ed. Joseph de Guignes (Paris: Didot l'aîné, 1772); Joseph-Marie Amiot, *Éloge de la ville de Mukden et de ses environs*, ed. Joseph de Guignes (Paris: N. M. Tilliard, 1770); Voltaire, D20040, in *Œuvres complètes, Correspondence and Related Documents*, ed. Thomas Besterman (Geneva: Institut et Musée Voltaire, 1968–1977).

12. Henri Bertin to Nicolas-Joseph Raux, December 25, 1787, MS 1524, "24 Lettres de Mr. Bertin aux Missionnaires en Chine, 1783–1788," Institut de France (IF), Paris, France; Joseph de Grammont to Henri Bertin, January 25, 1790, MS Henri Cordier, Archives jésuites de la province de France (AJPF), Vanves, France.

13. The standard reference is Hermans, "Biographie de Joseph-Marie Amiot." Two dissertations, one in Chinese (Long Yun, "Qian Deming yanjiu—18 shiji yi wei chu zai zhongfa wenhua jiaohuichu de chuanjiaoshi" [Research on Joseph-Marie Amiot—a missionary at the confluence of Chinese and French culture in the eighteenth century] [PhD diss., Peking University, 2010]) and one in Japanese (Nii Yoko, "Jūhasseiki ni okeru Chūgoku to Yōroppa no shisō kōryū: zaiKa Iezusu kaishi Amio no hōkoku wo chūshin ni 18" [Sino-European intellectual interactions in the eighteenth century: the case of Amiot, a Jesuit in China] [PhD diss., University of Tokyo, 2014]), were useful in writing this book, both of which have since been published.

14. Henri Cordier, "La Suppression de la Compagnie de Jésus et la mission de Peking," *T'oung Pao*, 2nd series, 17, no. 3 (1916): 271–347, at 283.

15. Prattle, *bavardage*; Joseph-Marie Amiot, "Mémoire sur le Thibet," MS 5409, Fonds Henri Cordier, "Notes sur la correspondance de Henri Bertin et sur les missions catholiques," IF.

16. On the last Jesuit missionaries in Beijing, see especially R. Po-chia Hsia, "Jesuit Survival and Restoration in China," in *Jesuit Survival and Restoration: A Global History, 1773–1900*, ed. Robert A. Maryks and Jonathan Wright (Leiden: Brill, 2014): 245–61; R. Po-chia Hsia, "The End of the Jesuit Mission in China," in *The Jesuit Sup-*

pression in Global Context, eds. Jeffrey D. Burson and Jonathan Wright (Cambridge: Cambridge University Press, 2015): 100–16; Joanna Waley-Cohen, "China and Western Technology in the Late Eighteenth Century," *American Historical Review* 98, no. 5 (1993): 1525–44; Kristina Kleutghen, *Imperial Illusions: Crossing Pictorial Boundaries in the Qing Palaces* (Seattle: University of Washington Press, 2015).

17. Two exceptions are Pierre Huard and Ming Wong, "Mesmer en Chine: Trois lettres médicales du R.P. Amiot, rédigées à Pékin, de 1783 à 1790," *Revue de Synthèse* series 3, 81, no. 17–18 (1960); and Joël Thoraval, "De la magie à la 'raison': Hegel et la religion chinoise," in *La Chine entre amour et haine: Actes du VIIIe Colloque de sinologie de Chantilly*, ed. Michel Cartier (Paris: Desclée de Brouwer, 1998): 111–41.

18. Hongwu 弘旿; as a member of the imperial clan, his full name was Aisin Gioro Hongwu 愛新覺羅・弘旿.

19. Needham compared it to the purges of Stalin and the extermination of the Knights Templar, with "untold anguish both physical and mental suffered by its most faithful supporters" (Joseph Needham, Ho Ping-Yü, and Lu Gwei-Djen, *Science and Civilisation in China: Volume V: Chemistry and Chemical Technology, Part III, Spagyrical Discovery and Invention* [Cambridge: Cambridge University Press, 1976]: 223).

20. Nicolas Standaert, ed., *Handbook of Christianity in China*, vol. 1 (Leiden: Brill, 2001): 384–85.

21. R. Hsia, "The End of the Jesuit Mission in China," 104.

22. *Dongtang* 東堂; *Nantang* 南堂; *Xitang* 西堂; *Beitang* 北堂.

23. Isabelle Vissière and Jean-Louis Vissière, "Un carrefour culturel: La mission française de Pékin au XVIIIe siècle," in *Actes du IIIe Colloque International de Sinologie: Appréciation par l'Europe de la tradition chinoise à partir du XVIIe siècle* (Paris: Les Belles Lettres, Cathasia, 1983): 211–22, at 211–13.

24. Noël Golvers, *Libraries of Western Learning for China: Circulation of Western Books between Europe and China in the Jesuit Mission (ca.1650–1750), ii: Formation of Jesuit Libraries* (Louvain: Ferdinand Verbiest Institute K. U. Leuven, 2013); Hubert Verhaeren, *Catalogue de la bibliothèque du Pé-t'ang* (Beijing: Imprimerie des Lazaristes, 1949).

25. Henri Bertin to François Bourgeois, January 24, 1779, MS 1523, "28 Lettres de Mr. Bertin aux Missionnaires en Chine, 1779–1782," IF.

26. François Bourgeois, "État de la Mission Française de Pékin," 1779, MS Brotier 135, AJPF.

27. R. Hsia, "The End of the Jesuit Mission in China," 104.

28. R. Hsia, "Jesuit Survival and Restoration in China," 245–61; R. Hsia, "The End of the Jesuit Mission in China," 100–16.

29. Henri Bertin to Pierre-Martial Cibot, December 31, 1780, MS 1523.

30. Joseph-Marie Amiot to Henri Bertin, October 16, 1790, MS 1517, "80 lettres du P. Amiot à Bertin," IF.

31. R. Hsia, "Jesuit Survival and Restoration in China," 245–46.

32. Amiot to Bertin, September 1, 1774, MS 1515, "80 lettres du P. Amiot à Bertin," IF.

33. Amiot to Bertin, October 1, 1774, MS 1515, IF.

34. Amiot to Bertin, September 15, 1776 [A], MS 1515, IF.

35. Amiot to Bertin, October 1, 1774, MS 1515, IF.

36. Amiot to Bertin, October 1, 1774, MS 1515, IF; Henri Bertin to Cardi-

nal Bernis, January 1780, MS 1526, "30 mémoires par divers savants adressés aux missionnaires," IF.

37. Mark C. Elliott, *The Manchu Way: The Eight Banners and Ethnic Identity in Late Imperial China* (Stanford, CA: Stanford University Press, 2001): 100.

38. Jean-Paul Collas to Henri Bertin, November 27, 1780, MS 1520, "Lettres des PP. Kô et Yang et de M. Brisson," IF.

39. Collas to Bertin, December 2, 1780, MS 1520, IF.

40. Bertin to Cardinal Bernis, January 1780, MS 1526, IF.

41. Henri Bertin to Joseph-Marie Amiot, October 1776, MS 1522, "36 Lettres de Mr. Bertin aux Missionnaires en Chine, 1773–1778," IF.

42. "Pièces concernant l'institution d'un Préfet Apostolique à Pékin," MS Brotier 134–135, AJPF.

43. The story of the internecine squabbles that ensued after the suppression has been told many times; see especially R. Hsia, "Jesuit Survival and Restoration in China," and R. Hsia, "The End of the Jesuit Mission in China." For a version told from French archives, see Camille de Rochemonteix, *Joseph Amiot et les derniers survivants de la mission française à Pékin (1750–1795)* (Paris: Alphonse Picard et fils, 1915); for one told from Portuguese and Roman archives, see Joseph Krahl, *China Missions in Crisis: Bishop Laimbeckhoven and His Times, 1738–1787* (Rome: Gregorian University Press, 1964).

44. Amiot to Bertin, September 15, 1776 [B], MS 1515, IF.

45. "Ordonnance du Roi, Concernant les Missions françaises en Chine," MS Brotier 135, AJPF.

46. Amiot to Bertin, November 19, 1777, MS 1515, IF.

47. Rochemonteix, *Joseph Amiot*, 219–22.

48. Bertin to Amiot, December 15, 1779, MS 1523, IF.

49. Krahl, *China Missions in Crisis*, 263–68; R. Hsia, "Jesuit Survival and Restoration in China," 247–51.

50. Astronomical Bureau, *Qintian jian* 欽天監.

51. *Neiwufu* 内務府; François Bourgeois to Henri Bertin, May 25, 1781, MS 1519, "Lettres de plusieurs missionnaires," IF.

52. Bourgeois to Bertin, May 25, 1781, MS 1519, IF.

53. Bourgeois to Bertin, May 25, 1781, MS 1519, IF.

54. Amiot to Bertin, May 22, 1781, MS 1516, "80 lettres du P. Amiot à Bertin," IF.

55. Paris Foreign Missions Society, *Missions Étrangères de Paris*.

56. Rochemonteix, *Joseph Amiot*, 386–87.

57. "Copie de la lettre de Mr. Le Sup. Général, écrite à M. de Castries, ministre de la Marine, le 14 Novembre 1782," MS 166.I B, Congrégation de la Mission (CM), Paris, France.

58. Jacquier, "Circulaires des Supérieurs Générales," January 1, 1783, MS 171.II B, CM.

59. Jacquier, "Circulaires des Supérieurs Générales," January 1, 1786, MS 171.II B, CM.

60. Bourgeois to Bertin, November 20, 1785, MS 1519, IF.

61. Rochemonteix, *Joseph Amiot*, 404–09; R. Hsia, "Jesuit Survival and Restoration in China," 255.

62. Amiot to Bertin, August 20, 1790 / October 4, 1790, MS 1517, IF; Amiot to Bertin, January 25, 1787, MS 1516, IF.

63. Amiot to Bertin, November 16, 1789, MS 1517, IF.

64. Amiot to Bertin, September 1, 1788, MS 1517, IF; Amiot to Bertin, October 10, 1789, MS 1517, IF.

65. Amiot to Bertin, October 10, 1789, MS 1517, IF.

66. Amiot to Bertin, September 26, 1780, MS 1516, IF; Pierre-Martial Cibot to Henri Bertin, October 23, 1779, MS 1519, IF.

67. Louis Antoine de Poirot and Giuseppe Panzi, letter of September 27, 1804, MS 164.II B, CM.

68. Amiot to Bertin, August 20, 1790 / October 4, 1790, MS 1517, IF.

69. Joseph-Marie Amiot's will, November 15, 1791, MS 1517, IF.

70. Emperor emeritus, *Taishang huang* 太上皇.

71. Benjamin A. Elman, *On Their Own Terms: Science in China, 1550–1900* (Cambridge, MA: Harvard University Press, 2005): 148.

72. *Yuanmingyuan* 圓明園, also called the Old Summer Palace; Louis-François Delatour, *Essais sur l'architecture des Chinois, sur leurs jardins, leurs principes de médecine, et leurs mœurs et usages; avec des notes*, vol. 1 (Paris: de Clousier, 1803): 152; Kleutghen, *Imperial Illusions*, 182–88.

73. Michel Benoist, letter of September 12, 1764, in *Revue de l'Extrême-Orient*, vol. 3, ed. Henri C order (1887): 242–51, at 248.

74. Amiot to Bertin, September 28, 1777, MS 1515, IF.

75. Benoist, "Lettre du Père Benoist" [17 72], 212–18; Benoist, letter of September 12, 1764, 248.

76. Nathan Sivin, "Copernicus in China: or, Good Intentions Gone Astray" (1973), in *Science in Ancient China: Researches and Reflections* (Aldershot: Variorum, 1995): 63–122, at 46.

77. Benoist, "Lettre du Père Benoist" [1772], 215.

78. Michel Benoist, "Lettre du Père Benoist" [c.1772], in *Lettres édifiantes et curieuses*, vol. 4, ed. Louis Aimé-Martin (Paris: Société du Panthéon Littéraire, 1843): 217–25, at 223–25.

79. Jinchuan 金川.

80. Agui 阿桂 et al, *Pingding liang Jinchuan fanglüe* [Strategies of the pacification of the two Jinchuans] (Beijing: 1784): *juan* 100.

81. Waley-Cohen, "China and Western Technology," 1537–40.

82. Benoist, letter of September 12, 1764, 248.

83. Benoist, "Lettre du Père Benoist" [1772], 215.

84. Wang Lun 王倫; Naquin, *Shantung Rebellion*.

85. William T. Rowe, "Introduction: The Significance of the Qianlong-Jiaqing Transition in Qing History," *Late Imperial China* 32, no. 2 (2011): 74–88, at 82.

86. R. Kent Guy, *The Emperor's Four Treasuries: Scholars and the State in the Late Ch'ien-Lung Era* (Cambridge, MA: Harvard University Asia Center, 1987); Benjamin A. Elman, *From Philosophy to Philology: Intellectual and Social Aspects of Change in Late Imperial China* (Cambridge, MA: Council on East Asian Studies, Harvard University, 1984): 15.

87. Joseph-Marie Amiot, "Extrait d'une lettre écrite de Péking le 13 juillet, 1778," in Joseph-Marie Amiot, François Bourgeois, Pierre-Martial Cibot, Aloys Ko, Aloys de Poirot, Charles Batteux, M. de Bréquigny, A. I. Silvestre de Sacy, and Antoine Gaubil, *Mémoires concernant l'histoire, les sciences, les arts, les moeurs, les usages, &c. des Chinois* [*MCC*], vols. 1–15 (Paris: Nyon l'aîné, 1776–1791): vol. 15 (1791): 281–91, at 290.

88. Kleutghen, *Imperial Illusions*, 217.

89. Lars Peter Laamann, *Christian Heretics in Late Imperial China: Christian Inculturation and State Control, 1720–1850* (New York: Routledge, 2006).

90. R. Hsia, "Jesuit Survival and Restoration in China," 259.

91. Wu Boya, "Cong xin chuban de qingdai dang'an kan Tianzhujiao chuan hua shi" [The history of Catholicism in China, based on newly published Qing archival documents] (Beijing: Chinese Academy of Social Sciences, 2005).

92. Western cult, *Xiyang xiejiao* 西洋邪教; *Archives Concerning Western Catholic Missions*, February 20, 1786, no. 375, FHA.

93. *Da Qing Gaozong Chun Huangdi shilu* [Veritable Records of the Great Qing, Qianlong period]. Reprint ed. (Taipei: Huawen shuju, 1970): *juan* 1218, Qianlong year 49, month 11.

94. *Da Qing Gaozong Chun Huangdi shilu*, *juan* 1221, Qianlong year 49, month 12.

95. Fukanggan, Fu Kang'an 福康安; *Archives Concerning Western Catholic Missions*, February 20, 1785, no. 328, FHA.

96. Tristan G. Brown, "A Mountain of Saints and Sages: Muslims in the Landscape of Popular Religion in Late Imperial China," *T'oung Pao* 105, no. 3–4 (2019): 437–91, at 484–88.

97. Fulungga, Fu Long'an 福隆安; R. Hsia, "Jesuit Survival and Restoration in China," 253.

98. *Archives Concerning Western Catholic Missions*, November 2, 1774, no. 149; July 16, 1781, no. 159, FHA.

99. Waley-Cohen, "China and Western Technology," 1538.

100. Joseph-Marie Amiot, "Extrait d'une lettre écrite de Péking le 2 oct. 1784," in Amiot et al., *MCC*, vol. 11 (1786): 515–68, at 553.

101. Arthur W. Hummel, ed., *Eminent Chinese of the Ch'ing Period (1644–1912)*, vol. 1 (Washington, DC: United States Government Printing Office, 1943): 259.

102. Bourgeois to Bertin, May 25, 1781, MS 1519, IF.

103. Charles O. Hucker, *A Dictionary of Official Titles in Imperial China* (Stanford, CA: Stanford University Press, 1985): 169.

104. R. Hsia, "The End of the Jesuit Mission in China," 107.

105. Pfister, *Notices biographiques et bibliographiques*, vol. 2, 774–75.

106. Fu Daren 傅大人.

107. Bourgeois to Bertin, May 25, 1781, MS 1519, IF.

108. Joseph-Marie Amiot, "Extrait d'une lettre écrite de Péking le 15 oct. 1785," in Amiot et al., *MCC*, vol. 12 (1786): 509–30, at 517.

109. D'Ollières, "Extrait d'une lettre du Père d'Ollières," 282.

110. Wish-Fulfilling Studio, *Ruyiguan* 如意館; Kleutghen, *Imperial Illusions*, 52–57.

111. Amiot to Bertin, September 28, 1777, MS 1515, IF.

112. Nicolas-Joseph Raux to Henri Bertin, November 25, 1787, MS Bréquigny 2, "Mélanges sur la Chine et les Chinois," Bibliothèque nationale de France (BnF), Paris, France.

113. Amiot, "Extrait d'une lettre écrite de Péking le 15 oct. 1785," 516–24. In Chinese records of the event, the only description next to the name "Qian Deming" was "Westerner," suggesting that he held no other rank or title at the time (*Baxun wanshou shengdian* [Grand celebration of the eightieth birthday] [Beijing: 1792]: *juan* 25).

114. 法羅海島銅人像，巴必鸞城公樂場，遠志七奇傳國俗，虔依萬壽祝天皇，筵霑尊斝寰瀛福，身傍璣衡霄漢光，龍角開杓瞻北闕，紫微天座燦中央. *Qinding qian sou yan shi* [Imperially commissioned poems from the Feast of the Thousand Old Men], *Siku quanshu* (1785): *jibu, zongji lei, juan* 25.

115. R. Hsia, "Jesuit Survival and Restoration in China," 250–51.

116. Amiot to Bertin, November 15, 1784, MS 1516, IF.

117. Joseph-Marie Amiot to Louis-Georges de Bréquigny, June 26, 1789, MS Bréquigny 2, BnF; Amiot to Bertin, January 25, 1787, MS 1516, IF; Amiot, "Extrait d'une lettre écrite de Péking le 26 juin 1789," in Amiot et al., *MCC*, vol. 15 (1791): v–xv, at xv.

118. Bourgeois, "État de la Mission Française de Pékin," 1779, MS Brotier 135, AJPF.

119. Amiot, "Discours du traducteur," in *Art militaire des Chinois*, 8.

120. Joseph-Marie Amiot, "Extrait d'une lettre écrite de Péking le 20 sep. 1786," in Amiot et al., *MCC*, vol. 13 (1788): 507–10.

121. Yang Yagebo, possibly 楊雅各伯.

122. Joseph-Marie Amiot to Jérôme-Frédéric Bignon, December 5, 1771, MS Joseph Bernard Maître 69, AJPF.

123. Amiot, "Extrait d'une lettre écrite de Péking le 2 oct. 1784," 520–22.

124. Amiot to Bertin, October 2, 1784, MS 1516, IF.

125. Amiot's will, November 15, 1791, MS 1517, IF.

126. Xu Guangqi 徐光啓; Zhao Yi 趙翼, *Yanpu zaji* [Miscellaneous records of sunning on the eaves] (late eighteenth century), reprint ed. (Beijing: Zhonghua shuju, 1982): *juan* 2, 6.

127. Wang Chang 王昶, ed., *Huhaiwen zhuan* [Collected works of lakes and seas] (Jingxun tang: 1837): *juan* 24.

128. Tribunal of Rites, *Libu* 禮部.

129. Joseph-Marie Amiot, trans., "*Hymne mandchou chanté à l'occasion de la conquête du Jin-Chuan,*" MS Manchu 285, BnF.

130. Joseph-Marie Amiot, "Lettre écrite de Péking le 17 août 1781," in Amiot et al., *MCC*, vol. 9 (1783): 441–54, at 454.

131. Amiot to Bréquigny, June 26, 1789, MS 1517, IF.

132. Amiot to Bertin, January 25, 1787, MS 1516, IF.

133. Amiot to Bertin, September 1, 1788, MS 1517, IF.

134. *Daren* 大人; Amiot to Bertin, August 20, 1790 / October 4, 1790, MS 1517, IF.

135. Catherine Jami, *The Emperor's New Mathematics: Western Learning and Imperial Authority During the Kangxi Reign (1662–1722)* (Oxford: Oxford University Press, 2012).

136. John W. Witek, "Manchu Christians and the Sunu Family," in *Handbook of Christianity in China*, vol. 1, ed. Nicolas Standaert (Leiden: Brill, 2001): 444–48.

137. Yinlu 胤禄; Benoist, letter of September 12, 1764, 249.

138. Amiot to Bertin, October 1, 1786, MS 1516, IF.

139. Jonathan Schlesinger, *A World Trimmed with Fur: Wild Things, Pristine Places, and the Natural Fringes of Qing Rule* (Stanford, CA: Stanford University Press, 2017): 95–99.

140. Joseph-Marie Amiot, "Observations sur le livre de M. de Pauw, intitulé *Recherches philosophiques sur les Égyptiens et les Chinois*" (1777), in Amiot et al., *MCC*, vol. 6 (1780): 275–345, at 306.

141. Amiot to Bertin, October 1, 1786, MS 1516, IF.

142. Rehe 熱河, or Jehol; Elliott, *The Manchu Way*, 183–87.

143. Amiot to Bertin, November 16, 1789, MS 1517, IF.

144. On the contentious question of the Sinicization of the Manchu court, see Elliott, *The Manchu Way*, 1–35.

145. Amiot to Bertin, August 20, 1790 / October 4, 1790, MS 1517, IF.

146. Amiot to Bertin, October 10, 1789, MS 1517, IF.

147. Amiot to Bertin, January 25, 1787, MS 1516, IF.

148. Amiot to Bertin, August 20, 1790 / October 4, 1790, MS 1517, IF.

149. Yongrong 永瑢.

150. Li Shi 李湜, *Zijin danqing—qing gong huihua de chuangzuo yu shoucang* [Forbidden City painting—collecting and creating paintings in the Qing palace] (Beijing: Zhongguo guoji guangbo chubanshe, 2008): 19.

151. *Quanzong* 03-0406-056, Qianlong year 40, month 5, day 23, *Zongguan Neiwufu*, FHA.

152. *Archives Concerning Western Catholic Missions*, November 2, 1774, no. 143, FHA.

153. *Da Qing Gaozong Chun Huangdi shilu, juan* 1271, Qianlong year 51, month 12; *juan* 1290, Qianlong year 52, month 10.

154. Benoist, "Lettre du Père Benoist," [1772], 209–25.

155. Bourgeois to Bertin, May 25, 1781, MS 1519, IF.

156. Jean-Matthieu Ventavon, "Extrait d'une lettre de M. de Ventavon, missionnaire à Pékin, en date du 25 novembre 1784," in *Lettres édifiantes et curieuses*, vol. 4, ed. Aimé-Martin, 306–10, at 307.

157. Collas to Bertin, November 27, 1780, MS 1520, IF.

158. Bourgeois to Bertin, May 25, 1781, MS 1519, IF.

159. Amiot to Bertin, November 15, 1784, MS 1516, IF. Two authors mention the prince briefly without speculating as to which prince he was (Waley-Cohen, "China and Western Technology," 1537; Long Yun 龍雲, "Qian Deming yanjiu," 241).

160. The prince was mentioned in thirteen letters written between 1784 and 1790, all currently held in the archives of the Institut de France and the Bibliothèque nationale de France. Amiot mentioned him in eight letters and gave the name in one of them; Bourgeois mentioned him in three letters and gave the name in two of them; and Raux mentioned him by name in one letter.

161. François Bourgeois, "Extrait d'une lettre écrite de Péking le 19 nov. 1784," in Amiot et al., *MCC*, vol. 11 (1786): 577–79, at 578.

162. Each missionary had his own Romanization system—Amiot used the spelling "*Houng-ou-yé*."

163. The characters would thus be: Hongwu *ye* 弘旿爺.

164. Bourgeois to Bertin, November 19, 1784, MS 1520, IF.

165. The missionaries typically translated *wang* 王 as "régulo" and *gong* 公 as "count" (Amiot, trans., "*Hymne mandchou chanté à l'occasion de la conquête du Kin-Tchouen*," MS Manchu 285, BnF).

166. Yang Danxia, "Hongwu de shuhua jiaoliu yu chuangzuo" [The exchange and creation of Hongwu's painting and calligraphy], *Forbidden City* (Beijing: Palace Museum, 2005); "Hongwu ji qi 'jiang shan gong cui tu'" [Hongwu and his "Painting of Rivers and Mountains Surrounded by Emerald Green"], *Shoucangjia* (Beijing: Capital Museum, 2003).

167. Yunbi 允祕; Hongli 弘曆.

168. Yang Danxia, "Hongwu ji qi 'jiang shan gong cui tu,'" 3–4.

169. Anonymous, "Lettre d'un missionnaire de Chine, mort du père Benoist, à Paris, année 1775," in *Lettres édifiantes et curieuses*, vol. 4, ed. Aimé-Martin, 225–33, at 231.

170. Second-rank general, *er deng fuguo jiangjun* 二等輔國將軍; banner prince, *beizi* 貝子; *Da Qing Gaozong Chun Huangdi shilu*, *juan* 680, Qianlong year 28, month 2, day 9; *juan* 951, Qianlong year 39, month 1, day 28.

171. *Da Qing Gaozong Chun Huangdi shilu*, *juan* 1051, Qianlong year 43, month 2, day 27.

172. *Da Qing Gaozong Chun Huangdi shilu*, *juan* 1053, Qianlong year 43, month 3, day 30.

173. Quoted in Yang Danxia, "Hongwu de shuhua jiaoliu," 184.

174. *Da Qing Gaozong Chun Huangdi shilu*, *juan* 1081, Qianlong year 44, month 4, day 28.

175. Amiot to Bertin, September 1, 1788, MS 1517, IF.

176. Li Yu'an and Huang Zhengyu, eds., *Zhongguo cangshujia tongdian* [Encyclopedia of Chinese book collectors] (Hong Kong: China International Culture Press, 2005): 462–63.

177. Li Junzhi, *Qing huajia shishi*, 丁上 [A poetic history of Qing artists] (Li Shuzhi, 1906): 2b.

178. Yang Danxia, "Hongwu de shuhua jiaoliu," 190.

179. Yongzhong 永忠; Arthur W. Hummel, ed., *Eminent Chinese of the Ch'ing Period (1644–1912)*, vol. 2 (Washington, DC: United States Government Printing Office, 1943): 962.

180. Ji Yun 紀昀, *Yuewei caotang biji* [Notes from the Thatched Cottage of Close Observation] (1800), ed. Bei Yuan 北原 et al. (Beijing: Zhongguo huaqiao chubanshe, 1994): *juan* 8.

181. Yongzhong, *Yanfen shi gao*, MS 25573, National Library of China, Rare Books (NLC), Beijing, China.

182. Na Yancheng 那彥成, Peng Yunmei 彭芸楣.

183. Compare *Ka'erka gong xiang ma tu juan* 喀爾喀貢象馬圖卷 [Drawing of the Khalkha Mongols tribute of horses and elephants] with *Ka'erka gong xiang ma tu* 喀爾喀貢象馬圖 [Illustration of the Khalkha Mongols tribute of horses and elephants]; Yang Danxia, "Hongwu de shuhua jiaoliu," 193.

184. Yang Danxia, "Hongwu de shuhua jiaoliu," 185–86.

185. 太古不知，名素心能，獨守月明，風露寒自，養靜中壽; Hongwu, *Yaohua Daoren shichao* [Poetry collection of the Daoist of Illustrious Jade], *juan* 10, MS 93051 [edition A], NLC.

186. Hongwu, *Yaohua Daoren shichao* [Poetry collection of the Daoist of Illustrious Jade], *juan* 9, MS 93051 [edition A], NLC; Hongwu, *Yaohua Daoren shichao* [Poetry collection of the Daoist of Illustrious Jade], *ce* 2, *juan* 5, MS 25072 (edition B), NLC.

187. "Yu the Great Controlling the Waters," "Da Yu zhi shui tu" 大禹治水圖.

188. Yaohua Daoren 瑤華道人.

189. Elliott, *The Manchu Way*, 9, 240.

190. Witek, "Manchu Christians and the Sunu Family," 446–48.

191. Daqudeng Alley 大取燈衚衕; Zhang Zhidong et al., *Guangxu Shuntian fu zhi* [Guangxu period gazetteer of the Capital prefecture] (1886): *juan* 13, *jingshi zhi* 13.

192. Bourgeois to Bertin, October 10, 1788, MS Bréquigny 2, BnF.

193. Amiot to Bertin, November 16, 1789, MS 1517, IF.

194. Amiot to Bertin, October 16, 1790, MS 1517, IF.

195. Nicolas-Joseph Raux to François Desvoyes, November 12, 1789, MS 1518, "Lettres de divers correspondants," IF.

196. Amiot to Bertin, November 16, 1789, MS 1517, IF.

197. Amiot to Bertin, August 20, 1790 / October 4, 1790, MS 1517, IF.

198. Amiot to Bertin, September 1, 1788, MS 1517, IF.

199. Bourgeois to Bertin, October 10, 1788, MS Bréquigny 2, BnF; Amiot to Bertin, October 1, 1788, MS 1517, IF; Amiot to Bertin, October 16, 1790, MS 1517, IF.

200. Amiot to Bertin, November 11, 1788, MS Bréquigny 2, BnF.

201. Agui 阿桂; Amiot to Bertin, January 25, 1787, MS 1516, IF.

202. Heshen 和珅.

203. François Bourgeois, "Catalogue des Objets envoyé de Peking en Novembre 1788," MS Bréquigny 2, BnF.

204. Waley-Cohen, "China and Western Technology," 1541–42; Rowe, "Introduction: The Significance of the Qianlong-Jiaqing Transition in Qing History," 77.

205. Amiot to Bertin, November 16, 1789, MS 1517, IF.

206. Amiot to Bertin, November 16, 1789, MS 1517, IF.

207. Amiot to Bertin, August 20, 1790 / October 4, 1790, MS 1517, IF.

208. Illustrations of Qing victories, *Desheng tu* 得勝圖.

209. Amiot to Bertin, January 25, 1787, MS 1516, IF.

210. Amiot to Bertin, August 20, 1790 / October 4, 1790, MS 1517, IF

211. Joanna Waley-Cohen, "Commemorating War in Eighteenth-Century China," *Modern Asian Studies* 30, no. 4 (1996): 869–99, at 891.

212. *Shi Quan Laoren* 十全老人, Waley-Cohen, "Commemorating War in Eighteenth-Century China," 869, 896.

213. *Baxun wanshou shengdian* [Grand celebration of the eightieth birthday], *juan* 52.

214. Amiot to Bertin, August 20, 1790 / October 4, 1790, MS 1517, IF.

215. Amiot to Bertin, November 15, 1784, MS 1516, IF.

216. Amiot to Bertin, November 11, 1788, MS Bréquigny 2, BnF.

217. Elman, *On Their Own Terms*, xxviii; Sivin, "Copernicus in China," 66.

218. Bruno Belhoste, *Paris savant: parcours et rencontres au temps des Lumières* (Paris: A. Colin, 2011).

219. Henri Bertin to Jean-Paul Collas, December 31, 1780, MS 1523, IF.

220. Bertin to Amiot, December 21, 1785, MS 1524, IF, Bertin to Amiot, December 20, 1787, MS 1524, IF.

221. Golvers, *Libraries of Western Learning for China.*

222. Bertin to Amiot, December 21, 1785, MS 1524, IF.

223. Amiot to Bertin, October 1, 1774, MS 1515, IF.

224. Amiot to Bertin, October 16, 1790, MS 1517, IF.

225. Bourgeois to Bertin, November 19, 1784, MS 1520, IF.

226. Amiot to Bertin, November 11, 1788, MS Bréquigny 2, BnF.

227. Amiot to Bertin, January 25, 1787, MS 1516, IF.

228. Amiot to Bertin, November 11, 1788, MS Bréquigny 2, BnF.

229. Bourgeois to Bertin, November 13, 1786, MS 1519, IF.

230. Bourgeois to Bertin, November 13, 1786, IF, MS 1519.

231. Joseph-Marie Amiot to Roze, October 20, 1784, MS Vivier I, AJPF.

232. Amiot to Bertin, October 16, 1790, MS 1517, IF.

233. Bertin to Amiot, December 21, 1785, MS 1524, IF.

234. Amiot to Bertin, November 15, 1784, MS 1516, IF; extracts also published in Amiot, "Extrait d'une autre lettre écrite de Péking le 15 nov. 1784," in Amiot et al., MCC, vol. 11 (1786): 569–76.

235. Harriet T. Zurndorfer, "Comment la science et la technologie se vendaient à la Chine au XVIIIe siècle," *Études chinoises* 7, no. 2 (1988): 59–90.

236. Mary Terrall, *The Man Who Flattened the Earth: Maupertuis and the Sciences in the Enlightenment* (Chicago: University of Chicago Press, 2002).

237. Amiot to Bertin, November 15, 1784, MS 1516, IF.

238. Amiot, "Extrait d'une lettre écrite de Péking le 2 oct. 1784," 528.

239. Jacques Roger, *Buffon: A Life in Natural History*, trans. Sarah Lucille Bonnefoi (Ithaca, NY: Cornell University Press, 1997): 406–10.

240. Amiot to Bertin, November 15, 1784, MS 1516, IF.

241. Bertin to Amiot, December 21, 1785, MS 1524, IF.

242. Amiot to Bertin, January 25, 1787, MS 1516, IF

243. Amiot to Bertin, September 1, 1788, MS 1517, IF.

244. Mathurin Jacques Brisson to Henri Bertin, September 7, 1764, MS 1520, IF.

245. Henri Bertin to Étienne Yang, November 30, 1777, MS 1522, IF; Bertin to Collas, December 31, 1780, MS 1519, IF.

246. Raux to Bertin, November 27, 1786, MS Bréquigny 3, "Mélanges sur la Chine et les Chinois," BnF.

247. Raux to Bertin, November 17, 1786, MS 1518, IF.

248. Bourgeois to Bertin, November 13, 1786, MS 1519, IF.

249. Raux to Bertin, November 27, 1786, MS Bréquigny 3, BnF

250. Bourgeois to Bertin, November 19, 1784, MS 1520, IF.

251. Bertin to Bourgeois, December 25, 1787, MS 1524, IF.

252. Bertin to Amiot, December 21, 1785, MS 1524, IF.

253. Bertin to Amiot, December 20, 1787, MS 1524, IF.

254. Amiot to Bertin, October 10, 1789, MS 1517, IF.

255. Earl George Macartney and Sir John Barrow, *Some Account of the Public Life, and a Selection from the Unpublished Writings, of the Earl of Macartney. . . . Journal of an Embassy from the King of Great Britain to the Emperor of China* (London: T. Cadell and W. Davies, 1807): vol. 1, 349; Waley-Cohen, "China and Western Technology," 1537.

256. Henrietta Harrison, "Chinese and British Diplomatic Gifts in the Macartney Embassy of 1793," *English Historical Review* 133, no. 560 (2018): 65–97, at 82–85; Waley-Cohen, "China and Western Technology," 1541–44.

257. Macartney and Barrow, *Journal of an Embassy*, vol. 2, 484.

258. Elman, *On Their Own Terms*, xxxii; Jonathan Porter, "The Scientific Community in Early Modern China," *Isis* 73, no. 4 (1982): 529–44, at 532.

259. Ruan Yuan 阮元, ed., *Chouren zhuan* [Biographies of mathematicians and astronomers] (Yangzhou: Wenxuan lou, 1799); Joseph Needham, *Science and Civilisation in China: Volume I: Introductory Orientations* (Cambridge: Cambridge University Press, 1954): 50.

260. Ruan, *Chouren zhuan, juan* 49; Ori Sela, "Confucian Scientific Identity: Qian Daxin's (1728–1804) Ambivalence Toward Western Learning and its Adherents," *East Asian Science, Technology, and Society* 6 (2012): 147–66, at 161–62; Minghui Hu,

China's Transition to Modernity: The New Classical Vision of Dai Zhen (Seattle: University of Washington Press, 2015): 134.

261. Zhu Gui 朱珪; Betty Peh-T'i Wei, *Ruan Yuan, 1764–1849: The Life and Work of a Major Scholar-Official in Nineteenth-Century China before the Opium War* (Hong Kong: Hong Kong University Press, 2006): 33.

262. *Wanshou si* 萬壽寺.

263. The event is recounted in Shen Chu 沈初, *Lanyuntang shiwen ji* [Collection of poems and texts from the Hall of Orchid Rhyme] (Qianlong Period): *shiji, juan* 11; and Ruan Yuan, *Xiaocanglang bitan* [Brush talk of the Blue Wave Tributary] (Yangzhou: Wenxuan lou, 1842): *juan* 2.

264. Yang Zhongshi, *Xueqiao shihua* [Snow Bridge poetry talks] (Qiushuzhai congshu: Early twentieth century):*yuji, juan* 5.

265. Bourgeois to Bertin, November 19, 1784, MS 1520, IF.

266. The extent to which alchemy and Daoism were historically linked is controversial (Nathan Sivin, "Taoism and Science," in *Medicine, Philosophy and Religion in Ancient China: Researches and Reflections* [Aldershot: Variorum, 1995]: 1–73).

Chapter 3

1. Peter Gay, *The Enlightenment: An Interpretation*, vol. 2, reprint ed. (New York: W. W. Norton, 1995): 606–08.

2. Paul Hazard, *The Crisis of the European Mind, 1680–1715*, trans. J. Lewis May (New York: New York Review Books, 2013), xiii.

3. The classic study is Daniel Pickering Walker, *The Ancient Theology: Studies in Christian Platonism from the Fifteenth to the Eighteenth Century* (Ithaca, NY: Cornell University Press, 1972); see also Frances Amelia Yates, *Giordano Bruno and the Hermetic Tradition* (Chicago: University of Chicago Press, 1964).

4. Dmitri Levitin, *Ancient Wisdom in the Age of the New Science* (Cambridge: Cambridge University Press, 2015): 8–10.

5. Urs App, *The Birth of Orientalism: Encounters with Asia* (Philadelphia: University of Pennsylvania Press, 2010): 254–66; John Greville Agard Pocock, *Barbarism and Religion, Volume 4: Barbarians, Savages and Empires* (New York: Cambridge University Press, 2005): 99–132.

6. Daniel Stolzenberg, *Egyptian Oedipus: Athanasius Kircher and the Secrets of Antiquity* (Chicago: University of Chicago Press, 2013): 6–8; Levitin, *Ancient Wisdom in the Age of the New Science*, 166–70.

7. Simon Schaffer, "The Asiatic Enlightenments of British Astronomy," in *The Brokered World: Go-Betweens and Global Intelligence, 1770–1820*, eds. Simon Schaffer, Lissa Roberts, Kapil Raj, and James Delbourgo (Sagamore Beach, MA: Science History Publications, 2009): 49–104, at 84–86, 97.

8. Levitin, *Ancient Wisdom in the Age of the New Science*, 193–95.

9. The classic study is Carl L. Becker, *The Heavenly City of the Eighteenth-Century Philosophers* (1932), reprint ed. (New Haven, CT: Yale University Press, 2003); see also David Spadafora, *The Idea of Progress in Eighteenth-Century Britain* (New Haven, CT: Yale University Press, 1990): 85–132.

10. Peter Gay, *The Enlightenment: An Interpretation*, vol. 1, reprint ed. (New York: W. W. Norton, 1995): 33–34; Dan Edelstein, *The Enlightenment: A Genealogy* (Chicago: University of Chicago Press, 2010): 1–6.

11. Cornelius de Pauw, *Recherches philosophiques sur les égyptiens et les chinois* (Berlin: C. J. Decker, 1773–1774): vol. 1, 15.

12. *Comprehensive Mirror for the Aid of Government, Zizhi tongjian* 資治通鑑; Joseph-Anne-Marie de Moyriac de Mailla, trans., *Histoire générale de la Chine, ou annales de cet Empire, traduites du Tong-kien-kang-mou*, ed. Michel-Ange le Roux Deshauterayes (Paris: Pierres and Clousier, 1777–1785); Antoine Gaubil, trans., *Le Chou-king: Un des livres sacrés des Chinois, qui renferme les fondements de leur ancienne histoire, les principes de leur gouvernement & de leur morale*, ed. Joseph de Guignes (Paris: N. M. Tilliard, 1770).

13. Nicolas Standaert, "Jesuit Accounts of Chinese History and Chronology and Their Chinese Sources," *East Asian Science, Technology, and Medicine* 35 (2013): 11–88, at 12.

14. Pierre-Martial Cibot, Aloys Ko, and Étienne Yang, "Essai sur l'antiquité des Chinois," in Joseph-Marie Amiot, François Bourgeois, Pierre-Martial Cibot, Aloys Ko, Aloys de Poirot, Charles Batteux, M. de Bréquigny, A. I. Silvestre de Sacy, and Antoine Gaubil, *Mémoires concernant l'histoire, les sciences, les arts, les moeurs, les usages, &c. des Chinois* [*MCC*], vols. 1–15 (Paris: Nyon l'aîné, 1776–1791): vol. 1 (1776), 1–271; Joseph-Marie Amiot, "Antiquité des Chinois, prouvée par les monuments," in Amiot et al, *MCC*, vol. 2 (1776): 1–364.

15. Jean-Louis Coster, ed., *L'Ésprit des journaux, françois et étrangers*, vol. 4 (Paris: L'Imprimerie du Journal, April 1777): 33.

16. Henri Bertin to Pierre-Martial Cibot, November 30, 1777, MS 1522, "36 Lettres de Mr. Bertin aux Missionnaires en Chine, 1773–1778," Institut de France (IF), Paris, France; Henri Bertin to François Bourgeois, November 30, 1777, MS 1522, IF.

17. Wu Huiyi, *Traduire la Chine au XVIIIe siècle: Les jésuites traducteurs de textes chinois et le renouvellement des connaissances européennes sur la Chine (1687–ca. 1740)* (Paris: Honoré Champion, 2017): 35.

18. Voltaire, *La Philosophie de l'histoire* (1764), *Œuvres complètes* (Oxford: Voltaire Foundation, 1968): 152, ARTFL Encyclopédie Projet, ed. Robert Morrissey, http://encyclopedie.uchicago.edu/.

19. Bertin to Bourgeois, November 30, 1777, MS 1522, IF.

20. Cibot et al., "Essai sur l'antiquité des Chinois," 23; Amiot, "Antiquité des Chinois," 1.

21. Cibot et al., "Essai sur l'antiquité des Chinois," 157–58; Amiot, "Antiquité des Chinois," 195–245.

22. Joseph Dehergne, "Une grande collection: Mémoires concernant les Chinois (1776–1814)," *Bulletin de l'École française d'Extrême-Orient* 72, no. 1 (1983): 267–98, at 270.

23. Michel-Ange le Roux Deshauterayes, "Observations," in de Mailla, *Histoire générale de la Chine*, vol. 1, lxvi.

24. Standaert, "Jesuit Accounts of Chinese History," 66–70.

25. "Histoire générale de la Chine," MS NAF 2491, "Deshauterayes I, Papiers Intimes / l'Apocalypse / Dissertations," and MS NAF 2492, "Deshauterayes II, Histoire générale de la Chine," Bibliothèque nationale de France (BnF), Paris, France.

26. Gaubil, trans., *Le Chou-king*.

27. "Letter on the *Chou-king*, 1771 [?]," in MS 5401, Fonds Henri Cordier, "Collection des Autographes," IF.

28. Joseph-Marie Amiot to Henri Bertin, November 5, 1778, MS 1515, IF.

29. Amiot, "Antiquité des Chinois," 8.

30. Deshauterayes, "Observations," in de Mailla, *Histoire générale de la Chine*, vol. 1, l.

31. Cibot et al., "Essai sur l'antiquité des Chinois," 40–41.

32. Joseph de Guignes, "Preface," in *Le Chou-king: Un des livres sacrés des Chinois, qui renferme les fondements de leur ancienne histoire, les principes de leur gouvernement & de leur morale*, ed. Joseph de Guignes, trans. Antoine Gaubil (Paris: N. M. Tilliard, 1770): i–xliii; Gaubil, *Le Chou-king*, xxvi.

33. Joseph de Guignes, "Examen critique des annales chinoises, ou Mémoire sur l'incertitude des douze premiers siècles de ces annales, et de la chronologie chinoise," in *Mémoires de l'Académie des inscriptions*, vol. 36 (1769): 164–89, at 183.

34. Michel-Ange le Roux Deshauterayes, "Extraits des historiens chinois," in *De l'origine des lois, des arts, et des sciences, et de leurs progrès chez les anciens peoples*, vol. 3 (1758), ed. Antoine-Yves Goguet, reprint ed. (Paris: L. Haussman and d'Hautel, 1809): 258; Frederic Clark, *The First Pagan Historian: The Fortunes of a Fraud from Antiquity to the Enlightenment* (New York: Oxford University Press, 2020): 187.

35. Amiot, "Antiquité des Chinois," 9.

36. Sima Qian 司馬遷, *Shiji* 史記, *Records of the Grand Historian*; Cibot et al., "Essai sur l'antiquité des Chinois," 134–35.

37. Cibot et al., "Essai sur l'antiquité des Chinois," 127; Amiot, "Antiquité des Chinois," 315–16.

38. Nickolas Panayiotis Roubekas, *An Ancient Theory of Religion: Euhemerism from Antiquity to the Present* (New York: Routledge, 2017).

39. Michel-Ange le Roux Deshauterayes to Anquetil-Duperron, October 8, 1755, MS NAF 8872, "Correspondance d'Anquetil-Duperron," BnF.

40. Joseph-Marie Amiot to Jérôme-Frédéric Bignon, December 20, 1769, MS Bréquigny 7, BnF.

41. Michel Benoist, "Lettre du Père Benoist," [1772], in *Lettres édifiantes et curieuses*, vol. 4, ed. Louis Aimé-Martin (Paris: Société du Panthéon Littéraire, 1843): 209–17, at 215.

42. Pocock, *Barbarism and Religion*, vol. 4, 99–155; App, *Birth of Orientalism*, 188–253.

43. Anonymous, "Idée générale de la Chine et de ses relations avec l'Europe sur le nom de Chine," in Amiot et al., *MCC*, vol. 5 (1780): 1–68, at 53.

44. Cibot et al., "Essai sur l'antiquité des Chinois," 93, 215; Amiot, "Antiquité des Chinois," 6–22.

45. Henri Bertin to Aloys Ko and Étienne Yang, December 2, 1764, MS 1521, "24 Lettres de M. Bertin aux Missionnaires en Chine, 1764–1772," IF.

46. Florence Hsia, "Athanasius Kircher's *China Illustrata* (1667)," in *Athanasius Kircher: The Last Man Who Knew Everything*, ed. Paula Findlen (New York: Routledge, 2004): 383–404, at 385–90.

47. Janine Hartman, "Ideograms and Hieroglyphs: The Egypto-Chinese Origins Controversy in the Enlightenment," *Dalhousie French Studies* 43 (1998): 101–18.

48. App, *Birth of Orientalism*, 207.

49. Joseph de Guignes, *Mémoire dans lequel on prouve que les Chinois sont une colonie égyptienne* (Paris: Desaint & Saillant, 1759): 37–38.

50. Joseph de Guignes, "Essai historique sur l'étude de la philosophie chez les

anciens Chinois," in *Mémoires de l'Académie des inscriptions*, vol. 38 (1777): 269–311, at 279–81.

51. De Guignes, *Mémoire dans lequel on prouve que les Chinois sont une colonie égyptienne*, 77–78.

52. App, *Birth of Orientalism*, 207–13.

53. Joseph-Marie Amiot, "Lettre sur les caractères chinois," (1764), in Amiot et al., *MCC*, vol. 1 (1776): 275–324.

54. Voltaire, "Article 4," in *Fragment sur l'histoire générale* (1773), *Œuvres complètes*.

55. De Pauw, *Recherches philosophiques sur les égyptiens et les chinois*, vol. 2, 190–92.

56. De Pauw, *Recherches philosophiques sur les égyptiens et les chinois*, vol. 2, 243.

57. De Pauw, *Recherches philosophiques sur les égyptiens et les chinois*, vol. 1, 368.

58. Pierre-Martial Cibot and Aloys Ko, "Remarques sur un écrit de M. de Pauw intitulé, Recherches sur les Égyptiens et les Chinois," in Amiot et al., *MCC*, vol. 2 (1776), 365–574; Joseph-Marie Amiot, "Observations sur le livre de M. de Pauw, intitulé *Recherches philosophiques sur les Égyptiens et les Chinois*" (1777), in Amiot et al., *MCC*, vol. 6 (1780): 275–345.

59. Voltaire, "Article 4," in *Fragment sur l'histoire générale* (1773), *Œuvres complètes*.

60. Bruno Belhoste, *Paris savant: parcours et rencontres au temps des Lumières* (Paris: A. Colin, 2011): 103.

61. Anne-Marie Mercier-Faivre, *Un supplément à L'Encyclopédie: Le "Monde primitif" d'Antoine Court de Gébelin, suivi d'une édition du "Génie allégorique et symbolique de l'Antiquité" extrait du "Monde primitif" (1773)* (Paris: Honoré Champion, 1999): 20–35.

62. Mercier-Faivre, *Un supplément à L'Encyclopédie*, 102.

63. Antoine Court de Gébelin, *Monde primitif, analysé et comparé avec le monde moderne, considéré dans son génie allégorique, et dans les allégories auxquelles conduis it ce génie précédé du plan général* (Paris: Court de Gébelin, 1773): 2.

64. Court de Gébelin, *Monde primitif . . . considéré dans son génie allégorique*, 258.

65. Charles Porset, *Franc-Maçonnerie et religions dans l'Europe des Lumières* (Paris: H. Champion, 2006): 64.

66. Antoine Court de Gébelin, *Lettre de l'auteur du monde primitif, à Messieurs ses souscripteurs, sur le magnétisme animal* (Paris: Gastelier, 1784): 45.

67. Mercier-Faivre, *Un supplément à L'Encyclopédie*, 102–08.

68. Porset, *Franc -Maçonnerie et religions*, 65; Mercier-Faivre, *Un supplément à L'Encyclopédie*, 82–91.

69. Court de Gébelin, *Monde primitif . . . considéré dans son génie allégorique*, 94.

70. Court de Gébelin, *Monde primitif . . . considéré dans son génie allégorique*, 98–99.

71. Darrin McMahon, *Enemies of the Enlightenment: The French Counter-Enlightenment and the Making of Modernity* (Oxford: Oxford University Press, 2001): 1–53.

72. Mercier-Faivre, *Un supplément à L'Encyclopédie*, 102.

73. Louis Amiable, *La Loge des Neuf sœurs*, ed. Charles Porset (Paris: Edimaf, 1989): 245–353.

74. Belhoste, *Paris savant*, 121–25.

75. *Catalogue des Livres de Feu M. Court de Gébelin, de la Société Économique de Berne, de l'Académie Royale de la Rochelle, Président honoraire du Musée de Paris, etc.* (Chez Musier: maison de M. Didot l'aîné, imprimeur, 1786), MS DELTA 11947, BnF.

76. Antoine Court de Gébelin to Macquer, February 20, 1782, MS Française 12305, "Correspondance Macquer," BnF; Antoine Court de Gébelin to Louis-Georges de Bréquigny, December 22, 1783, MS Bréquigny 158, "Correspondance de Bréquigny I," BnF.

77. Auguste Viatte, *Les sources occultes du Romantisme*, vol. 1 (Paris: Honoré Champion, 1928): 186.

78. Jean-Paul Rabaut Saint-Étienne to Antoine Court de Gébelin, July 26, 1782, MS 318, "Papiers Paul Rabaut, Lettres à Court de Gébelin," Bibliothèque de la Société de l'Histoire du Protestantisme français (BSHPF), Paris, France.

79. Amiable, *La Loge des Neuf sœurs*, 101.

80. "Lettre de M. l'Abbé de Beaulieu," in MS 367, "Lettres de Rabaut-Saint-Étienne à Court de Gébelin," BSHPF.

81. "Correspondance de Bréquigny I," MS Bréquigny 158, BnF; "Papiers Court de Gébelin," MS 361, BSHPF.

82. Bertin to Ko and Yang, December 31, 1774, MS 1522, IF.

83. Antoine Court de Gébelin to Henri Bertin, 1779, MS 1518, "Lettres de divers correspondants," IF.

84. Amiable, *La Loge des Neuf sœurs*, 273; Court de Gébelin, *Monde primitif . . . considéré dans son génie allégorique*, 98.

85. Bertin to Cibot, December 20, 1778, MS 1522, IF.

86. Pierre-Martial Cibot to Henri Bertin, November 10, 1777, MS 1519, "Lettres de plusieurs missionnaires," IF.

87. *Shou* 壽; Amiot to Bertin, September 28, 1777, MS 1515, "80 lettres du P. Amiot à Bertin," IF.

88. *Yu bei* 禹碑; also called the *Goulou feng bei* 岣嶁峰碑 after the peak on which it was placed; "Copie du monument que le grand yu éleva sur la Montagne hengchan . . . , 1777," MS Chinois 1170, BnF.

89. *Hengshan* 衡山.

90. Amiot to Bertin, September 28, 1777, MS 1515, IF.

91. *Kedou wen* 蝌蚪文.

92. Jonathan Chaves, "Still Hidden by Spirits and Immortals: The Quest for the Elusive 'Stele of Yu the Great,'" *Asia Major* 3rd series 26, no. 1 (2013): 1–22; Bruce Rusk, *Goulou Feng Bei: Stele of Goulou Peak, Ink Squeeze of Mao Huijian's Stele of King Yu of 1666* (Melbourne: Quirin Press, 2016).

93. "Yu gong" 禹貢.

94. Qiong Zhang, *Making the New World Their Own: Chinese Encounters with Jesuit Science in the Age of Discovery* (Leiden: Brill, 2015): 203–63.

95. Wang Chang 王昶; *kaozheng* 考證. Arthur W. Hummel, ed., *Eminent Chinese of the Ch'ing Period (1644–1912)*, vol. 2 (Washington, DC: United States Government Printing Office, 1943): 805–07.

96. Wang Chang, *Jinshi cuibian* [Compendium of seals and stones] (Jingxun tang: 1805): *juan* 2.

97. Yang Danxia, *Hongwu de shuhua jiaoliu yu chuangzuo* [The exchange and creation of Hongwu's painting and calligraphy], *Forbidden City* (Beijing: Palace Museum, 2005): 184. For Wang's poems mentioning Hongwu, see Wang Chang 王昶, *Chunrongtang ji* [Collection of the Hall of the Spring Thaw] (Shunan shushe: 1808): *juan* 22.

98. Hummel, ed., *Eminent Chinese of the Ch'ing Period*, vol. 2, 806.

99. Jinchuan 金川; Joseph-Marie Amiot, “Lettre sur la réduction des Miao-tsée en 1775,” in Amiot et al., *MCC*, vol. 3 (1778): 387–411.

100. Amiot to Bertin, September 28, 1777, MS 1515, IF.

101. Mao Huijian 毛会建.

102. Xi’an Stele Forest, Xi’an *beilin* 西安碑林.

103. Amiot to Bertin, September 28, 1777, MS 1515, IF.

104. Amiot to Bertin, September 28, 1777, MS 1515, IF.

105. Henri Bertin to Joseph-Marie Amiot, December 15, 1779, MS 1523, “28 Lettres de Mr. Bertin aux Missionnaires en Chine, 1779–1782,” IF.

106. Henri Bertin to Chompré [1779?], MS 1518, IF.

107. Ronald Decker, Thierry Depaulis, and Michael Dummett, *A Wicked Pack of Cards: The Origins of the Occult Tarot* (London: Duckworth, 1996): 23–51.

108. Decker, Depaulis, and Dummett, *A Wicked Pack of Cards*, 52–73.

109. Antoine Court de Gébelin, *Monde primitif, analysé et comparé avec le monde moderne, considéré dans divers objets concernant l’histoire, le blason, les monnoies, les jeux, les voyages des Phéniciens autour du monde, les langues Américaines, &c. ou dissertations mêlées* (Paris: Court de Gébelin, 1781): 365.

110. Court de Gébelin, *Monde primitif . . . Dissertations mêlées*, 394.

111. Court de Gébelin, *Monde primitif . . . Dissertations mêlées*, 367.

112. Louis-Raphaël-Lucrèce de Fayolle, comte de Mellet, “Recherches sur les Tarots, et sur la divination par les cartes des Tarots” (1781), in *Monde primitif . . . Dissertations mêlées*, ed. Court de Gébelin (Paris: Court de Gébelin, 1781): 395–410, at 396–400.

113. Court de Gébelin, *Monde primitif . . . Dissertations mêlées*, 383–84.

114. Frank Palmeri, *State of Nature, Stages of Society: Enlightenment Conjectural History and Modern Social Discourse* (New York: Columbia University Press, 2016): 1.

115. Court de Gébelin, *Monde primitif . . . Dissertations mêlées*, 394.

116. Mellet, “Recherches sur les Tarots,” 404.

117. Mellet, “Recherches sur les Tarots,” 406–07.

118. Mellet, “Recherches sur les Tarots,” 404.

119. Mellet, “Recherches sur les Tarots,” 405.

120. Mellet, “Recherches sur les Tarots,” 408.

121. Court de Gébelin, *Monde primitif . . . Dissertations mêlées*, 379–80.

122. Court de Gébelin, *Monde primitif . . . Dissertations mêlées*, 367; Antoine Court de Gébelin to Joseph-Marie Amiot, January 1, 1780, MS 1518, IF.

123. Court de Gébelin, *Monde primitif . . . Dissertations mêlées*, 380.

124. In fact, the French word *tarot* is likely derived from the Italian *tarocchi*, which probably came from *trionfi*, or “triumph,” the source of the English word “trump” (Decker, Depaulis, and Dummett, *A Wicked Pack of Cards*, 41).

125. Court de Gébelin, *Monde primitif . . . Dissertations mêlées*, 387–88.

126. Court de Gébelin to Amiot, January 1, 1780, MS 1518, IF.

127. Antoine Court de Gébelin to Jean-Paul Rabaut Saint-Étienne, January 6, 1780, MS 316, “Papiers Paul Rabaut, Lettres 1777–1783,” BSHPF.

128. MS NAF 279, “De Guignes, papiers divers,” BnF.

129. Court de Gébelin to Amiot, January 1, 1780, MS 1518, IF.

130. Bertin to Chompré [1779?], MS 1518, IF.

131. Antoine Court de Gébelin to Jean-Paul Rabaut Saint-Étienne, January 6, 1780, MS 316, “Papiers Paul Rabaut, Lettres 1777–1783,” BSHPF.

132. The stele that Mao Huijian inscribed and placed in the Xi'an Stele Forest seems to be the only one arranged in six columns of multiples of seven (Chen Zhongkai et al. eds., *Xi'an beilin bowuguan cang beike zongmu tiyao* [Annotated catalogue of the stele inscriptions in the collection of the Xi'an Stele Forest Museum] [Beijing: Xi'an zhuang shuju, 2006]: 36).

133. Amiot to Bertin, August 17, 1781, MS 1516, "80 lettres du P. Amiot à Bertin," IF.

134. Court de Gébelin, *Monde primitif . . . Dissertations mêlées*, 387–88.

135. Court de Gébelin to Rabaut Saint-Étienne, January 6, 1780, MS 316, BSHPF.

136. Amiot to Bertin, August 17, 1781, MS 1516, IF.

137. Amiot to Bertin, October 20, 1782, MS 438, "Chine, Lettres, 1780–1787," Archives des Missions Étrangères de Paris, France.

138. Amiot to Bertin, November 15, 1784, MS 1516, IF.

139. Amiot to Bertin, October 20, 1782," MS 438, Archives des Missions Étrangères de Paris.

140. Dan Edelstein, "Jean-Sylvain Bailly (1736–1793)," *The Super-Enlightenment: A Digital Archive,* accessed July 18, 2022, https://exhibits.stanford.edu/super-e/feature/jean-sylvain-bailly-1736-1793; Dan Edelstein, "Introduction to the Super-Enlightenment," in *The Super-Enlightenment: Daring to Know Too Much,* ed. Dan Edelstein (Oxford: Voltaire Foundation, 2010): 1–34, at 26–27.

141. Jean-Antoine-Nicolas de Caritat, marquis de Condorcet, "À Voltaire" (1777), in *Œuvres complètes de Condorcet*, eds. F. Condorcet O'Connor and F. Arago (Paris: Firmin Didot, 1847–1849): vol. 1, 147–49, at 148.

142. Jean-Sylvain Bailly, *Lettres sur l'Atlantide de Platon et sur l'ancienne histoire de l'Asie: pour servir de suite aux Lettres sur l'origine des sciences, adressées à M. de Voltaire par M. Bailly* (London: M. Elmesly; Paris: les frères de Bure, 1779): 28.

143. Jean-Sylvain Bailly and Voltaire, *Lettres sur l'origine des sciences et sur celle des peuples de l'Asie: Adressées à M. de Voltaire Par M. Bailly, & précédées de quelques lettres de M. de Voltaire à l'auteur* (London: M. Elmesly; Paris: de Bure l'aîné, 1777): 4; Bailly, *Lettres sur l'Atlantide de Platon, Avertissement.*

144. Bailly and Voltaire, *Lettres sur l'origine des sciences*, 93–97.

145. Bailly and Voltaire, *Lettres sur l'origine des sciences*, 187–88.

146. Bailly and Voltaire, *Lettres sur l'origine des sciences*, 156–63.

147. Bailly, *Lettres sur l'Atlantide de Platon*, 403–12.

148. Bailly may have coined the geographical use of the word *plateau* in this context: "*Nous n'avons point de mot pour nommer ces grandes terres, élevées sur d'autres terres comme nos collines sur nos petites plaines; Permettez-moi de désigner ces terres sous le nom de* plateau" (Bailly, *Lettres sur l'Atlantide de Platon*, 218).

149. Bailly, *Lettres sur l'Atlantide de Platon*, 413.

150. Bailly, *Lettres sur l'Atlantide de Platon*, 25.

151. Bailly and Voltaire, *Lettres sur l'origine des sciences*, 205.

152. Bailly, *Lettres sur l'Atlantide de Platon*, 417.

153. Dan Edelstein, "Hyperborean Atlantis: Jean-Sylvain Bailly, Madame Blavatsky, and the Nazi Myth," *Studies in Eighteenth-Century Culture* 35 (2006): 267–91; David Allen Harvey, "The Lost Caucasian Civilization: Jean-Sylvain Bailly and the Roots of the Aryan Myth," *Modern Intellectual History* (2014): 279–306.

154. Amiable, *La Loge des Neuf sœurs*, 248.

155. Joseph-Marie Amiot to Louis-Georges de Bréquigny, June 26, 1789, MS Bréquigny 2, "Mélanges sur la Chine et les Chinois," BnF; "Lettres de Rabaut-Saint-

Étienne à Court de Gébelin," MS 367, BSHPF; Jean-Paul Rabaut Saint-Étienne, *Lettres sur l'histoire primitive de la Grèce* (Paris: de Bure l'aîné, 1787): 51; Jean-Paul Rabaut-Saint-Étienne, *Lettres à Monsieur Bailly sur l'histoire primitive de la Grèce* (Paris: de Bure l'aîné, 1787).

156. Bailly and Voltaire, *Lettres sur l'origine des sciences*, 232.

157. Bailly, *Lettres sur l'Atlantide de Platon*, 70, 272.

158. Bailly and Voltaire, *Lettres sur l'origine des sciences*, 273.

159. Jean-Sylvain Bailly, *Traité de l'astronomie indienne et orientale* (Paris: de Bure l'aîné, 1787): cx. His writings on the topic, over five hundred pages mostly dating to the 1780s, were published in two posthumous volumes, *Essai sur les fables et sur leur histoire* (Paris: de Bure l'aîné, 1799).

160. "*Gin-hoang;*" *Ren huang* 人皇; Bailly, *Traité de l'astronomie indienne et orientale*, cxiii.

161. The definitive biography of Condorcet remains Keith Michael Baker, *Condorcet: From Natural Philosophy to Social Mathematics* (Chicago: University of Chicago Press, 1975).

162. Baker, *Condorcet*, 35–47.

163. Charles Coutel, "Utopie et Perfectibilité: Significations de l'Atlantide chez Condorcet," in *Condorcet: Homme des lumières et de la revolution*, eds. Anne-Marie Chouillet and Pierre Crépel (Fontenat-aux-Roses: ENS Éditions, 1997): 99–107, at 103.

164. Charles Coulston Gillispie, *Science and Polity in France: The Revolutionary and Napoleonic Years* (Princeton, NJ: Princeton University Press, 2004): 67–77.

165. Baker, *Condorcet*, 344–47; Morris Ginsberg, "Progress in the Modern Era," in *Dictionary of the History of Ideas*, vol. 3, ed. Philip P. Wiener (New York: Charles Scribner's Sons, 1973): 664; John Bagnell Bury, *The Idea of Progress: An Inquiry into Its Origin and Growth* (1932), reprint ed. (New York: Dover Publications, 1955): 209.

166. I. Bernard Cohen, "The Eighteenth-Century Origins of the Concept of Scientific Revolution," *Journal of the History of Ideas* 37, no. 2 (1976): 257–88, at 275–79.

167. Bailly and Voltaire, *Lettres sur l'origine des sciences*, 205–06.

168. Bailly and Voltaire, *Lettres sur l'origine des sciences*, 206.

169. Jean-Antoine-Nicolas de Caritat, marquis de Condorcet, Review of Bailly, *Histoire de l'astronomie ancienne* (17 75), *Histoire de l'Académie royale des sciences* (1775): 44–53, at 47–48.

170. Jean-Antoine-Nicolas de Caritat, marquis de Condorcet, "Discours . . . en réponse à celui de M. Bailly" (1784), in *Œuvres complètes de Condorcet*, vol. 10, eds. Condorcet O'Connor and Arago: 429–34, at 431.

171. Condorcet, "Discours . . . en réponse à celui de M. Bailly," 431.

172. Jean-Antoine-Nicolas de Caritat, marquis de Condorcet, *Esquisse d'un tableau historique des progrès de l'esprit humain* (1794), in *Œuvres complètes d e Condorcet*, vol. 6, eds. Condorcet O'Connor and Arago: 1–513, at 58.

173. Bury, *The Idea of Progress*, 8; Charles Frankel, *The Faith of Reason: The Idea of Progress in the French Enlightenment* (New York: Columbia University Press, 1948): 149–51.

174. Bailly and Voltaire, *Lettres sur l'origine des sciences*, 194; Bailly, *Lettres sur l'Atlantide de Platon*, 428.

175. Coutel, "Utopie et Perfectibilité," 100–06.

176. Jean-Antoine-Nicolas de Caritat, marquis de Condorcet, "Fragment sur l'Atlantide, ou efforts combinés de l'espèce humaine pour le progrès des sciences" (1794), in *Œuvres complètes d e Condorcet*, vol. 6, eds. Condorcet O'Connor and Arago: 597–660, at 660.

177. Antoine Lilti, "La civilisation est-elle européenne? Écrire l'histoire de l'Europe au XVIIIe siècle," in *Penser l'Europe au XVIIIe siècle: Commerce, civilisation, empire*, eds. Antoine Lilti and Céline Spector (Oxford: Voltaire Foundation, 2014): 161–66.

178. Bailly, *Lettres sur l'Atlantide de Platon*, 239.

179. Jean-Sylvain Bailly, *Histoire de l'astronomie ancienne: depuis son origine jusqu'à l'établissement de l'école d'Alexandrie* (Paris: les frères de Bure, 1775): 118.

180. Ba illy, *Traité de l'astronomie indienne et orientale*, cx; "Listes des souscripteurs à l'Histoire de la Chine du P. de Mailla en 1777–1783," MS Joseph Bernard Maître 84, Archives jésuites de la province de France (AJPF).

181. Bailly, *Traité de l'astronomie indienne et orientale*, lxxviii.

182. Jean-Sylvain Bailly, *Éloge de Leibnitz, qui a remporté le prix à l'Académie royale des sciences et belles-lettres* (Berlin: Haude and Spener, 1768): 11.

183. Bailly and Voltaire, *Lettres sur l'origine des sciences*, 39.

184. Bailly and Voltaire, *Lettres sur l'origine des sciences*, 25

185. Bailly and Voltaire, *Lettres sur l'origine des sciences*, 30.

186. Bailly and Voltaire, *Lettres sur l'origine des sciences*, 27–28.

187. Dhruv Raina, "Betwixt Jesuit and Enlightenment Historiography: Jean-Sylvain Bailly's History of Indian Astronomy," *Revue d'histoire des mathématiques* 9 (2003): 253–306, at 267–82.

188. enlightened, *éclairé*; Bailly and Voltaire, *Lettres sur l'origine des sciences*, 189–91.

189. Bailly and Voltaire, *Lettres sur l'origine des sciences*, 33–35.

190. Bailly, *Lettres sur l'Atlantide de Platon*, 436.

191. Bailly, *Histoire de l'astronomie ancienne*, 119.

192. Amiot to Bertin, January 25, 1787, MS 1516, IF.

193. Amiot to Bignon, December 20, 1769, MS Bréquigny 7, BnF; this section was mostly omitted from the published version of the letter, which appeared as the "Discours préliminaire" to Joseph-Marie Amiot, "Abrégé chronologique de l'histoire universelle de l'Empire Chinois" (1769), in Amiot et al., *MCC*, vol. 13 (1788): 74–308.

194. Amiot to Bréquigny, June 26, 1789, MS Bréquigny 2, BnF.

195. Amiot to Bertin, January 25, 1787, MS 1516, IF.

196. Joseph-Marie Amiot, "Suite du mémoire sur les danses religieuses, politiques, et civiles des anciens chinois" (1789), in *Les Danses rituelles chinoises d'après Joseph-Marie Amiot: Aux sources de l'ethnochorégraphie*, eds. Yves Lenoir and Nicolas Standaert (Namur: Presses Universitaires de Namur, 2005): 243–59, at 250.

197. Joseph-Marie Amiot to Louis-Raphaël-Lucrèce de Fayolle, comte de Mellet, October 18, 1784, MS 2446, "Copies de lettres à sujet médical rédigés à Pékin adressées à M. Desvoyes pseudonyme de l'abbé Louis Augustin Bertin" [mislabeled], Bibliothèque interuniversitaire de Santé (BIS), Paris, France.

198. Amiot to Bréquigny, June 26, 1789, MS Bréquigny 2, BnF.

199. *Kaozheng* 考證.

200. Qian Daxin 錢大昕; Ruan Yuan 阮元; Benjamin A. Elman, *From Philosophy to Philology: Intellectual and Social Aspects of Change in Late Imperial China* (Cam-

bridge, MA: Council on East Asian Studies, Harvard University, 1984): 27–42; Ori Sela, *China's Philological Turn: Scholars, Textualism, and the Dao in the Eighteenth Century* (New York: Columbia University Press, 2018): 101–17.

201. William Theodore de Bary, "Neo-Confucian Cultivation and the Seventeenth-Century 'Enlightenment,'" in *The Unfolding of Neo-Confucianism*, ed. William Theodore de Bary (New York: Columbia University Press, 2019): 141–216, at 141–45.

202. Liu Dun, "Qingchu lisuan dashi Mei Wending" [Mei Wending, early Qing master of astronomical calculation], *Ziran bianzhengfa tongxun* 1 (1986): 60–63.

203. *Xixue Zhongyuan* 西學中源; John B. Henderson, "Ch'ing Scholars' Views of Western Astronomy," *Harvard Journal of Asiatic Studies* 46, no. 1 (1986): 121–48; Elman, *From Philosophy to Philology*, 80–83; Sela, *China's Philological Turn*, 154–61.

204. Ruan Yuan, *Yanjingshi ji* [Collection of the Study of Classics Research], ed. Zhang Yuanji et al. (Shanghai: Commercial Press 1919): *ji* 3, *juan* 5.

205. Han Qi, "Baijin de 'Yijing' yanjiu he Kangxi shidai de 'xixue zhongyuan' shuo" [Bouvet's *Yijing* studies and the Kangxi Reign-period 'Chinese origin of Western studies' theory], *Hanxue yanjiu* 16, no. 1 (1998): 185–201.

206. Sela, *China's Philological Turn*, 154–55.

207. *Jue xue* 絕學; Qiong Zhang, *Making the New World Their Own: Chinese Encounters with Jesuit Science in the Age of Discovery* (Leiden: Brill, 2015): 250; Sela, *China's Philological Turn*, 148.

208. Wang Chang, ed., *Huhaiwen zhuan* [Collected works of lakes and seas] (Jingxun tang: 1837): *juan* 43.

209. Amiot to Bertin, January 25, 1787, MS 1516, IF.

210. Bertin to Amiot, December 21, 1785, MS 1524, IF

211. Bailly, *Traité de l'astronomie indienne et orientale*, 244.

212. Amiot, "Extrait d'une lettre écrite de Péking le 2 oct. 1784," in Amiot et al., *MCC*, vol. 11 (1786): 515–68, at 537, 528.

213. This necessity has been a major focus of the "New Qing History"; see especially Peter C. Perdue, *China Marches West: The Qing Conquest of Central Eurasia* (Cambridge: Belknap Press, 2010).

214. Matthew Mosca, "Empire and the Circulation of Frontier Intelligence: Qing Conceptions of the Ottomans," *Harvard Journal of Asiatic Studies*, 70, no. 1 (2010): 147–207, at 147–93; Gregory Afinogenov, *Spies and Scholars: Chinese Secrets and Imperial Russia's Quest for World Power* (Cambridge, MA: Harvard University Press, 2020): 139–58.

215. Wu Huiyi, "'The Observations We Did in the Indies and in China': The Shaping of French Jesuits' Knowledge of China by Other Non-Western Regions," *East Asian Science, Technology and Medicine* 46 (2018): 47–88, at 67–77.

216. Mark C. Elliott, "The Limits of Tartary: Manchuria in Imperial and National Geographies," *Journal of Asian Studies* 69, no. 3 (2000): 603–46, at 624–26.

217. Mark C. Elliott, "Abel-Rémusat, la langue mandchoue et la sinologie," at *Jean-Pierre Abel-Rémusat et ses successeurs*, symposium, June 11–13, 2014, Collège de France, Paris, https://www.college-de-france.fr/site/pierre-etienne-will/symposium-2013-2014.htm.

218. Amiot to Bertin, October 9, 1767, MS 1515, IF.

219. Amiot to Bertin, October 1, 1786, MS 1516, IF.

220. Amiot to Bréquigny, September 25, 1786, MS Bréquigny 2, BnF.

221. Jiang Fan 江繁, *Siyiguan kao* [Investigations of the Translation Bureau]

(1696), reprint ed. (Beijing: *Beijing tushuguan guji zhenben congkan*, vol. 59, 2000). Louis Pfister gave no title for the original Chinese text and incorrectly identified the author as a different Jiang Fan 江藩, who was active in the late eighteenth century (*Notices biographiques et bibliographiques sur l es jésuites de l'ancienne mission de Chine*, vol. 1 [Shanghai: Imprimerie de la Mission Catholique, 1932]: 12). My thanks to Wesley Cheney for helping to identify the Chinese source.

222. Siyiguan 四譯館.

223. Joseph-Marie Amiot, "Introduction à la connaissance des peuples qui ont été ou qui sont actuellement tributaires de la Chine," in Amiot et al., *MCC*, vol. 14 (1789): 1–238, at 7.

224. Amiot to Bréquigny, June 26, 1789, MS Bréquigny 2, BnF.

225. Bailly, *Traité de l'astronomie indienne et orientale*, cx–cxiv.

226. Bailly, *Traité de l'astronomie indienne et orientale*, clxxx.

227. Jean-Sylvain Bailly, *Recueil de pièces intéressantes sur les arts, les sciences, et la littérature: Ouvrage posthume de Sylvain Bailly* (Paris: Ainé et Jeune, 1810): 148–49.

228. Arhats, *luohan* 羅漢. Consulting the Chinese original reveals that the mountain is called Lingshan 靈山, or Soul Mountain.

229. Amiot, "Introduction à la connaissance des peuples," 21–22.

230. Amiot to Bréquigny, June 26, 1789, MS Bréquigny 2, BnF.

231. Amiot, "Suite du mémoire sur les danses," 246.

Chapter 4

1. Simon Schaffer, "The Accomplishment of Facts at the End of the Enlightenment" (forthcoming).

2. Lawrence M. Principe, "The End of Alchemy? The Repudiation and Persistence of Chrysopoeia at the Académie Royale des Sciences in the Eighteenth Century," *Osiris* 29, no. 1 (2014): 96–116, at 100–15; Simon Schaffer, "The Astrological Roots of Mesmerism," *Studies in History and Philosophy of Biological and Biomedical Sciences* 42, no. 2 (2010): 158–68, at 158–61.

3. Michael Adas, *Machines as the Measure of Men: Science, Technology, and Ideologies of Western Dominance* (1989), reprint ed. (Ithaca, NY: Cornell University Press, 2015).

4. Auguste Viatte, *Les sources occultes du Romantism*, vol. 1 (Paris: Honoré Champion, 1928); Dan Edelstein, "Introduction to the Super-Enlightenment," in *The Super-Enlightenment: Daring to Know Too Much*, ed. Dan Edelstein (Oxford: Voltaire Foundation, 2010): 1–34.

5. Peter Reill, "The Hermetic Imagination in the High and Late Enlightenment," in *The Super-Enlightenment*, ed. Edelstein, 43–51; Riskin, "Mr. Machine and the Imperial Me," in *The Super-Enlightenment*, ed. Edelstein,93–94.

6. Edelstein, "Introduction to the Super-Enlightenment," 27; Simon Schaffer, "Late Enlightenment Crises of Facts: Mesmerism and Meteorites," *Configurations* 26, no. 2: 119–48, at 137–41.

7. David Bates, "Super-Epistemology," in *The Super-Enlightenment*, ed. Edelstein, 53–74; Gaukroger, *The Collapse of Mechanism and the Rise of Sensibility: Science and the Shaping of Modernity, 1680–1760* (New York: Oxford University Press, 2010): 444–52.

8. Simon Schaffer, Lissa Roberts, Kapil Raj, and James Delbourgo, "Introduction,"

in *The Brokered World: Go-Betweens and Global Intelligence, 1770–1820*, eds. Schaffer, Roberts, Raj, and Delbourgo (Sagamore Beach, MA: Science History Publications, 2009): ix–xxxviii.

9. Simon Schaffer, "Newton on the Beach: The Information Order of *Principia Mathematica*," *History of Science* 47 (2009): 243–76.

10. Kapil Raj, *Relocating Modern Science: Circulation and the Construction of Knowledge in South Asia and Europe, 1650–1900* (New York: Palgrave Macmillan, 2007): 8–12.

11. Fa-ti Fan, "The Global Turn in the History of Science," *East Asian Science, Technology and Society* 6 (2012): 249–58, at 249–51.

12. Raj, *Relocating Modern Science*, 62–82; Londa Schiebinger, *Plants and Empire: Colonial Bioprospecting in the Atlantic World* (Cambridge, MA: Harvard University Press, 2007): 150–93.

13. The two most detailed studies are based mostly on just three of Amiot's letters: Pierre Huard and Ming Wong, "Mesmer en Chine: Trois lettres médicales du R.P. Amiot, rédigées à Pékin, de 1783 à 1790," *Revue de Synthèse* series 3, 81, no. 17–18 (1960); Linda L. Barnes, *Needles, Herbs, Gods, and Ghosts: China, Healing, and the West to 1848* (Cambridge, MA: Harvard University Press, 2009): 194–211. Another important reference is Nii Yoko, *Iezusu kaishi to fuhen no teikoku: ZaiKa senkyōshi niyoru bunmei no honyaku* [Jesuit missionaries and universal empire, translation of civilization by missionaries in China] (Nagoya: Nagoya daigaku shuppankai, 2017), which identifies many of their French and Chinese sources; I have not been able to make use of it as much as I would have liked, because it was published fairly recently, and is in Japanese.

14. Richard Smith and Danny Wynn Ye Kwo, eds., *Cosmology, Ontology, and Human Efficacy: Essays in Chinese Thought* (Honolulu: University of Hawaii Press, 1993): vii–xiii; Michael Strickmann, *Chinese Magical Medicine*, ed. Bernard Faure (Stanford, CA: Stanford University Press, 2002).

15. Mellet's main surviving archive consists of about twenty-five notes and letters occupying forty-five folio pages of a bound volume, MS Bréquigny 163, "Correspondance de Bréquigny VI," in the Bibliothèque nationale de France (BnF) in Paris. All these manuscripts were composed between 1784 and 1787 or 1788, and most were addressed to Bréquigny. Due to problems with organization and legibility, more precise information is hard to assign for many of the letters. Names and dates given in the text can be established with some confidence; additional information in the footnotes is often tentative.

16. Louis-Raphaël-Lucrèce de Fayolle, comte de Mellet to [Bréquigny?], February 19 [1785?], MS Bréquigny 163, BnF.

17. Louis-Raphaël-Lucrèce de Fayolle, comte de Mellet to [Barthélemy?], December 23 [1784?], MS Bréquigny 163, BnF.

18. Joseph-Marie Amiot to Louis-Raphaël-Lucrèce de Fayolle, comte de Mellet, November 14, 1789, MS 1517, Institut de France (IF), Paris, France.

19. René Louis de Roussel, *État militaire de France pour l'année 1772* (Paris: Guillyn, 1772): 52.

20. Antoine Court de Gébelin, *Monde primitif, analysé et comparé avec le monde moderne, considéré dans l'histoire civile, religieuse et allégorique du calendrier ou almanac* (Paris: Court de Gébelin, 1776): xvii; Jean-Marie Lhôte, *Le Tarot* (Paris: Berg, 1983): 144.

21. Mellet to [Bréquigny?], February 19 [1785?], MS Bréquigny 163, BnF.

22. Brian Vickers, *Occult Scientific Mentalities in the Renaissance* (Cambridge: Cambridge University Press, 1986): 1–4.

23. Mellet to Bréquigny, [1785?], MS Bréquigny 163, BnF.

24. Anthony Grafton, *Defenders of the Text: The Traditions of Scholarship in an Age of Science, 1450–1800* (Cambridge, MA: Harvard University Press, 1994): 145–61.

25. Mellet to Bréquigny, [1785?], MS Bréquigny 163, BnF.

26. Francis Bacon, *Novum Organum* (1620), in *Francis Bacon: Selected Philosophical Works*, ed. Rose-Mary Sargent (Indianapolis: Hackett Publishing Co., 1999): 119.

27. Mellet to [Barthélemy?], December 23 [1784?], MS Bréquigny 163, BnF.

28. Louis-Raphaël-Lucrèce de Fayolle, comte de Mellet, "Première lettre, de M. le comte de Mellet, maréchal de camp, Paris, ce 8 Octobre 1787," in Armand-Marie-Jacques Chastenet de Puységur, *Du magnétisme animal: considéré dans ses rapports avec diverses branches de la physique générale* (Paris: Cellot, 1807): 383–387, at 383–84.

29. The classic work is Robert Darnton, *Mesmerism and the End of the Enlightenment in France* (Cambridge, MA: Harvard University Press, 1968).

30. Franz Anton Mesmer, *Mémoire sur la découverte du magnétisme animal* (Geneva: Didot le jeune, 1779): 83.

31. Darnton, *Mesmerism*, 8.

32. Jessica Riskin, *Science in the Age of Sensibility: The Sentimental Empiricists of the French Enlightenment* (Chicago: University of Chicago Press, 2002): 189–92.

33. Adam Crabtree, *Animal Magnetism, Early Hypnotism, and Psychical Research, 1766–1925: An Annotated Bibliography* (Millwood, NY: Kraus International Publications, 1988): 1–38.

34. Antoine Court de Gébelin, *Lettre de l'auteur du monde primitif, à Messieurs ses souscripteurs, sur le magnétisme animal* (Paris: Gastelier, 1784): 22, 45; Darnton, *Mesmerism*, 116–17.

35. Jean-Sylvain Bailly, *Exposé des expériences qui ont été faites pour l'examen du magnétisme animal* (Paris: Mourard, 1784): 8; Riskin, *Science in the Age of Sensibility*, 215–225; Schaffer, "The Accomplishment of Facts at the End of the Enlightenment," 1–15.

36. *Gongfu* 功夫, literally "skill," "labor," or "achievement," can refer to various therapeutic and martial arts.

37. Pierre-Martial Cibot, "Notice du cong-fou, des bonzes Tao-sée," in Joseph-Marie Amiot, François Bourgeois, Pierre-Martial Cibot, Aloys Ko, Aloys de Poirot, Charles Batteux, M. de Bréquigny, A. I. Silvestre de Sacy, and Antoine Gaubil, *Mémoires concernant l'histoire, les sciences, les arts, les moeurs, les usages, &c. des Chinois* [*MCC*], vols. 1–15 (Paris: Nyon l'aîné, 1776–1791): vol. 4 (1779), 441–51, at 448.

38. Cibot, "Notice du cong-fou," 449.

39. *Daoshi* 道士.

40. Cibot, "Notice du cong-fou," 441–42.

41. Cibot, "Notice du cong-fou," 451.

42. Joseph Dehergne, "Les historiens Jésuites du Taoïsme," in *La Mission française de Pékin aux XVIIe et XVIIIe siècles: La Chine au temps des Lumières* 2, 59–67 (Paris: Les Belles Lettres, Cathasia, 1976): 65–66.

43. Mellet, "Première lettre, de M. le comte de Mellet," 384.

44. Mellet was not the last person who mistakenly thought that the essay on "Cong-fou" was by Amiot—modern historians, including Huard and Wong, also

considered this authorship likely ("Mesmer en Chine," 62). But it was certainly by Cibot, as Dehergne argued, and as Amiot's first letter to Mellet proves: "I kindly admit to you that I am still very new to the article on *Kong fou*" (Amiot to Mellet, October 1, 1783, MS 2446, "Copies de lettres à sujet médical rédigés à Pékin adressées à M. Desvoyes pseudonyme de l'abbé Louis Augustin Bertin" [mislabeled], Bibliothèque interuniversitaire de Santé [BIS], Paris, France).

45. Mellet, "Première lettre, de M. le comte de Mellet," 384–85.

46. Amiot to Mellet, October 1, 1783, MS 2446, BIS; Amiot to Mellet, October 18, 1784, MS 2446, BIS.

47. In total, at least five letters from Amiot to Mellet survive, but confusion about the addressee of the first two along with virtual ignorance of the last two has made it impossible for historians to place them in proper context. The only significant study of the relevant archival material was conducted in 1960, when Pierre Huard and Ming Wong annotated and published "three medical letters of the Reverend Father Amiot," two of which had been unknown even to Amiot's usually thorough Jesuit bibliographers (Huard and Wong, "Mesmer en Chine"; Louis Pfister, *Notices biographiques et bibliographiques sur les jésuites de l'ancienne mission de Chine* [Shanghai: Imprimerie de la Mission Catholique, 1932]). All later historians have followed Huard and Wong in the conclusion that the first two letters were sent to one "Desvoyes," whom they took to be Bertin's brother, and the third to Mellet, about whom they were quite ignorant. But in fact, all three of the letters that they published, plus at least two more of which they were unaware, were sent to Mellet.

The reasons for the misidentification are clear enough. The three letters studied by Huard and Wong were not originals, but eighteenth-century copies, held at the Faculté de médecine de Paris and catalogued under the title "Copies de lettres à sujet médical rédigés à Pékin adressées à M. Desvoyes pseudonyme de l'abbé Louis-Augustin Bertin." Since the metadata is undated, it is impossible to say whether it reflects their research, or vice versa. The only relevant evidence on the documents themselves is the name "M. Desvoyes" scrawled across the top of one letter in a different hand from that of the copyist. Who was this Desvoyes? Most historians have followed Huard and Wong in asserting that the name was the pseudonym of the abbé Louis-Augustin Bertin, brother of Henri-Leonard. In fact, François Étienne Desvoyes, a Paris lawyer, was Henri Bertin's secretary (Michel Hermans, "Biographie de Joseph-Marie Amiot," in *Les Danses rituelles chinoises d'après Joseph-Marie Amiot: Aux sources de l'ethnochorégraphie*, eds. Yves Lenoir and Nicolas Standaert [Namur: Presses Universitaires de Namur, 2005]: 11–77, at 66; Bruno Belhoste, David Armando, and Stéphane Lamassé, "Harmonia Universalis," LabEx Hastec project, with IHMC, Centre Koyré, CARE, ISPF, C.R.I.S.E.S., and AHRHA (2016), https://harmoniauniversalis.univ-paris1.fr/#/.) Here, too, the original source of confusion is apparent: a speculative comment by Bourgeois in 1783 ("M. Desvoyes could well be Monsieur the abbé Bertin. Going from there, I am in an unknown land" [François Bourgeois to an unknown recipient, November 20, 1783, MS 1519, "Lettres de plusieurs missionnaires," IF]). But the other missionaries knew or soon learned exactly who Desvoyes was (Joseph-Marie Amiot to Henri Bertin, October 20, 1790, MS 1517, IF). During the Revolution, when Bertin's Chinese cabinet was catalogued for auction, the Committee for Public Safety delegated the task to him, since he "had been attached to this correspondence" ("Desvoyes," MS F^17, 1047, Archives nationales [AN], Pierrefitte-sur-Seine, France).

Whoever he was, Desvoyes was certainly not the recipient of any of Amiot's known letters on animal magnetism. Huard and Wong were apparently unaware that there exists another eighteenth-century copy of one of the letters they published. On that one, the name Desvoyes does not appear, and the addressee is instead given only as "Monseigneur," suggesting a higher-ranking member of the First or Second Estate (Amiot to Mellet, 1783, MS Bréquigny 5, BnF). On the copy they did know, there is a note next to Desvoyes's name: "*pour copier*." One possible interpretation is that the letters were copied for Desvoyes, since, as a founding member of the mesmerist "Société de l'harmonie universelle," he might have been interested. A more likely one is that Desvoyes himself was the copyist. In 1788, Mellet asked someone else if Desvoyes could make a copy of a letter to be sent to China, so it makes sense that he would have made copies of letters going the other way, too (Mellet to Bréquigny, April 19 [1778?], MS Bréquigny 163, BnF). In any case, based on the contents of the unpublished letters and the published attribution of one of them in 1807, there is no question that the original recipient of all three was Mellet.

48. Amiot to Mellet, October 1, 1783, MS 2446, BIS.

49. Universal figure, *Chiffre universel*; *Taiji tu* 太極圖; Zhou Dunyi 周敦頤.

50. *Liushu benyi* 六書本義 [Foundational Meaning of the Six Books]; Nii Yoko, *Iezusu kaishi to fuhen no teikoku*, 135–36.

51. *Li* 理; Amiot insisted on the Latin term *fomes* for lack of a French equivalent.

52. Amiot to Mellet, October 1, 1783, MS 2446, BIS.

53. Joseph Needham with Wang Ling, *Science and Civilisation in China: Volume II: History of Scientific Thought* (Cambridge: Cambridge University Press, 1956): 273–78, 455–84.

54. Amiot to Mellet, October 18, 1784, MS 2446, BIS.

55. Amiot to Mellet, October 18, 1784, MS 2446, BIS.

56. Amiot to Mellet, October 1, 1783, MS 2446, BIS.

57. Amiot to Bertin, October 2, 1784, MS 1516, "80 lettres du P. Amiot à Bertin," IF.

58. *Longyan* 龍煙; *Hanming* 含明.

59. Amiot to Mellet, October 18, 1784, MS 2446, BIS.

60. Amiot to Mellet, September 29, 1786, MS Bréquigny 1, "Mélanges sur la Chine et les Chinois," BnF.

61. *Zangfu bu* 臟腑部 [Organs section], *Gujin tushu jicheng* [Complete collection of ancient and modern figures and texts], *Minglun huibian, Renshi dian, ce* 386, *juan* 21 (1726), reprint ed. (Shanghai: Zhonghua Publishing Company, 1934). The only identifying information Amiot provided was that the volume had been included in a recent "imperial collection." But in one archival folio next to a letter by Mellet, a short note by a later bibliography reads: "kou-kin-tou-chou-tsie-tching, collection of ancient and modern figures; in the first book, concerning man, which mentions their functions, their rapport with the elements, their union with metals, planets, etc." (Mellet to [Bréquigny?], August [1785?], MS Bréquigny 163, BnF).

62. *Bencao gangmu* 本草綱目; *Huangdi neijing* 黃帝內經.

63. *Huainanzi* 淮南子; *Zihuazi* 子華子.

64. Paul U. Unschuld, *Medicine in China: A History of Ideas* (Berkeley: University of California Press, 1985): 51.

65. *Shen* 神.

66. *Sancai Tuhui* 三才圖會 [Collected Illustrations of the Three Realms]; "Organs Section," 48.

67. Mesmer, *Mémoire*, 77.

68. "Organs Section," 51; Mesmer, *Mémoire*, 7.

69. Darnton, *Mesmerism*, 18.

70. *he* 和, *diao* 調, "Organs Section," 48–53; Belhoste, Armando, and Lamassé, "Harmonia Universalis."

71. Unschuld, *Medicine in China*, 72.

72. "Organs Section," 48–52.

73. Joseph-Marie Amiot, "Antiquité des Chinois, prouvée par les monuments," in Amiot et al., *MCC*, vol. 2 (1776), 1–364, at 152.

74. *wu xing* 五行; Carla Nappi, *The Monkey and the Inkpot: Natural History and its Transformations in Early Modern China* (Cambridge, MA: Harvard University Press, 2009): 71–72.

75. "Organs Section," 54.

76. Amiot to Mellet, October 18, 1784, MS 2446, BIS.

77. Amiot to Mellet, October 18, 1784, MS 2446, BIS.

78. Joseph-Marie Amiot to Louis-Georges de Bréquigny, June 26, 1789, MS Bréquigny 2, "Mélanges sur la Chine et les Chinois," BnF.

79. Amiot to Mellet, October 1, 1783, MS 2446, BIS.

80. Amiot to Mellet, October 18, 1784, MS 2446, BIS.

81. Amiot to Mellet, October 1783, MS 2446, BIS.

82. *L'Ordre du Chiffre Universel.*

83. It is not clear if Amiot knew that the "Unknown Philosopher" was the illuminist, mystic, and occultist Louis-Claude de Saint-Martin, an acquaintance of Court de Gébelin and influence on Mellet (David Bates, "The Mystery of Truth: Louis-Claude de Saint-Martin's Enlightened Mysticism," *Journal of the History of Ideas* 61, no. 4 [2000]: 635–55).

84. One can only speculate as to Amiot's condescension toward the Swiss.

85. Amiot to Mellet, October 1, 1783, MS 2446, BIS.

86. Riskin, *Science in the Age of Sensibility*, 226–89; Schaffer, "Late Enlightenment Crises of Facts," 119–37.

87. Bruno Belhoste and Nicole Edelman, eds., *Mesmer et mesmérismes: Le magnétisme animal en contexte* (Paris: Omniscience, 2015).

88. Crabtree, *Animal Magnetism*.

89. Underlining in the original; Mellet to [Barthélemy?], December 23 [1784?], MS Bréquigny 163, BnF.

90. Mellet to [Bréquigny?], April [1785?], MS Bréquigny 163, BnF.

91. Mellet, "Première lettre, de M. le comte de Mellet," 383.

92. Mellet to [Barthélemy?], December 23 [1784?], MS Bréquigny 163, BnF.

93. Mellet, "Première lettre, de M. le comte de Mellet," 386.

94. Mellet to [Barthélemy?], December 23 [1784?], MS Bréquigny 163, BnF.

95. The dates on the relevant manuscripts are particularly confusing; it is possible that the exchange between Mellet and Le Roy took place one year later.

96. Benjamin Franklin, *The Writings of Benjamin Franklin, Vol. X (1789–1790)*, ed. Albert Henry Smyth (New York: Macmillan, 1907): 69.

97. Riskin, *Science in the Age of Sensibility*, 201.

98. Jean-Baptiste Le Roy to Louis-Raphaël-Lucrèce de Fayolle, comte de Mellet, October 20 [1784?], MS Bréquigny 163, BnF.

99. Mellet to [Bréquigny?], February 19 [1785?], MS Bréquigny 163, BnF.

100. Mellet to [Bréquigny?], June 26, 1785, MS Bréquigny 163, BnF.

101. [Bailly?] to Mellet, May 1785, MS Bréquigny 163, BnF.

102. Mellet to [Bréquigny?], June 26, 1785, MS Bréquigny 163, BnF.

103. Darrin McMahon, *Enemies of the Enlightenment: The French Counter-Enlightenment and the Making of Modernity* (Oxford: Oxford University Press, 2001): 86; Christophe Félix Louis Galart de Montjoye, *Lettre sur le magnétisme animal adressée à M. Bailly* (Paris: Pierre-J. Duplain, 1784): 3.

104. Jean-Sylvain Bailly, *Traité de l'astronomie indienne et orientale* (Paris: de Bure l'aîné, 1787): xxiv.

105. Mellet to [Bréquigny?], April [1785?], MS Bréquigny 163, BnF.

106. Mellet to [Bréquigny?], June 14 [1787?], MS Bréquigny 163, BnF

107. Mellet to [Bréquigny?], December 14, 1787, MS Bréquigny 163, BnF.

108. Mellet to [Bréquigny?], April [1785?], MS Bréquigny 163, BnF.

109. Mellet, "Première lettre, de M. le comte de Mellet," 385.

110. Mellet, "Première lettre, de M. le comte de Mellet"; Amiot to Mellet, September 29, 1786, MS Bréquigny 1, BnF.

111. Amiot to Bertin, August 20, 1790 / October 4, 1790, MS 1517, IF.

112. Mellet to Bréquigny, December 1787, MS Bréquigny 1, BnF.

113. Amiot to Mellet, September 29, 1786, MS Bréquigny 1, BnF.

114. Pierre-Martial Cibot, "Notice du sang de Cerf, employé comme remède" (1781), in Amiot et al., *MCC*, vol. 8 (1782): 271–74.

115. Amiot to Mellet, September 29, 1786, MS Bréquigny 1, BnF.

116. Nappi, *The Monkey and the Inkpot.*

117. *Lei huo zhen* 雷火針.

118. Amiot to Mellet, September 29, 1786, MS Bréquigny 1, BnF.

119. Mellet to [Bréquigny?], June 14 [1787?], MS Bréquigny 163, BnF

120. Mellet to [Bréquigny?], April [1785?], MS Bréquigny 163, BnF.

121. Michael R. Lynn, "Divining the Enlightenment: Public Opinion and Popular Science in Old Regime France," *Isis* 92, no. 1 (2001): 31–53.

122. Mellet to [Bréquigny?], December 14, 1787, MS Bréquigny 163, BnF.

123. Michel-Ange le Roux Deshauterayes to Louis-Georges de Bréquigny, May 6, 1790, MS Bréquigny 160, "Correspondance de Bréquigny III," BnF.

124. Mellet to [Bréquigny?], December 14, 1787, MS Bréquigny 163, BnF.

125. Jean-Baptiste Sirey and Le Moine Devilleneuve, *Recueil général des lois et des arrêts, en matière civile, criminelle, commercial et de Droit public*, vol. 29 (Paris: Lachevardière, 1829): 99.

126. This distinction is contentious among scholars today (Fabrizio Pregadio, "Religious Daoism" (2016), *Stanford Encyclopedia of Philosophy* (Fall 2020), ed. Edward N. Zalta, https://plato.stanford.edu/archives/fall2020/entries/daoism-religion/); in any case, the Daoist classics of the fourth century BCE and the Daoist cannon of the second century CE were not the same (Nathan Sivin, "Taoism and Science," in *Medicine, Philosophy and Religion in Ancient China: Researches and Reflections* [Aldershot: Variorum, 1995]: 1–73).

127. Amiot to Bertin, August 20, 1790 / October 4, 1790, MS 1517, IF.

128. Joseph-Marie Amiot, "Extrait d'une lettre écrite de Péking le 16 oct. 1787 (Traité sur la secte des *Tao-sée*)," in Amiot et al., *MCC*, vol. 15 (1791): 208–59, at 208.

129. Amiot to Bertin, August 20, 1790 / October 4, 1790, MS 1517, IF.

130. Unfortunately, none of the books sent to Beijing during this period appear

in the bibliography of the Beitang library (Hubert Verhaeren, *Catalogue de la bibliothèque du Pé-t'ang* [Beijing: Imprimerie des Lazaristes, 1949]).

131. Amiot to Bertin, November 15, 1784, MS 1516, IF.

132. Bertin to Amiot, December 31, 1784, MS 1524, IF.

133. *yuanqi* 元氣.

134. Amiot to Mellet, September 29, 1786, MS Bréquigny 1, BnF.

135. Amiot to Bertin, October 2, 1784, MS 1516, IF.

136. Jacques-François-Maxime de Chastenet de Puységur et al., *État actuel de l'art et de la science militaire à la Chine: Tiré des livres militaires des Chinois. Avec diverses observations sur l'étendue & les bornes des connoissances militaires chez les Européens* (Paris: Didot l'aîné, 1773).

137. Louis-Raphaël-Lucrèce de Fayolle, comte de Mellet to Marius-Jean-Baptiste-Nicolas d'Aine, February 12, 1793, MS 291 AP/1, "Papiers de Marius-Jean-Baptiste-Nicolas d'AINE," AN.

138. Amiot to Mellet, November 14, 1789, MS 1517, IF.

139. A. A. Tardy de Montravel, *Essai sur la théorie du Somnambulisme magnétique* (London [Paris?]: 1786): 27–28.

140. Tardy de Montravel, *Essai sur la théorie du Somnambulisme magnétique*, 7.

141. The "*sixième sens*" was already a theme in French spiritualism, though apparently not in English.

142. Tardy de Montravel, *Essai sur la théorie du Somnambulisme magnétique*, 50.

143. Amiot to Mellet, October 18, 1784, MS 2446, BIS.

144. Amiot to Mellet, November 14, 1789, MS 1517, IF.

145. The sources for Amiot's discussion of Daoist practices are particularly uncertain. He mentioned having "drawn from some of their books," but gave no quotes or other identifying information. Probably, his account reflected some combination of scattered reading, personal conversations, and accumulated knowledge.

146. Amiot to Mellet, November 14, 1789, MS 1517, IF.

147. Amiot to Mellet, November 14, 1789, MS 1517, IF.

148. *titou* 剃頭.

149. Amiot to Mellet, September 24, 1790, MS 1517, IF.

150. Amiot to Mellet, September 24, 1790, MS 1517, IF.

151. *bienséance*; Amiot to Mellet, November 14, 1789, MS 1517, IF.

152. Amiot to Mellet, November 14, 1789, MS 1517, IF.

153. Amiot to Mellet, November 14, 1789, MS 1517, IF.

154. Amiot to Mellet, September 24, 1790, MS 1517, IF.

155. Amiot to Mellet, November 14, 1789, MS 1517, IF.

156. Amiot to Mellet, September 24, 1790, MS 1517, IF.

157. Amiot to Bertin, August 20, 1790 / October 4, 1790, MS 1517, IF.

158. Amiot to Mellet, November 14, 1789, MS 1517, IF.

159. Amiot to Mellet, November 14, 1789, MS 1517, IF.

160. Amiot to Mellet, November 14, 1789, MS 1517, IF.

161. Amiot to Bertin, October 16, 1787, MS Bréquigny 2, BnF.

162. Amiot to Mellet, November 14, 1789, MS 1517, IF.

163. Amiot to Bertin, October 10, 1789, MS 1517, IF.

164. Amiot to Mellet, September 24, 1790, MS 1517, IF.

165. Amiot to Mellet, September 24, 1790, MS 1517, IF.

166. Robert de Lo-Looz, *Recherches physiques et métaphysiques sur les influences célestes, sur le magnétisme universel, et sur le magnétisme animal: dont on trouve la pratique de temps immémorial chez les Chinois* (London: Couturier, 1788): 135; Jean-Baptiste Montmignon, ed., *Choix des lettres édifiantes: écrites des missions étrangères; avec des additions, des notes critiques, et des observations pour la plus grande intelligence de ces lettres* (Paris: Maradan, 1808): 110; Puységur, *Du magnétisme animal*, 396.

167. Amiot to Bréquigny, September 20, 1786, MS Bréquigny 2, BnF; Joseph-Marie Amiot, "Extrait d'une lettre écrite de Péking le 20 sep. 1786," in Amiot et al., *MCC*, vol. 13 (1788): 507–10.

168. Louis-Georges de Bréquigny to Joseph-Marie Amiot, December 15, 1787, MS Bréquigny 2, BnF.

169. Amiot to Bréquigny, June 26, 1789, MS Bréquigny 2, BnF.

170. Lo-Looz, *Recherches physiques et métaphysiques*, 127.

171. Jean-Louis Alibert, *Nouveaux éléments des thérapeutiques et de matière médicale*, 5th ed. (Paris: Béchet, 1826): 507–08.

172. Antoine François Delandine, *De la Philosophie corpusculaire: Ou des connoissances et des procédés magnétiques chez les divers peuples* (Paris: Cuchet, 1785): 170.

173. Jacques Cambry, *Traces du magnétisme* (Le Hague, 1784): 37.

174. Cambry, *Traces du magnétisme*, 9–11.

175. Cambry, *Traces du magnétisme*, 26.

176. Cambry, *Traces du magnétisme*, 46–47.

177. Pierre Jean Baptiste Nougaret, *Tableau mouvant de Paris, ou variétés amusantes, ouvrage enrichi de notes historique & critiques, & mis au jour*, vol. 2. (London: Thomas Hookham, Libraire, la Veuve Duchesne, 1787): 239–42.

178. "Mémoire pour servir à l'histoire du somnambulisme," MS 2270, BIS.

179. Lo-Looz, *Recherches physiques et métaphysiques*.

180. Antoine-Gabriel de Becdelièvre-Hamal, *Biographie liégeoise* (Liège: Jeunehomme, 1837): 486.

181. Robert de Lo-Looz, *Les Militaires au-delà du Gange* (Paris: Bailly, 1770).

182. Lo-Looz, *Recherches physiques et métaphysiques*, 127, 135.

183. Lo-Looz, *Recherches physiques et métaphysiques*, 127–48.

184. Lo-Looz, *Recherches physiques et métaphysiques*, 136.

185. Lo-Looz, *Recherches physiques et métaphysiques*, 144, 140.

186. Lo-Looz, *Recherches physiques et métaphysiques*, 127.

187. Lo-Looz, *Recherches physiques et métaphysiques*, 139.

188. Lo-Looz, *Recherches physiques et métaphysiques*, 144.

189. Lo-Looz, *Recherches physiques et métaphysiques*, 136–38.

190. Lo-Looz, *Recherches physiques et métaphysiques*, vol. 2, 138.

191. Lo-Looz, *Recherches physiques et métaphysiques*, 138.

192. Frédéric Barbier et al., *Dictionnaire des imprimeurs, libraires et gens du livre à Paris: A–C* (Geneva: Droz, 2007): 567.

193. Anonymous ("Un Amateur de la Vérité"), *Rapport du Rapport de MM. les commissaires nommés par le Roi pour examiner la pratique de M. Deslon sur le magnétisme animal* (Couturier: Beijing, 1784); Charles Hervier, *Lettre sur la découverte du magnétisme animal à Court de Gébelin* (Couturier: Beijing, 1784).

194. Emil Ottokar Weller, *Die falschen und fingierten Druckorte: Repertorium der seit Erfindung der Buchdruckerkunst unter falscher Firma erschienenen deutschen,*

lateinischen und französischen Schriften (Leipzig: Wilhelm Engelmann, 1864): 220–24.

195. Crabtree, *Animal Magnetism*, 26; Riskin, *Science in the Age of Sensibility*, 206–09.

196. Crabtree, *Animal Magnetism*, 60.

197. Puységur, *Du magnétisme animal*, 383–96.

198. Puységur, *Du magnétisme animal*, 396–97.

199. Helena Petrovna Blavatsky, ed., *The Theosophist*, vol. 1 (Madras: Theosophical Society, 1879–1880): 30–31.

200. Montmignon, *Choix des lettres édifiantes*, 264–67.

201. Montmignon, *Choix des lettres édifiantes*, xvii.

202. Montmignon, *Choix des lettres édifiantes*, 264.

203. Montmignon, *Choix des lettres édifiantes*, vii, 264.

Chapter 5

1. Anonymous, "Notice historique sur la vie et les ouvrages de M. de Bréquigny," in *Mémoires de l'Académie des Inscriptions et Belles-Lettres*, vol. 50 (Paris: Imprimerie Impériale, 1808): 721; Anonymous, "Note sur la vie et les ouvrages de M. de Guignes," in *Mémoires de l'Académie des Inscriptions et Belles-Lettres*, vol. 48 (Paris: Imprimerie Impériale, 1808): 772; Antoine-Isaac Silvestre de Sacy, "Deshautesrayes, Michel-Ange-André Le Roux," in *Biographie universelle* (Paris: Imprimeur du Roi, 1814): 180.

2. John Finlay, *Henri Bertin and the Representation of China in Eighteenth-Century France* (New York: Routledge, 2020): 148.

3. MS 291 AP/1, "Papiers de Marius-Jean-Baptiste-Nicolas d'AINE," Archives nationales (AN), Pierrefitte-sur-Seine, France; Jérôme Delandine de Saint-Esprit, *Vie de S. A. R. Charles Ferdinand d'Artois, duc de Berry* (Paris L.-E. Herhan, 1820): 152.

4. Nicolas Viton de Saint-Allais et al., *Nobiliaire de France*, vol. 11, part 1 (Paris: Bachelin-Deflorenne, 1876): 150.

5. Joseph-Marie Amiot to François Desvoyes, November 2, 1792, MS 1517, "80 lettres du P. Amiot à Bertin," Institut de France (IF), Paris, France.

6. Haidian海淀.

7. Joseph-Marie Amiot to Louis-Raphaël-Lucrèce de Fayolle, comte de Mellet, September 24, 1790, MS 1517, IF.

8. Joseph-Marie Amiot to Pierre Jules Amyot, September 20, 1792, MS 1717, IF.

9. Joseph-Marie Amiot to Marie Julie Victoire Amiot, May 26, 1793, MS 1517, IF.

10. Michel Hermans, "Biographie de Joseph-Marie Amiot," in *Les Danses rituelles chinoises d'après Joseph-Marie Amiot: Aux sources de l'ethnochorégraphie*, eds. Yves Lenoir and Nicolas Standaert (Namur: Presses Universitaires de Namur, 2005): 11–77, at 17.

11. Nicolas-Joseph Raux, note on Amiot's will, 1793, MS 1517, IF.

12. Earl George Macartney and Sir John Barrow, *Some Account of the Public Life, and a Selection from the Unpublished Writings, of the Earl of Macartney. . . . Journal of an Embassy from the King of Great Britain to the Emperor of China* (London: T. Cadell and W. Davies, 1807): vol. 2, 225, 306.

13. Recent historians argue that the Qianlong emperor was not the ignorant dupe that the British made him out to be, but a savvy player in an emerging international

system. For the traditional interpretation, see Alain Peyrefitte, *The Immobile Empire*, trans. Jon Rothschild (New York: Knopf, 1992); for the revisionist one, see James L. Hevia, *Cherishing Men from Afar: Qing Guest Ritual and the Macartney Embassy of 1793* (Durham, NC: Duke University Press, 1995).

14. Quoted in Chen Pei-Kei and Michael Lestz, with Jonathan Spence, *The Search for Modern China: A Documentary Collection* (New York: W. W. Norton, 1999): 105.

15. Macartney and Barrow, *Journal of an Embassy*, vol. 2, 304.

16. Niall Ferguson, *Empire: The Rise and Demise of the British World Order and the Lessons for Global Power* (New York: Basic Books, 2002): 29.

17. Klaus Mülhan, *Making China Modern: From the Great Qing to Xi Jinping* (Cambridge, MA: Belknap Press, 2019): 85–93.

18. Matthew Mosca, *From Frontier Policy to Foreign Policy: The Question of India and the Transformation of Geopolitics in Qing China* (Stanford, CA: Stanford University Press, 2013): 163–98.

19. Robert and Eliza Morrison, *Memoirs of the Life and Labours of Robert Morrison*, vol. 1 (London: Longman, Orme, Brown, and Longmans, 1839): 187–94.

20. P. J. Marshall, "Introduction," in *The Oxford History of the British Empire: Volume II: The Eighteenth Century*, eds. P. J. Marshall and Alaine Low (Oxford: Oxford University Press, 1998): 1–27, at 7–8; Muriel Détrie, "L'évolution de l''Europe Chinoise' de la fin du XVIIIe siècle au début du XXe siècle," in *Idées de la Chine au XIXe siècle*, eds. Marie Dollé and Geneviève Espagne (Paris: Les Indes Savants, 2014): 19–38.

21. Jean-Pierre Abel-Rémusat, *Programme du cours de langue et de littérature chinoises et de tartare-mandchou; précédé du discours prononcé à la première séance de ce cours, dans l'une des salles du Collège royal de France, le 16 janvier 1815* (Paris: Charles, 1815): 10.

22. Ashley Eva Millar, "Revisiting the Sinophilia/Sinophobia Dichotomy in the European Enlightenment through Adam Smith's 'Duties of Government,'" *Asian Journal of Social Science* 38 (2010): 716–37, at 716–18; Zhang Chunjie, "From Sinophilia to Sinophobia: China, History, and Recognition," *Colloquia Germanica* 41, no. 2 (2008): 97–110.

23. Jean-Pierre Abel-Rémusat, *Mélanges posthumes d'histoire et de littérature orientales* (Paris: Imprimerie Royale, 1843): 334.

24. Willy Vande Walle and Noël Golvers, *The History of the Relations Between the Low Countries and China in the Qing Era (1644–1911)* (Leuven: Leuven University Press, 2003): 17–18.

25. Edward Said, *Orientalism* (New York: Pantheon Books, 1978): 2–3.

26. Said, *Orientalism*, 121.

27. Jürgen Osterhammel, *Unfabling the East: The Enlightenment's Encounter with Asia* (Princeton, NJ: Princeton University Press, 2018): 5–11; Alexander Bevilacqua, *The Republic of Arabic Letters* (Cambridge, MA: Harvard University Press, 2019): 199–203.

28. Said, *Orientalism*, 125.

29. Said, *Orientalism*, 26.

30. Jean-Pierre Abel-Rémusat, "Lettre au Rédacteur sur l'état et les progrès de la littérature chinoise en Europe," *Journal Asiatique* 1 (1822): 279–91, at 287.

31. Henri Cordier, "Les études chinoises sous la Révolution et l'Empire," *T'oung Pao* 2nd series, 19, no. 2 (May 1918): 59–103, at 101–03; Pierre-Étienne Will, "Abel-

Rémusat, l'Orientaliste," at *Jean-Pierre Abel-Rémusat et ses successeurs*, symposium, June 11–13, 2014, Collège de France, Paris, https://www.college-de-france.fr/site/pierre-etienne-will/symposium-2013-2014.htm.

32. Bevilacqua, *The Republic of Arabic Letters*, 5; Suzanne L. Marchand, *German Orientalism in the Age of Empire: Religion, Race, and Scholarship* (Cambridge: Cambridge University Press, 2010): 1.

33. Anonymous, "Notice Historique sur la vie et les ouvrages de M. Abel Rémusat," in *Mémoires de l'Institut royal de France: Académie des inscriptions et belles-lettres*, vol. 12 (Paris: Imprimerie Royale, 1839): 375–400, at 379.

34. Isabelle Landry-Deron, "Les outils de l'apprentissage du chinois en France en 1814 et les efforts d'Abel-Rémusat pour les améliorer," at *Jean-Pierre Abel-Rémusat et ses successeurs*, Symposium, June 11–13, 2014, Collège de France, Paris; Mark C. Elliott, "Abel-Rémusat, la langue mandchoue et la sinologie," at *Jean-Pierre Abel-Rémusat et ses successeurs*, symposium, June 11–13, 2014.

35. Jean-Pierre Abel-Rémusat, *Dissertatio de glossosemeiotice sive de signis morborum qua è linguâ sumuntur praesertim apud Sinenses* (Paris: Didot, 1813).

36. Knud Lundbaek, "Notes on Abel Rémusat," in *Actes du VIIe Colloque International de Sinologie de Chantilly: Échanges culturels et religieux entre la Chine et l'Occident*, eds. Edward Malatesta, Yves Raguin, and Adrianus C. Dudink (Paris: Ricci Institute, 1995): 207–221.

37. MS 14 CDF art 65a, "Sinologie, Chaire de Langue et littérature chinoises et tartare-mandchou, 1814–1832," Collège de France (CF), Paris, France.

38. Abel-Rémusat, "Lettre au Rédacteur," 279–91.

39. *Jean-Pierre Abel-Rémusat et ses successeurs*, symposium, June 11–13, 2014.

40. Collège de France, "Intellectual History of China," accessed May 10, 2022, https://www.college-de-france.fr/site/en-intellectual-history-china/index.htm.

41. Joël Thoraval, "De la magie à la 'raison': Hegel et la religion chinoise," in *La Chine entre amour et haine: Actes du VIIIe Colloque de sinologie de Chantilly*, ed. Michel Cartier (Paris: Desclée de Brouwer, 1998): 111–41, at" 133–37.

42. M. Henri Wallon, "Notice historique sur la vie et les travaux d'Aignan-Stanislas Julien," *Mémoires de l'Institut de France, Académie des inscriptions et belles-lettres* 31 (1884): 409–58, at 421–22.

43. Louis-Alexandre-Marguerite Bourgeat, "No. XIV, Langues Orientales," *Mercure Étrangère, ou Annales de la littérature Étrangère*, vol. 3 (Paris: D. Collas, 1814): 73.

44. Académie Française, *Supplément au Dictionnaire de l'Académie Française* (Paris: Gustave Barba, 1836): 733, ARTFL Encyclopédie Projet, edited by Robert Morrissey, http://encyclopedie.uchicago.edu/.

45. Académie Française, *Dictionnaire de l'Académie Française, Cinquième Édition* (1798); *Sixième Édition* (1835) (Paris), ARTFL.

46. Cordier, "Les études chinoises sous la Révolution et l'Empire," 101–03.

47. Harriet T. Zurndorfer, "Orientalism, Sinology, and Public Policy: Baron Antoine Isaac Silvestre de Sacy and the Foundation of Chinese Studies in Post-Revolutionary France," in *Images de La Chine: Le Contexte Occidental de La Sinologie Naissante*, eds. Edward Malatesta and Yves Raguin (San Francisco: Ricci Institute for Chinese-Western Cultural History, 1995): 175–92, at 191.

48. Landry-Deron, "Les outils de l'apprentissage du chinois en France en 1814."

49. Abel-Rémusat, "Lettre au Rédacteur," 286.

50. Abel-Rémusat, *Mélanges posthumes*, 162–63.

51. Jean-Pierre Abel-Rémusat, *Essai sur la langue et la littérature chinoises: avec cinq planches, contenant des textes chinois, accompagnés de traductions, de remarques et d'un commentaire littéraire et grammatical. Suivi de notes et d'une table alphabétique des mots chinois* (Paris: Treuttel et Wurtz, 1811): 124; Jean-Pierre Abel-Rémusat, *Nouveaux mélanges asiatiques: ou, Recueil de morceaux de critique et de mémoires, relatifs aux religions, aux sciences, aux coutumes, à l'histoire et la géographie des nations orientales* (Paris: Schubart et Heideloff, 1829): vol. 2, 143.

52. Abel-Rémusat, *Programme du cours*, 9–10.

53. Marie Dollé and Geneviève Espagne, "Introduction," in *Idées de la Chine au XIXe siècle*, eds. Dollé and Espagne (Paris: Les Indes Savants, 2014): 7–14.

54. Gregory Afinogenov, *Spies and Scholars: Chinese Secrets and Imperial Russia's Quest for World Power* (Cambridge, MA: Harvard University Press, 2020): 211–13; Koos Kuiper, *The Early Dutch Sinologists (1854-1900): Training in Holland and China, Functions in the Netherlands Indies* (Leiden: Brill, 2017): vol. 1, 1–13.

55. Henri Cordier, "Un orientaliste Allemand: Jules Klaproth," *Comptes rendus des séances de l'Académie des inscriptions et Belles-Lettres* 61, no. 4 (1917): 297–308, at 299–300.

56. P. F. Kornicki, "Review: Julius Klaproth and His Works," *Monumenta Nipponica* 55, no. 4 (2000): 579–91, at 580–81.

57. Norman T. Girardot, "James Legge and the Strange Saga of British Sinology and the Comparative Science of Religions in the Nineteenth Century," *Journal of the Royal Asiatic Society*, 3rd series, 12, no. 2 (2002): 155–65, at 155–57.

58. Henrietta Harrison, *The Perils of Interpreting: China and the Rise of the British Empire* (Princeton, NJ: Princeton University Press, forthcoming): 179–82.

59. Timothy Barrett, "A Bicentenary in Robert Morrison's Scholarship on China and His Significance for Today," *Journal of the Royal Asiatic Society*, 3rd series, 25, no. 4 (2015): 705–16.

60. Norman T. Girardot, *The Victorian Translation of China: James Legge's Oriental Pilgrimage* (Berkeley: University of California Press, 2002): 164.

61. Arthur Wright, "The Study of Chinese Civilization," *Journal of the History of Ideas* 21, no. 2 (1960): 233–55, at 241–43.

62. Académie française, *Le Dictionnaire de l'Académie française, Sixième Édition*.

63. Marchand, *German Orientalism*, xxvi–xxix.

64. Peter K. J. Park, *Africa, Asia, and the History of Philosophy: Racism in the Formation of the Philosophical Canon, 1780–1830* (Albany: State University of New York Press, 2013): 1–9.

65. Anne Cheng, "'Y a-t-il une philosophie chinoise?': Est-ce une bonne question?," *Extrême-Orient, Extrême-Occident* 27, no. 27 (2005): 5–12, at 5.

66. Johann Jakob Brucker, *Historia Critica Philosophiae a Tempore Resuscitatarum in Occidente Literarum ad Nostra Tempora* (Leipzig: Breitkopf, 1744): 846–906.

67. Anne-Lise Dyck, "La Chine hors de la philosophie: essai de généalogie à partir des traditions sinologique et philosophique françaises au XIXe siècle," *Extrême-Orient, Extrême-Occident* 27, no. 27 (2005): 13–47, at 20–26.

68. Park, *Africa, Asia, and the History of Philosophy*, 3–4.

69. Anne Cheng, "Philosophy and the French Invention of Sinology: Mapping Academic Disciplines in Nineteenth Century Europe," *China Report* 50, no. 1 (2014): 11–30.

70. Abel-Rémusat, *Mélanges posthumes*, 161.

71. Abel-Rémusat, "Lettre au Rédacteur," 288, 290.

72. Abel-Rémusat, *Mélanges asiatiques: ou, choix de morceaux de critique et de mémoires, relatifs aux religions, aux sciences, aux coutumes, à l'histoire et à la géographie des nations orientales* (Paris: Dondey-Dupré père et fils, 1825–1826): vol. 1, 330; *Mélanges posthumes*, 231; *Mélanges asiatiques*, vol. 1, 88.

73. Abel-Rémusat, *Nouveaux mélanges asiatiques*, vol. 1, 305.

74. Abel-Rémusat, *Mélanges asiatiques*, vol. 1, 330, 132, 88–89.

75. Exceptions include Cordier, "Les études chinoises sous la Révolution et l'Empire"; Landry-Deron, "Le *Dictionnaire Chinois, Français et Latin* de 1813," *T'oung Pao* 101, no. 4–5 (2015): 407–40; Zurndorfer, "Orientalism, Sinology, and Public Policy."

76. Abel-Rémusat, *Essai sur la langue et la littérature chinoises*, 1–3; Paul Demiéville, "Aperçu historique des études sinologiques en France," *Acta Asiatica* (September 1966): 56–100, at 72; Landry-Deron, "Le *Dictionnaire Chinois, Français et Latin*."

77. David Porter, *Ideographia: The Chinese Cipher in Early Modern Europe* (Stanford, CA: Stanford University Press, 2001): 15–76; Haun Saussy, *Great Walls of Discourse and Other Adventures in Cultural China* (Cambridge, MA: Harvard University Asia Center, 2001): 35–52.

78. Jean Chrétien Ferdinand Hoefer, ed., *Nouvelle biographie générale*, vol. 23 (Paris: Firmin Didot, 1852–1856): 95–96.

79. William Huttmann, "A Notice of Several Chinese-European Dictionaries," *Asiatic Journal and Monthly Register for British India and its Dependencies* 12 (1821): 240–44, at 242.

80. Cordier, "Les études chinoises sous la Révolution et l'Empire," 72.

81. Joseph Hager, *An Explanation of the Elementary Characters of the Chinese: With an Analysis of Their Ancient Symbols and Hieroglyphics* (London: Richard Phillips, 1801): ii–xv.

82. Henry McAnally, "Antonio Montucci," *Modern Language Quarterly* 7, no. 1 (1946): 65–81.

83. H. Harrison, *The Perils of Interpreting*, 68–78.

84. Antonio Montucci, *The Title Page Reviewed: The Characteristic Merits of the Chinese Language* (London: Montucci, Spilsbury, and Snowhill, 1801): 2.

85. Antonio Montucci, *Letters to the Editor of the Universal Magazine, on Chinese Literature: Including Strictures on Dr. Hager's Two Works, and the Reviewers' Opinions Concerning Them* (London: Knight and Compton, 1804): 2.

86. "Desvoyes," F^17, 1047, AN; "Commission Temporaire des Arts," F^17, 1188, AN.

87. "Notice des articles curieux composant le Cabinet Chinois de feu M. Bertin," MS Joseph Bernard Maître 50–52, Archives jésuites de la province de France (AJPF), Vanves, France.

88. Finlay, *Henri Bertin*, 147–51.

89. Hoefer, *Nouvelle biographie générale*, vol. 23, 95–96; Hager, *An Explanation of the Elementary Characters of the Chinese*; Joseph Hager, *Description des médailles chinoises du Cabinet Impérial de France: précédée d'un essai de numismatique chinoise* (Paris: Imprimerie Impériale, 1805).

90. Joseph Hager, *Monument de Yu, Ou la plus ancienne inscription de la Chine* (Paris: Treuttel et Wurtz, 1802).

91. Hager, *An Explanation of the Elementary Characters of the Chinese*, xxxvi; Pierre-Martial Cibot, "Essai sur la langue et les caractères des Chinois I," in Joseph-Marie Amiot, François Bourgeois, Pierre-Martial Cibot, Aloys Ko, Aloys de Poirot, Charles Batteux, M. de Bréquigny, A. I. Silvestre de Sacy, and Antoine Gaubil, *Mémoires concernant l'histoire, les sciences, les arts, les moeurs, les usages, &c. des Chinois* [*MCC*], vols. 1–15 (Paris: Nyon l'aîné, 1776–1791): vol. 8 (1782): 133–266, at 192.

92. Hager, *Monument de Yu*, 1–2.

93. Hager, *Monument de Yu*, 9.

94. Montucci, *Letters to the Editor*, 4.

95. Montucci, *Letters to the Editor*, 23.

96. Montucci, *Letters to the Editor*, 22.

97. Antonio Montucci, *Remarques philologiques sur les voyages en Chine de M. de Guignes* (Berlin: Montucci, 1809): 15.

98. Hoefer, *Nouvelle biographie générale*, vol. 23, 96.

99. Hoefer, *Nouvelle biographie générale*, vol. 36, 405.

100. Landry-Deron, "Le *Dictionnaire Chinois, Français et Latin*," 427.

101. Joseph de Guignes to Henri Bertin, September 30, 1785, MS 5401, Fonds Henri Cordier, "Collection des Autographes," IF; Joseph de Guignes to the Académie des inscriptions, October 4, 1785, MS 5401, IF.

102. Chrétien-Louis-Joseph de Guignes, *Voyages à Pékin, Manille, et L'Îsle de France* (Paris: Imprimerie Impériale, 1808): vol. 1, 256.

103. Landry-Deron, "Le *Dictionnaire Chinois, Français et Latin*," 426–28.

104. Joseph-Marie Amiot to Henri Bertin, October 10, 1789, MS 1517, IF.

105. Amiot to Desvoyes, November 17, 1789, MS 1517, IF.

106. Amiot to Bertin, August 20, 1790 / October 4, 1790, MS 1517, IF.

107. Landry-Deron, "Le *Dictionnaire Chinois, Français et Latin*," 433–35.

108. Cordier, "Un orientaliste allemand: Jules Klaproth"; Helmut Walravens, "Julius Klaproth: His Life and Works with Special Emphasis on Japan," *Japonica Humboldtiana* 10 (2006): 177–91.

109. Anonymous, *Asiatic Journal and Monthly Register for British and Foreign India, China and Australasia* XIX (1836): 70.

110. Antonio Montucci, *Réponse à une lettre imprimée et signée Julius v. Klaproth* (Berlin, 1810).

111. Julius Klaproth, *Leichenstein auf dem Grabe der chinesischen Gelehrsamkeit des Herrn Joseph Hager* (Berlin: Waisenhausdruckerei, 1811).

112. Helmut Walravens, "Julius Klaproth, Stanislas Julien, et les débuts de la sinologie européenne," in *Idées de la Chine au XIXe siècle*, eds. Dollé and Espagne, 145–56, at 149.

113. *Shen Yu bei zheng yi* 神禹碑正義.

114. *Records of the Unification of the Great Ming, Daming yitong zhi* 大明一統志; *Huguang Gazetteer, Huguang tongzhi* 湖廣通志.

115. Jules Klaproth, *Inschrift des Yü* (Halle: Waisenhausbuchhandlung, 1811): 19–30.

116. Abel-Rémusat, *Mélanges asiatiques*, vol. 2, 276.

117. Abel-Rémusat, *Mélanges asiatiques*, vol. 2, 272.

118. Bruce Rusk, *Goulou Feng Bei: Stele of Goulou Peak, Ink Squeeze of Mao Huijian's Stele of King Yu of 1666* (Melbourne: Quirin Press, 2016).

119. Abel-Rémusat, "Lettre au Rédacteur," 290; Stanislas Julien, *Le Livre de la voie et de la vertu: Composé dans le VIe siècle avant l'ère chrétienne* (Paris: Imprimerie Royale, 1842), ii.

120. Cheng, "Philosophy and the French Invention of Sinology," 26; Détrie, "L'évolution de l'"Europe Chinoise,'" 29.

121. Jean-Pierre Abel-Rémusat, *Le livre des récompenses et des peines* (Paris: Antoine-Augustin Renouard, 1816): 51.

122. Frédéric Obringer, "Jean-Pierre Abel-Rémusat, médecin et sinologue," at *Jean-Pierre Abel-Rémusat et ses successeurs,* symposium, June 11–13, 2014.

123. Jean-Pierre Abel-Rémusat, *Mémoire sur la vie et les opinions de Lao-Tseu, philosophe chinois du VIe siècle avant notre ère: qui a professé les opinions communément attribuées à Pythagore, à Platon et à leurs disciples* (Paris: Imprimerie Royale, 1823): 1–2.

124. Dehergne, "Les historiens Jésuites du Taoïsme," in *La Mission française de Pékin aux XVIIe et XVIIIe siècles: La Chine au temps des Lumières 2* (Paris: Les Belles Lettres, Cathasia, 1976): 59–67; Knut Walf, "Fascination and Misunderstanding: The Ambivalent Western Reception of Daoism," *Monumenta Serica* 53 (2005): 273–86, at 276–78.

125. Abel-Rémusat, *Mélanges posthumes,* 164.

126. Abel-Rémusat, *Mélanges posthumes,* 240–41.

127. Julien, *Le Livre de la voie et de la vertu,* ii.

128. *Daodejing* 道德經.

129. Abel-Rémusat, *Mélanges asiatiques,* vol. 1, 91.

130. Abel-Rémusat, *Mémoire sur la vie et les opinions de Lao-Tseu,* 2.

131. Anne Cheng, "Abel-Rémusat et Hegel: sinologie et philosophe dans l'Europe du XIXe siècle," at *Jean-Pierre Abel-Rémusat et ses successeurs,* symposium, June 11–13, 2014.

132. Abel-Rémusat, *Mémoire sur la vie et les opinions de Lao-Tseu,* 35.

133. Abel-Rémusat, *Mémoire sur la vie et les opinions de Lao-Tseu,* 19.

134. Abel-Rémusat, *Mélanges posthumes,* 176.

135. Abel-Rémusat, *Mémoire sur la vie et les opinions de Lao-Tseu,* 20.

136. Abel-Rémusat, *Mélanges asiatiques,* vol. 2, 95.

137. Julien, *Le Livre de la voie et de la vertu,* 5.

138. Abel-Rémusat, *Mémoire sur la vie et les opinions de Lao-Tseu,* 1.

139. Abel-Rémusat, *Mémoire sur la vie et les opinions de Lao-Tseu,* 49.

140. Abel-Rémusat, *Mémoire sur la vie et les opinions de Lao-Tseu,* 49.

141. Abel-Rémusat, *Mémoire sur la vie et les opinions de Lao-Tseu,* 51–52.

142. Abel-Rémusat, *Mémoire sur la vie et les opinions de Lao-Tseu,* 24–25.

143. Abel-Rémusat, *Mémoire sur la vie et les opinions de Lao-Tseu,* 52.

144. Abel-Rémusat, *Mémoire sur la vie et les opinions de Lao-Tseu,* 53-54.

145. Walf, "Fascination and Misunderstanding," 278.

146. 道生一，一生二，二生三，三生萬物. 萬物負陰而抱陽，沖氣以為和; *La raison a produit un; un a produit deux; deux a produit trois; trois a produit toutes choses. Toutes choses reposent sur la matière, et sont enveloppées par l'éther; une vapeur ou un souffle qui les unit entretient en eux l'harmonie* (Abel-Rémusat, *Mémoire sur la vie et les opinions de Lao-Tseu,* 31–32).

147. G. E. R. Lloyd and Nathan Sivin, *The Way and the Word: Science and Medicine in Early China and Greece* (New Haven, CT: Yale University Press, 2003).

148. 視之不見名曰夷，聽之不聞名曰希，搏之不得名曰微。此三者不可致詰，故混而爲一; *Celui que vous regardez et que vous ne voyez pas, se nomme I; celui que vous écoutez et que vous n'entendez pas, se nomme Hi; celui que votre main cherche et qu'elle ne peut saisir, se nomme Wei. Ce sont trois êtres qu'on ne peut comprendre, et qui, confondus, n'en font qu'un* (Abel-Rémusat, *Mémoire sur la vie et les opinions de Lao-Tseu*, 40).

149. Dehergne, "Les historiens Jésuites du Taoïsm," 60–61; Abel-Rémusat, *Mémoire sur la vie et les opinions de Lao-Tseu*, 2.

150. Abel-Rémusat, *Mémoire sur la vie et les opinions de Lao-Tseu*, 47.

151. Abel-Rémusat, *Mémoire sur la vie et les opinions de Lao-Tseu*, 40.

152. Abel-Rémusat, *Mélanges posthumes*, 175–76.

153. Abel-Rémusat, *Mémoire sur la vie et les opinions de Lao-Tseu*, 46–47.

154. Abel-Rémusat, *Mélanges posthumes*, 176.

155. Abel-Rémusat, *Mélanges posthumes*, 182.

156. Abel-Rémusat, *Mélanges asiatiques*, vol. 1, 98.

157. *Une vive lumière éclairait la haute antiquité; mais à peine quelques rayons sont venus jusqu'à nous. Il nous semble que les anciens étaient dans les ténèbres, parce que nous les voyons à travers les nuages épais dont nous venons de sortir. L'homme est un enfant né à minuit; quand il voit lever le soleil, il croit qu'hier n'a jamais existé*; Abel-Rémusat, *Mélanges asiatiques*, vol. 1, 99.

158. Abel-Rémusat, *Mélanges posthumes*, 175.

159. Abel-Rémusat, *Mélanges posthumes*, 182; Cheng, "Philosophy and the French Invention of Sinology," 24.

160. Thoraval, "De la magie à la 'raison,'" 114–19.

161. Cheng, "Abel-Rémusat et Hegel."

162. Georg Wilhelm Friedrich Hegel, *Hegel: The Letters*, trans. Clark Butler and Christiane Seiler (Bloomington: Indiana University Press, 1984): 655.

163. Personal communication from Mireille Lamarque, archivist of the Institut de France, 2018.

164. "Trois lettres d'Abel Rémusat à Victor Cousin," in MS VC 214, "Correspondance générale de Victor Cousin," Bibliothèque interuniversitaire de la Sorbonne (BIU), Paris, France.

165. Julien, *Le Livre de la voie et de la vertu*, xvi.

166. "Cinq lettres de Georg Wilhelm Friedrich Hegel à Victor Cousin," in MS VC 232, "Cinq lettres de Georg Wilhelm Friedrich Hegel à Victor Cousin," BIU; Clark Butler in *Hegel: The Letters*, trans. Butler and Seiler, 631.

167. Cheng, "'Y a-t-il une philosophie chinoise?,'" 5.

168. Park, *Africa, Asia, and the History of Philosophy*; Cheng, "Philosophy and the French Invention of Sinology"; Dyck, "La Chine hors de la philosophie"; Fang Weigui, "Transferts de savoirs et représentations de la Chine en France et en Allemagne au XIXe siècle," in *Idées de la Chine au XIXe siècle*, eds. Dollé and Espagne, 39–58.

169. Ernst Rose, "China as a Symbol of Reaction in Germany, 1830–1880," *Comparative Literature* 3, no. 1 (1951): 57–76, at 58.

170. Park, *Africa, Asia, and the History of Philosophy*, 130.

171. Park, *Africa, Asia, and the History of Philosophy*, 9; Young Kun Kim, "Hegel's Criticism of Chinese Philosophy," *Philosophy East and West* 28, no. 2 (1978): 173–80; Wong Kwok Kui, "Hegel's Criticism of Laozi and its Implications," *Philosophy East and West* 61, no. 1 (2011): 56–79.

172. Daniel Carey and Sven Trakulhun, "Universalism, Diversity, and the Postcolonial Enlightenment," in *The Postcolonial Enlightenment: Eighteenth-Century Colonialism and Postcolonial Theory*, eds. Daniel Carey and Lynn Festa (Oxford: Oxford University Press, 2013): 254–89, at 254–78.

173. Dipesh Chakrabarty, *Provincializing Europe: Postcolonial Thought and Historical Difference* (Princeton, NJ: Princeton University Press, 2000); Gayatri Spivak, *A Critique of Postcolonial Reason: Toward a History of the Vanishing Present* (Cambridge, MA: Harvard University Press, 1999).

174. Isaiah Berlin, "The Counter-Enlightenment," in *Dictionary of the History of Ideas*, vol. II, ed. Philip Wiener (New York: Scribner, 1973): 100–12, at 100–01.

175. Mark Lilla, "What is Counter-Enlightenment?," in *Isaiah Berlin's Counter-Enlightenment*, eds. Joseph Mali and Robert Wokler (Philadelphia: American Philosophical Society, 2003): 1–11, at 7–10; Paul Redding, "Hegel," 1997; revised 2015, *Stanford Encyclopedia of Philosophy*, ed. Edward N. Zalta, https://plato.stanford.edu/entries/hegel; George Dennis O'Brien, *Hegel on Reason and History* (Chicago: University of Chicago Press, 1975): 32–35; John E. Toews, "Berlin's Marx: Enlightenment, Counter-Enlightenment, and the Historical Construction of Cultural Identities," in *Isaiah Berlin's Counter-Enlightenment*, eds. Mali and Wokler, 163–76.

176. Georg Wilhelm Friedrich Hegel, *The Philosophy of History*, trans. J. Sibree, rev. ed. (New York: Colonial Press, 1900): 72; Georg Wilhelm Friedrich Hegel, *Elements of the Philosophy of Right*, ed. Allen W. Wood, trans. H. B. Nisbet (Cambridge: Cambridge University Press, 1991): 21.

177. Thoraval, "De la magie à la 'raison,'" 123.

178. Joseph-Marie Amiot, "Extrait d'une lettre écrite de Péking le 16 oct. 1787 (Traité sur la secte des *Tao-sée*)," in Amiot et al., *MCC*, vol. 15 (1791): 208–59, at 210, 227.

179. Amiot, "Traité sur la secte des *Tao-sée*," 208.

180. Jiang Ziya姜子牙; King Wu of Zhou, Zhou Wu Wang 周武王; Amiot, "Traité sur la secte des *Tao-sée*," 233.

181. Amiot, "Traité sur la secte des *Tao-sée*," 241.

182. Amiot, "Traité sur la secte des *Tao-sée*," 215.

183. Amiot, "Traité sur la secte des *Tao-sée*," 258.

184. *Investiture of the Gods, Feng shen yanyi* 封神演義;.

185. Thoraval, "De la magie à la 'raison,'" 124–36.

186. Amiot, "Traité sur la secte des *Tao-sée*," 254–56.

187. Peter C. Hodgson in Georg Wilhelm Friedrich Hegel, *Lectures on the Philosophy of Religion: Determinate Religion*, ed. Peter C. Hodgson, trans. R. F. Brown, Peter C. Hodgson, and J. M. Stewart, with H. S. Harris (Berkeley: University of California Press, 1987): vol. 2, 5.

188. Hegel, *Lectures on the Philosophy of Religion*, vol. 2, 301.

189. Hegel, *Lectures on the Philosophy of Religion*, vol. 2, 299.

190. Hegel, *Lectures on the Philosophy of Religion*, vol. 2, 301–03.

191. Thoraval, "De la magie à la 'raison,'" 126–36.

192. Hodgson, "Editorial Introduction," in Hegel, *Lectures on the Philosophy of Religion*, 1–90, at 58–60; Thoraval, "De la magie à la 'raison,'" 118–20.

193. Hegel, *Lectures on the Philosophy of Religion*, vol. 2, 557–58.

194. Hodgson in Hegel, *Lectures on the Philosophy of Religion*, vol. 2, 530.

195. Hegel, *Lectures on the Philosophy of Religion*, vol. 2, 559.

196. *Dao* 道; *Tian* 天; Joseph-Marie Amiot, "Antiquité des Chinois, prouvée par les monuments," in Amiot et al., *MCC*, vol. 2 (1776): 1–364, at 22. The modern editors of the *Lectures on the Philosophy of Religion* understandably failed to see how the comment could apply to the character *Dao* and were unable to locate its original source (Hodgson in Hegel, *Lectures on the Philosophy of Religion*, vol. 2, 559).

197. Hegel, *Lectures on the Philosophy of Religion*, vol. 2, 558.

198. Amiot, "Traité sur la secte des *Tao-sée*," 228.

199. Hegel, *Lectures on the Philosophy of Religion*, vol. 2, 561.

200. The text was compiled from notes taken from 1822 to 1831, but context and references date the relevant section to later in that decade: for example, Hegel mentioned the later influence of Ferdinand Eckstein's *Le Catholique*, which began publication in 1826, as well as Friedrich Schlegel's *Philosophy of History*, which was published in 1829.

201. Hegel, *The Philosophy of History*, 58–59.

202. Hegel, *The Philosophy of History*, 120.

203. Redding, "Hegel," section 3.2.1.

204. Hegel, *The Philosophy of History*, 133–38.

205. Hegel, *The Philosophy of History*, 135.

206. Cheng, "Philosophy and the French Invention of Sinology," 24–25.

207. Robert Nisbet, "Idea of Progress: A Bibliographical Essay," *Liberty Fund*, 1979, https://oll4.libertyfund.org/page/idea-of-progress-a-bibliographical-essay-by-robert-nisbet.

208. Muriel Détrie, "L'image du Chinois dans la littérature occidentale au XIXe siècle," in *La Chine entre amour et haine: Actes du VIIIe Colloque de sinologie de Chantilly*, ed. Cartier, 403–29, at 411.

209. Hegel, *The Philosophy of History*, 105.

210. Hegel, *The Philosophy of History*, 103.

211. Hegel, *The Philosophy of History*, 116.

212. Hegel, *The Philosophy of History*, 113, 105.

213. Georg Wilhelm Friedrich Hegel, *Lectures on the History of Philosophy*, trans. E. S. Haldane (London: Kegan Paul, Trench, Trübner, 1892): 121.

214. Joseph Needham, *Science and Civilisation in China: Volume I: Introductory Orientations* (Cambridge: Cambridge University Press, 1954): 38–39; Abel-Rémusat, *Programme du cours*, 21–23. One significant difference is that in the original, Abel-Rémusat referred to the "progress of these sciences" in Europe not as "recent," but rather as taking place "in the last 200 years."

215. Needham, *Science and Civilisation in China: Volume I*, 9.

216. Needham, *Science and Civilisation in China: Volume I*, 4; Joseph Needham and Kenneth Girdwood Robinson, *Science and Civilisation in China: Volume VII, The Social Background, Part 2, General Conclusions and Reflections* (Cambridge: Cambridge University Press, 2004): 1.

217. "Why did modern science, the mathematization of hypotheses about Nature, with all its implications for advanced technology, take its meteoric rise *only* in the West at the time of Galileo?" See Joseph Needham, *The Grand Titration: Science and Society in East and West* (1969), reprint ed. (Oxford: Routledge, 2005): 16; Nathan Sivin, "Why the Scientific Revolution Did Not Take Place in China—or Didn't It?," *Chinese Science* 5 (1982): 45–66, at 45.

218. Mark Elvin, "Vale Atque Ave," in Needham and Robinson, *Science and Civili-*

sation in China: Volume VII, Part II, xxiv–xliii, at xl–xli; Timothy Brook, "The Sinology of Joseph Needham," *Modern China* 22, no. 3 (1996): 340–48, at 342–45.

219. Needham and Robinson, Science and Civilisation in China: Volume VII, Part II, 210.

220. Needham and Robinson, *Science and Civilisation in China: Volume VII, Part II*, xliv–xlv.

221. Needham, *Science and Civilisation in China: Volume I*, 3.

222. Needham and Robinson, *Science and Civilisation in China: Volume VII, Part II*, 24–25.

223. Needham, *Science and Civilisation in China: Volume I*, 3.

224. Abel-Rémusat, *Programme du cours*, 21–23.

225. It was posthumously published twice: first in 1833, and again with significant revisions in 1843; Abel-Rémusat, *Mélanges posthumes*, 219.

226. Jean-Pierre Abel-Rémusat, "Observations sur l'état des sciences naturelles chez les peuples de l'Asie orientale," *Mémoires de l'Institut Royale de France, Académie des inscriptions et belles-lettres*, vol. 10 (Paris: Imprimerie Royale, 1833): 130.

227. Abel-Rémusat, "Observations sur l'état des sciences naturelles chez les peuples de l'Asie orientale," x, 130; Abel-Rémusat, *Mélanges posthumes*, 219–20.

228. Abel-Rémusat, *Mélanges posthumes*, 207.

229. Abel-Rémusat, *Mélanges posthumes*, 214.

230. Abel-Rémusat, *Mélanges posthumes*, 211; Wilhelm von Humboldt, *Lettre à M. Abel-Rémusat* (Paris: Dondey-Dupré, 1827); the idea is still discussed by some today (Joseph Needham and Christoph Harbsmeier, *Science and Civilisation in China, Volume VII: The Social Background, Part I: Language and Logic* (Cambridge: Cambridge University Press, 1998): 25–26.

231. Abel-Rémusat, *Mélanges posthumes*, 217–18, 199.

232. Joseph Needham with Wang Ling, *Science and Civilisation in China: Volume II: History of Scientific Thought* (Cambridge: Cambridge University Press, 1956): 457-58.

233. Abel-Rémusat, *Mélanges posthumes*, 207.

234. Abel-Rémusat, *Mélanges posthumes*, 224.

235. Abel-Rémusat, *Mélanges posthumes*, 211; Bruce Mazlish, *Civilization and Its Contents* (Stanford, CA: Stanford University Press, 2004): 1–19.

236. Needham, *Science and Civilisation in China: Volume I*, 19; Needham and Robinson, *Science and Civilisation in China: Volume VII, Part II*, 19.

237. Abel-Rémusat, *Mélanges asiatiques*, vol. 1, 242.

238. Abel-Rémusat, *Mélanges asiatiques*, vol. 1, 241.

239. Abel-Rémusat, *Mélanges posthumes*, 224.

240. Abel-Rémusat, *Mélanges posthumes*, 265–66.

241. Abel-Rémusat, *Mélanges posthumes*, 230.

242. Abel-Rémusat, *Mélanges posthumes*, 225.

243. Abel-Rémusat, *Mélanges posthumes*, 277.

244. Abel-Rémusat, *Mélanges posthumes*, 253.

245. Markus Messling, "Repräsentation und Macht: Selbstkritik der Philologie in Zeiten ihrer Ermächtigung," in *Sprachgrenzen—Sprachkontakte—kulturelle Vermittler: Kommunikation zwischen Europäern und Außereuropäern (16.-20. Jahrhundert)*, eds. Mark Häberlin and Alexander Keese (Stuttgart: Franz Steiner Verlag, 2010): 247–60.

246. Michael Adas, *Machines as the Measure of Men: Science, Technology, and Ideologies of Western Dominance* (1989), reprint ed. (Ithaca, NY: Cornell University Press, 2015): 199–219.

247. Abel-Rémusat, *Mélanges posthumes*, 231.

248. Abel-Rémusat, *Mélanges posthumes*, 220.

249. Abel-Rémusat, *Mélanges posthumes*, 252.

250. Abel-Rémusat, *Mélanges posthumes*, 245.

251. Abel-Rémusat, *Mélanges posthumes*, 177.

252. Abel-Rémusat, *Mélanges posthumes*, 196.

253. Abel-Rémusat, *Mélanges posthumes*, 234.

254. Said, *Orientalism*, 26.

255. Abel-Rémusat, *Mélanges posthumes*, 231.

256. John King Fairbank, *Trade and Diplomacy on the China Coast: The Opening of the Treaty Ports, 1842–1854* (Cambridge, MA: Harvard University Press, 1953); Julia Lowell, *The Opium War: Drugs, Dreams, and the Making of Modern China* (London: Picador, 2011).

257. Détrie, "L'image du Chinois," 407; Paul A. Cohen, *Between Tradition and Modernity: Wang T'ao and Reform in Late Ch'ing China* (Cambridge, MA: Harvard University Press, 1974); Wright, "The Study of Chinese Civilization," 233–45.

258. Raymond Schwab, *La Renaissance orientale* (Paris: Payot, 1950): xx.

259. Marchand, *German Orientalism*, 53–74.

260. Schwab, *La Renaissance orientale*, 20; John Bagnell Bury, *The Idea of Progress: An Inquiry into Its Origin and Growth* (1932), reprint ed. (New York: Dover Publications, 1955): 258.

261. Urs App, *Arthur Schopenhauer and China*, ed. Victor H. Mair (Philadelphia: Sino-Platonic Papers, 2010): vi.

262. Said, *Orientalism*, 115; Georgios Varouxakis, "The Godfather of 'Occidentality': Auguste Comte and the Idea of 'The West,'" *Modern Intellectual History* (October 11, 2017): 1–31, at 1–11, https://doi.org/10.1017/S1479244317000415.

263. Said, *Orientalism*, 125.

264. Auguste Viatte, *Les sources occultes du Romantisme* (Paris: Honoré Champion, 1928): 168.

Conclusion

1. Immanuel Kant, *Practical Philosophy*, ed. Mary J. Gregor (Cambridge: Cambridge University Press, 1996): 11.

2. Michel Foucault, *The Foucault Reader*, ed. Paul Rabinow (New York: Pantheon Books, 1984): 34.

3. Peter J. Bowler and Iwan Rhys Morus, *Making Modern Science: A Historical Survey* (Chicago: University of Chicago Press, 2005): 329–37; Peter Harrison, *The Fall of Man and the Foundations of Science* (Cambridge: Cambridge University Press, 2007): 245–58.

4. Steven Press, *Rogue Empires: Contract and Conmen in Europe's Scramble for Africa* (Cambridge, MA: Harvard University Press, 2017): 166–218.

5. Alison Butler, *Victorian Occultism and the Making of Modern Magic: Invoking Tradition* (Basingstoke: Palgrave McMillan, 2011): viii–xiii, 1–16.

6. Edward Said, *Orientalism* (New York: Pantheon Books, 1978): 166–97.

7. Said, *Orientalism*, 125–26, 137–38.

8. Joscelyn Godwin, *Theosophical Enlightenment* (Albany: State University of New York Press, 1994): 277–331.

9. Henrik Bogdan and Martin P. Starr, eds., *Aleister Crowley and Western Esotericism* (Oxford: Oxford University Press, 2012); Aleister Crowley, cited in Jason Ananda Josephson-Storm, *The Myth of Disenchantment: Magic, Modernity, and the Birth of the Human Sciences* (Chicago: University of Chicago Press, 2017): 174.

10. Josephson-Storm, *The Myth of Disenchantment*, 1–21.

11. Max Weber, *The Vocation Lectures*, eds. David Owen and Tracy B. Strong (Indianapolis, IN: Hackett, 2004): 13; scare quotes in the original.

12. Franz Boas, *A Franz Boas Reader: The Shaping of American Anthropology, 1883–1911*, ed. George W. Stocking Jr. (Chicago: University of Chicago Press, 1974): 67–71.

13. Sigmund Freud, *Civilization and Its Discontents*, ed. James Strachey (New York: W. W. Norton, 1962): 33–45.

14. Marie-Jean-Léon Lecoq, marquis d'Hervey de Saint-Denys, *Les rêves et les moyens de les diriger: Observations pratiques* (Paris: Amyot, 1867); Carl Gustav Jung, "Foreword to the Second German Edition," in *The Secret of the Golden Flower: A Chinese Book of Life*, ed. Richard Wilhelm (New York: Harcourt, Brace & World, 1962): xi–xiv.

15. Kang Youwei 康有為; Thomas Fröhlich, "Introduction: Progress, History, and Time in Chinese Discourses after the 1890s," in Thomas Fröhlich and Axel Schneider, *Chinese Visions of Progress, 1895 to 1949* (Leiden: Brill, 2020): 1–40, at 1–8.

16. Li Qiang, "The Idea of Progress in Modern China: The Case of Yan Fu," in Fröhlich and Schneider, *Chinese Visions of Progress, 1895 to 1949*, 103–20.

17. Peter Zarrow, "An Anatomy of the Utopian Impulse in Modern Chinese Political Thought, 1890–1940," in Fröhlich and Schneider, *Chinese Visions of Progress, 1895 to 1949*, 165–205, at 183–91.

18. Yan Fu, *Yan Fu ji* (Selected works of Yan Fu), ed. Wang Shi (Beijing: Zhonghua shuju, 1986): vol. 1, 210–18.

19. Chen Duxiu, *Duxiu wencun* (Collected works of Chen Duxiu) (1922), reprint ed. (Hefei: Anhui renmin chubanshe, 1987): vol. 1, 10.

20. Vera Schwarcz, *The Chinese Enlightenment: Intellectuals and the Legacy of the May Fourth Movement of 1919* (Berkeley: University of California Press, 1986).

21. Liang Qichao 梁啓超, quoted in William Theodore de Bary and Richard Lurfrano, eds., *Sources of Chinese Tradition: Volume II: From 1600 Through the Twentieth Century*, 2nd ed. (New York: Columbia University Press, 2000): 378.

22. G. William Skinner, "What the Study of China Can Do for Social Science," *Journal of Asian Studies* 23, no. 4 (1964): 517–22, at 522.

23. Frederick W. Mote, "The Case for the Integrity of Sinology," *Journal of Asian Studies* 23, no. 4 (1964): 531–33.

24. Skinner, "What the Study of China Can Do for Social Science," 519–20.

25. Francis Fukuyama, *The End of History and the Last Man* (London: Penguin, 1992).

26. Barack Obama, "Remarks by President Obama in Address to the United Nations General Assembly," New York, September 24, 2014, https://obamawhitehouse.archives.gov.

27. Donald J. Trump, "Inaugural Address," Washington, DC, January 20, 2017, https://www.govinfo.gov/features/presidential-inaugural-addresses.

28. Steven Pinker, *Enlightenment Now: The Case for Reason, Science, Humanism, and Progress* (New York: Viking, 2018).

29. Bill Gates, "My New Favorite Book of All Time," *GatesNotes: The Blog of Bill Gates*, January 26, 2018, https://www.gatesnotes.com/books/enlightenment-now.

30. Caroline Winterer, "Buck Up, Everyone! We Are Riding along the Enlightenment's Long Path of Progress," *Washington Post*, February 23, 2018; David Bell, "The PowerPoint Philosophe: Waiting for Steven Pinker's Enlightenment," *Nation*, April 2, 2018; Jessica Riskin, "Pinker's Pollyannish Philosophy and Its Perfidious Politics," *Los Angeles Review of Books*, December 15, 2019.

31. Pinker, *Enlightenment Now*, 39.

32. Ross Douthat, "The Edges of Reason," *New York Times*, February 25, 2018.

33. Amanda Svachula, "Like a Virgo: How the Times Covers Astrology," *New York Times*, August 31, 2019.

34. Sarah Pulliam Bailey, "Tarot Cards Are Having a Moment with Help from the Pandemic," *Washington Post*, December 10, 2021; MIT Libraries, "Distinctive Collections," https://libraries.mit.edu/distinctive-collections/collections/visual-collections.

35. Mary B. Campbell, Lorraine Daston, Arnold Ira Davidson, John Forrester, and Simon Goldhill, "Enlightenment Now: Concluding Reflections on Knowledge and Belief," *Common Knowledge* 13, no. 2–3 (2007): 429–50.

36. Larry Laudan, "The Demise of the Demarcation Problem," in *Physics, Philosophy and Psychoanalysis*, eds. Robert S. Cohen and Larry Laudan (Dordrecht: D. Reidel, 1983): 120–25.

37. Thomas Kuhn, *The Structure of Scientific Revolutions* (Chicago: University of Chicago Press, 1962); Bruno Latour and Steve Woolgar, *Laboratory Life: The Construction of Scientific Facts* (Beverly Hills: Sage Publications, 1979).

38. Michael Gordin, *On the Fringe: Where Science Meets Pseudoscience* (New York: Oxford University Press, 2021): 1–28.

39. Johns Hopkins Medicine, "Acupuncture," accessed May 13, 2022, https://www.hopkinsmedicine.org/health/wellness-and-prevention/acupuncture; William S. Waldren, "Buddhism in Psychology and Psychotherapy," *Oxford Bibliographies*, last modified September 13, 2010, https://www.oxfordbibliographies.com/view/document/obo-9780195393521/obo-9780195393521-0130.xml.

40. Bruno Latour, *Facing Gaia: Eight Lectures on the New Climatic Regime*, trans. Catherine Porter (Cambridge, MA: Polity Press, 2017); Carolyn Merchant, *The Anthropocene and the Humanities: From Climate Change to a New Age of Sustainability* (New Haven, CT: Yale University Press, 2020); Jim Robbins, "Native Knowledge: What Ecologists Are Learning from Indigenous People," Yale School of the Environment: *Yale Environment 360*, April 26, 2018, https://e360.yale.edu/features/native-knowledge-what-ecologists-are-learning-from-indigenous-people; James C. Scott, *Seeing Like a State: How Certain Schemes to Improve the Human Condition Have Failed* (New Haven, CT: Yale University Press, 1998); Julie Cruikshank, "Are Glaciers 'Good to Think With'? Recognising Indigenous Environmental Knowledge," *Anthropological Forum* 22, no. 3 (2012): 239–50.

41. Hu Jintao, "Full text of Hu's report at the 18th Party Congress," Beijing, November 8, 2012, posted on *China Daily*, November 18, 2012, https://www.chinadaily.com.cn/china/19thcpcnationalcongress/2012-11/18/content_29578562.htm.

42. Bridie Andrews, *The Making of Modern Chinese Medicine, 1850-1960* (Honolulu: University of Hawaii Press, 2015): 1–14; Richard J. Smith, *Fathoming the Cosmos*

and Ordering the World: The Yijing (I Ching, or Classic of Changes) and Its Evolution in China (Charlottesville: University of Virginia Press, 2008): 218–19.

43. Xi Jinping, "Full Text of Xi Jinping's Report at 19th CPC National Congress," Beijing, October 18, 2017, posted on *China Daily*, updated November 4, 2017, https://www.chinadaily.com.cn/china/19thcpcnationalcongress/2017-11/04/content_34115212.htm.

44. Joseph-Marie Amiot to Louis-Raphaël-Lucrèce de Fayolle, comte de Mellet, November 14, 1789, MS 1517, "80 lettres du P. Amiot à Bertin," Institut de France (IF), Paris, France.

Bibliography

Published Sources Written before 1850

Abel-Rémusat, Jean-Pierre. *Dissertatio de glossosemeiotice sive de signis morborum qua è linguâ sumuntur praesertim apud Sinenses*. Paris: Didot, 1813.

Abel-Rémusat, Jean-Pierre. *Essai sur la langue et la littérature chinoises: avec cinq planches, contenant des textes chinois, accompagnés de traductions, de remarques et d'un commentaire littéraire et grammatical. Suivi de notes et d'une table alphabétique des mots chinois*. Paris: Treuttel et Wurtz, 1811.

Abel-Rémusat, Jean-Pierre. *Le livre des récompenses et des peines*. Paris: Antoine-Augustin Renouard, 1816.

Abel-Rémusat, Jean-Pierre. "Lettre au Rédacteur, sur l'état et les progrès de la littérature chinoise en Europe." *Journal Asiatique* 1 (1822): 279–91.

Abel-Rémusat, Jean-Pierre. *Mélanges asiatiques, ou, choix de morceaux de critique et de mémoires, relatifs aux religions, aux sciences, aux coutumes, à l'histoire et à la géographie des nations orientales*. Vols. 1–2. Paris: Dondey-Dupré père et fils, 1825–1826.

Abel-Rémusat, Jean-Pierre. *Mélanges posthumes d'histoire et de littérature orientales*. Paris: Imprimerie Royale, 1843.

Abel-Rémusat, Jean-Pierre. *Mémoire sur la vie et les opinions de Lao-Tseu, philosophe chinois du VIe siècle avant notre ère: Qui a professé les opinions communément attribuées à Pythagore, à Platon et à leurs disciples*. Paris: Imprimerie Royale, 1823.

Abel-Rémusat, Jean-Pierre. *Nouveaux mélanges asiatiques: ou, Recueil de morceaux de critique et de mémoires, relatifs aux religions, aux sciences, aux coutumes, à l'histoire et la géographie des nations orientales*. Paris: Schubart et Heideloff, 1829.

Abel-Rémusat, Jean-Pierre. "Observations sur l'état des sciences naturelles chez les peuples de l'Asie orientale." In *Mémoires de l'Institut Royale de France, Académie des inscriptions et belles-lettres*, vol. 10, 116–67. Paris: Imprimerie Royale, 1833.

Abel-Rémusat, Jean-Pierre. *Programme du cours de langue et de littérature chinoises et de tartare-mandchou; précédé du discours prononcé à la première séance de ce cours, dans l'une des salles du Collège royal de France, le 16 janvier 1815*. Paris: Charles, 1815.

Académie française. *Le Dictionnaire de l'Académie Française, Cinquième Édition*. Paris: 1798. ARTFL Encyclopédie Projet, edited by Robert Morrissey.

Académie française. *Le Dictionnaire de l'Académie Française, Sixième Édition*. Paris: 1835. ARTFL Encyclopédie Projet, edited by Robert Morrissey.

Académie française. *Supplément au Dictionnaire de l'Académie Française.* Paris: Gustave Barba, 1836. ARTFL Encyclopédie Projet, edited by Robert Morrissey.

Agui 阿桂 et al. *Pingding liang Jinchuan fanglüe* 平定兩金川方略 [Strategy of the pacification of the two Jinchuans]. Beijing: 1784.

Aimé-Martin, Louis, ed. *Lettres édifiantes et curieuses, concernant l'Asie, l'Afrique, et l'Amérique.* Vols. 3–4. Paris: Société du Panthéon Littéraire, 1843.

Alibert, Jean-Louis. *Nouveaux éléments des thérapeutiques et de matière médicale.* 5th ed. Paris: Béchet, 1826.

Amiot, Joseph-Marie. "Abrégé chronologique de l'histoire universelle de l'Empire Chinois" (1769). In Amiot et al., *MCC*, vol. 13 (1788): 74–308.

Amiot, Joseph-Marie. "Antiquité des Chinois, prouvée par les monuments." In Amiot et al., *MCC*, vol. 2 (1776): 1–364.

Amiot, Joseph-Marie. *Art militaire des Chinois, ou, Recueil d'anciens traits sur la guerre, composés avant l'ère chrétienne, par différents généraux Chinois.* Edited by Joseph de Guignes. Paris: Didot l'aîné, 1772.

Amiot, Joseph-Marie. *Éloge de la ville de Mukden et de ses environs.* Edited by Joseph de Guignes. Paris: N. M. Tilliard, 1770.

Amiot, Joseph-Marie. "Extrait d'une lettre écrite de Péking le 15 oct. 1773." In Amiot et al., *MCC*, vol. 1 (1776): 419–27.

Amiot, Joseph-Marie. "Extrait d'une lettre écrite de Péking le 13 juillet, 1778." In Amiot et al., *MCC*, vol. 15 (1791): 281–91.

Amiot, Joseph-Marie. "Extrait d'une lettre écrite de Péking le 13 août 1780." In Amiot et al., *MCC*, vol. 9 (1783): 6–24.

Amiot, Joseph-Marie. "Extrait d'une lettre écrite de Péking le 26 sep. 1780." In Amiot et al., *MCC*, vol. 9 (1783): 45–59.

Amiot, Joseph-Marie. "Extrait d'une lettre écrite de Péking le 17 août 1781." In Amiot et al., *MCC*, vol. 9 (1783): 441–54.

Amiot, Joseph-Marie. "Extrait d'une lettre écrite de Péking le 2 oct. 1784." In Amiot et al., *MCC*, vol. 11 (1786): 515–68.

Amiot, Joseph-Marie. "Extrait d'une autre lettre écrite de Péking le 15 nov. 1784." In Amiot et al., *MCC*, vol. 11 (1786): 569–76.

Amiot, Joseph-Marie. "Extrait d'une lettre écrite de Péking le 15 oct. 1785." In Amiot et al., *MCC*, vol. 12 (1786): 509–30.

Amiot, Joseph-Marie. "Extrait d'une lettre écrite de Péking le 20 mai 1786." In Amiot et al., *MCC*, vol. 13 (1788): 417–58.

Amiot, Joseph-Marie. "Extrait d'une lettre écrite de Péking le 20 sep. 1786." In Amiot et al., *MCC*, vol. 13 (1788): 507–10.

Amiot, Joseph-Marie. "Extrait d'une lettre écrite de Péking le 16 oct. 1787 (Traité sur la secte des *Tao-sée*)." In Amiot et al., *MCC*, vol. 15 (1791): 208–59.

Amiot, Joseph-Marie. "Extrait d'une lettre écrite de Péking le 19 nov. 1787." In Amiot et al., *MCC*, vol. 14 (1789): 536–81.

Amiot, Joseph-Marie. "Extrait d'une lettre écrite de Péking le 26 juin 1789." In Amiot et al., *MCC*, vol. 15 (1791): v–xv.

Amiot, Joseph-Marie. "Introduction à la connaissance des peuples qui ont été ou qui sont actuellement tributaires de la Chine." In Amiot et al., *MCC*, vol. 14 (1789): 1–238.

Amiot, Joseph-Marie. "Lettre du Père Amiot au Père Allart . . . le 20 octobre 1752."

In *Lettres édifiantes et curieuses*, vol. 3, edited by Louis Aimé-Martin, 832–39. Paris: Société du Panthéon Littéraire, 1843.

Amiot, Joseph-Marie. "Lettre sur la réduction des Miao-tsée en 1775" (1776). In Amiot et al., *MCC*, vol. 3 (1778): 387–411.

Amiot, Joseph-Marie. "Lettre sur les caractères chinois" (1764). In Amiot et al., *MCC*, vol. 1 (1776): 275–324.

Amiot, Joseph-Marie. "Observations sur le livre de M. de Pauw, intitulé *Recherches philosophiques sur les Égyptiens et les Chinois*" (1777). In Amiot et al., *MCC*, vol. 6 (1780): 275–345.

Amiot, Joseph-Marie. "Suite du mémoire sur les danses religieuses, politiques, et civiles des anciens chinois" (1789). In *Les Danses rituelles chinoises d'après Joseph-Marie Amiot: Aux sources de l'ethnochorégraphie*, edited by Yves Lenoir and Nicolas Standaert, 243–59. Namur: Presses Universitaires de Namur, 2005.

Amiot, Joseph-Marie, François Bourgeois, Pierre-Martial Cibot, Aloys Ko, Aloys de Poirot, Charles Batteux, M. de Bréquigny, A. I. Silvestre de Sacy, and Antoine Gaubil. *Mémoires concernant l'histoire, les sciences, les arts, les moeurs, les usages, &c. des Chinois* [*MCC*]. Vols. 1–15. Paris: Nyon l'aîné, 1776–1791.

Anonymous. *Asiatic Journal and Monthly Register for British and Foreign India, China and Australasia* XIX (1836).

Anonymous. "Idée générale de la Chine et de ses relations avec l'Europe sur le nom de Chine." In Amiot et al., *MCC*, vol. 5 (1780): 1–68.

Anonymous. "Lettre d'un missionnaire de Chine, mort du père Benoist, à Paris, année 1775." In *Lettres édifiantes et curieuses*, vol. 4, edited by Louis Aimé-Martin, 225–33. Paris: Société du Panthéon Littéraire, 1843.

Anonymous. "Note sur la vie et les ouvrages de M. de Guignes." In *Mémoires de l'Académie des Inscriptions et Belles-Lettres*, vol. 48, 770–72. Paris: Imprimerie Impériale, 1808.

Anonymous. "Notice Historique sur la vie et les ouvrages de M. Abel Rémusat." In *Mémoires de l'Institut royal de France: Académie des inscriptions et belles-lettres*, vol. 12, 375–400. Paris: Imprimerie Royale, 1839.

Anonymous. "Notice historique sur la vie et les ouvrages de M. de Bréquigny." In *Mémoires de l'Académie des Inscriptions et Belles-Lettres*, vol. 50, 719–21. Paris: Imprimerie Impériale, 1808.

Anonymous. "Notice Historique sur l'Académie des inscriptions et Belles-Lettres." *Comptes rendus des séances de l'Académie des inscriptions et Belles-Lettres* (1857): 1–43.

Anonymous. "Préface." In Amiot et al., *MCC*, vol. 1 (1776): i–xv.

Anonymous ("Un Amateur de la Vérité"). *Rapport du Rapport de MM. les commissaires nommés par le Roi pour examiner la pratique de M. Deslon sur le magnétisme animal.* Couturier: Beijing [Paris?], 1784.

Bacon, Francis. *Novum Organum* (1620). In *Francis Bacon: Selected Philosophical Works*, edited by Rose-Mary Sargent. Indianapolis: Hackett Publishing Co., 1999.

Bailly, Jean-Sylvain. *Éloge de Leibnitz, qui a remporté le prix à l'Académie royale des sciences et belles-lettres.* Berlin: Haude and Spener, 1768.

Bailly, Jean-Sylvain. *Essai sur les fables et sur leur histoire.* Vols. 1–2. Paris: de Bure l'aîné, 1799.

Bailly, Jean-Sylvain. *Exposé des expériences qui ont été faites pour l'examen du magnétisme animal*. Paris: Mourard, 1784.

Bailly, Jean-Sylvain. *Histoire de l'astronomie ancienne: depuis son origine jusqu'à l'établissement de l'école d'Alexandrie*. Paris: les frères de Bure, 1775.

Bailly, Jean-Sylvain. *Lettres sur l'Atlantide de Platon et sur l'ancienne histoire de l'Asie: pour servir de suite aux Lettres sur l'origine des sciences, adressées à M. de Voltaire par M. Bailly*. London: M. Elmesly; Paris: les frères de Bure, 1779.

Bailly, Jean-Sylvain. *Recueil de pièces intéressantes sur les arts, les sciences et la littérature: Ouvrage posthume de Sylvain Bailly*. Paris: Ainé et Jeune, 1810.

Bailly, Jean-Sylvain. *Traité de l'astronomie indienne et orientale*. Paris: de Bure l'aîné, 1787.

Bailly, Jean-Sylvain, and Voltaire. *Lettres sur l'origine des sciences et sur celle des peuples de l'Asie: Adressées à M. de Voltaire Par M. Bailly, & précédées de quelques lettres de M. de Voltaire à l'auteur*. London: M. Elmesly; Paris: de Bure l'aîné, 1777.

Barrett, Timothy H. "A Bicentenary in Robert Morrison's Scholarship on China and His Significance for Today." *Journal of the Royal Asiatic Society*, 3rd series, 25, no. 4 (2015): 705–16.

Baudeau, Nicolas. "Le Chou-King." In *Éphémérides du citoyen, ou Bibliothèque raisonnée des sciences morales et politiques*, vol. 7, 139–69. Paris: Lacombe, 1770.

Baxun wanshou shengdian 八旬萬壽盛典 [Grand celebration of the eightieth birthday]. Beijing: 1792.

Becdelièvre-Hamal, Antoine-Gabriel de. *Biographie liégeoise*. Liège: Jeunehomme, 1837.

Benoist, Michel. Letter of September 12, 1764. In *Revue de l'Extrême-Orient*, vol. 3, edited by Henri Corder, 242–51. 1887.

Benoist, Michel. "Lettre du Père Benoist" [1772]. In *Lettres édifiantes et curieuses*, vol. 4, edited by Louis Aimé-Martin, 209–17. Paris: Société du Panthéon Littéraire, 1843.

Benoist, Michel. "Lettre du Père Benoist" [c.1772]. In *Lettres édifiantes et curieuses*, vol. 4, edited by Louis Aimé-Martin, 217–25. Paris: Société du Panthéon Littéraire, 1843.

Bourgeat, Louis-Alexandre-Marguerite. "No. XIV, Langues Orientales." *Mercure Étrangère, ou Annales de la littérature Étrangère*. Vol. 3. Paris: D. Collas, 1814.

Bourgeois, François. "Extrait d'une lettre écrite de Péking le 19 nov. 1784." In Amiot et al., *MCC*, vol. 11 (1786). 577–79.

Breton de la Martinière, Jean Baptiste Joseph. *China: its Costume, Arts, Manufactures, &c. Edited Principally from the Originals in the Cabinet of the Late M. Bertin*. 2nd ed. Translated. London: J. J. Stockdale, 1812.

Brucker, Johann Jakob. *Historia Critica Philosophiae a Tempore Resuscitatarum in Occidente Literarum ad Nostra Tempora*. Leipzig: Breitkopf, 1744.

Cambry, Jacques. *Traces du magnétisme*. Le Hague, 1784.

Cibot, Pierre-Martial. "Essai sur la langue et les caractères des Chinois I." In Amiot et al., *MCC*, vol. 8 (1782): 133–266.

Cibot, Pierre-Martial. "Essai sur la langue et les caractères des Chinois II." In Amiot et al., *MCC*, vol. 9 (1783): 282–430.

Cibot, Pierre-Martial. "Notice du cong-fou des bonzes Tao-sée." In Amiot et al., *MCC*, vol. 4 (1779): 441–51.

Cibot, Pierre-Martial. "Notice du sang de Cerf, employé comme remède." (1781). In Amiot et al., *MCC*, vol. 8 (1782): 271–74.

Cibot, Pierre-Martial. "Observations de Physique et d'Histoire naturelle de l'empereur K'ang Hi" (1771). In Amiot et al., *MCC*, vol. 4 (1779): 452–83.

Cibot, Pierre-Martial. "Requête à l'Empereur pour la cérémonie du Labourage" (1767). In Amiot et al., *MCC*, vol. 3 (1778): 499–504.

Cibot, Pierre-Martial. "Tchong-yong, ou Juste Milieu." In Amiot et al., *MCC*, vol. 1 (1776): 459–89.

Cibot, Pierre-Martial, and Aloys Ko. "Remarques sur un écrit de M. de Pauw intitulé, Recherches sur les Égyptiens et les Chinois." In Amiot et al., *MCC*, vol. 2 (1776): 365–574.

Cibot, Pierre-Martial, Aloys Ko, and Étienne Yang. "Essai sur l'antiquité des Chinois." In Amiot et al., *MCC*, vol. 1 (1776): 1–271.

Condorcet, Jean-Antoine-Nicolas de Caritat, marquis de. *Œuvres complètes de Condorcet*. Edited by F. Condorcet O'Connor and F. Arago. Paris: Firmin Didot, 1847–1849.

"Discours . . . en réponse à celui de M. Bailly" (1784). Vol. 10: 429–34.

L'Esquisse d'un tableau historique des progrès de l'esprit humain (1794). Vol. 6: 1–513.

"Fragment sur l'Atlantide, ou efforts combinés de l'espèce humaine pour le progrès des sciences" (1794). Vol. 6: 597–660.

"À Voltaire" (1777). Vol. 1: 147–49.

Condorcet, Jean-Antoine-Nicolas de Caritat, marquis de. Review of Bailly, *Histoire de l'astronomie ancienne* (1775). *Histoire de l'Académie royale des sciences* (1775): 44–53.

Coster, Jean-Louis, ed. *L'Ésprit des journaux, françois et étrangers*. Vol. 4. Paris: L'Imprimerie du Journal, April 1777.

Couplet, Philippe, ed. *Confucius sinarum philosophus, sive scientia sinensis: latine exposita*. Paris: Daniel Horthemels, 1687.

Court de Gébelin, Antoine. *Lettre de l'auteur du monde primitif, à Messieurs ses souscripteurs, sur le magnétisme animal*. Paris: Gastelier, 1784.

Court de Gébelin, Antoine. *Monde primitif, analysé et comparé avec le monde moderne, considéré dans divers objets concernant l'histoire, le blason, les monnoies, les jeux, les voyages des Phéniciens autour du monde, les langues Américaines, &c. ou dissertations mêlées*. Paris: Court de Gébelin, 1781.

Court de Gébelin, Antoine. *Monde primitif, analysé et comparé avec le monde moderne, considéré dans l'histoire civile, religieuse et allégorique du calendrier ou almanac*. Paris: Court de Gébelin, 1776.

Court de Gébelin, Antoine. *Monde primitif, analysé et comparé avec le monde moderne, considéré dans son génie allégorique et dans les allégories auxquelles conduisit ce génie précédé du plan général*. Paris: Court de Gébelin, 1773.

Da Qing Gaozong Chun Huangdi shilu 大清高宗純皇帝實錄 [Veritable records of the Great Qing, Qianlong period]. Reprint ed. Taipei: Huawen shuju, 1970.

De Guignes, Chrétien-Louis-Joseph. *Voyages à Péking, Manille, et L'Îsle de France*. Paris: Imprimerie Impériale, 1808.

De Guignes, Joseph. "Essai historique sur l'étude de la philosophie chez les anciens Chinois." In *Mémoires de l'Académie des inscriptions*, vol. 38 (1777): 269–311.

De Guignes, Joseph. "Examen critique des annales chinoises, ou Mémoire sur l'incertitude des douze premiers siècles de ces annales, et de la chronologie chinoise." In *Mémoires de l'Académie des inscriptions*, vol. 36 (1769): 164–89.

De Guignes, Joseph. *Histoire générale des Huns, des Turcs, des Mogols, et des autres Tartares occidentaux*. Paris: Desaint & Saillant: 1756–1768.

De Guignes, Joseph. "Idée de la littérature Chinoise en général, et particulièrement des historiens et de l'étude de l'histoire à la Chine." In *Mémoires de l'Académie des inscriptions*, vol. 36 (1769): 190–238.

De Guignes, Joseph. "Lettre de M. De Guignes, interprète du roy." *Journal des sçavans* (January 1752): 812–14.

De Guignes, Joseph. *Mémoire dans lequel on prouve que les Chinois sont une colonie égyptienne*. Paris: Desaint & Saillant, 1759.

De Guignes, Joseph. Preface to *Le Chou-king: Un des livres sacrés des Chinois, qui renferme les fondements de leur ancienne histoire, les principes de leur gouvernement & de leur morale*, translated by Antoine Gaubil, edited by Joseph de Guignes, i–xliii. Paris: N. M. Tilliard, 1770.

Delandine, Antoine François. *De la philosophie corpusculaire: Ou des connoissances et des procédés magnétiques chez les divers peuples*. Paris: Cuchet, 1785.

Delatour, Louis-François. *Essais sur l'architecture des Chinois, sur leurs jardins, leurs principes de médecine, et leurs moeurs et usages; avec des notes*. Paris: de Clousier, 1803.

De Mailla, Joseph-Anne-Marie de Moyriac, trans. *Histoire générale de la Chine, ou annales de cet Empire, traduites du Tong-kien-kang-mou*. Edited by Deshauterayes. Paris: Pierres and Clousier, 1777–1785.

De Pauw, Cornelius. *Recherches philosophiques sur les égyptiens et les chinois*. Berlin: C. J. Decker, 1773–1774.

Deshauterayes, Michel-Ange le Roux. "Extraits des historiens chinois." In *De l'origine des lois, des arts, et des sciences, et de leurs progrès chez les anciens peoples*, vol. 3 (1758), edited by Antoine-Yves Goguet. Reprint ed. Paris: L. Haussman and d'Hautel, 1809.

Diderot, Denis. "Chine, la." In *Encyclopédie, ou dictionnaire raisonné des sciences, des arts et des métiers*, edited by Denis Diderot and Jean le Rond d'Alembert. 1753. ARTFL Encyclopédie Projet, edited by Robert Morrissey.

Diderot, Denis. "Chinois, Philosophie des." In *Encyclopédie, ou dictionnaire raisonné des sciences, des arts et des métiers*, edited by Denis Diderot and Jean le Rond d'Alembert. 1753. ARTFL Encyclopédie Projet, edited by Robert Morrissey.

Diderot, Denis, and Jean le Rond d'Alembert, eds. *Encyclopédie, ou dictionnaire raisonné des sciences, des arts et des métiers*. 1753. ARTFL Encyclopédie Projet, edited by Robert Morrissey.

D'Ollières, Jacques-François. "Extrait d'une lettre du Père d'Ollières, à son frère, curé de Lexie, près Longwi, 1780." In *Lettres édifiantes et curieuses*, vol. 4, edited by Louis Aimé-Martin, 275–82. Paris: Société du Panthéon Littéraire, 1843.

Du Halde, Jean-Baptiste. *A Description of the Empire of China and Chinese-Tartary, together with the Kingdoms of Korea, and Tibet*. Translated. London: Edward Cave, 1738–1741.

Du Halde, Jean-Baptiste. *Description géographique, historique, chronologique, politique, et physique de l'empire de la Chine et de la Tartarie chinoise*. Vols. 1–3. Paris: P. G. Le Mercier, 1735.

Felice, Fortunato de. "Chinois (de la Littérature des)." In *Supplément à l'Encyclopédie. Encyclopédie, ou dictionnaire raisonné des sciences, des arts et des métiers*. Vol. 2. Amsterdam: M. Rey, 1776.

Franklin, Benjamin. *The Writings of Benjamin Franklin, Vol. X (1789–1790)*. Edited by Albert Henry Smyth. New York: Macmillan, 1907.

Galart de Montjoye, Christophe Félix Louis. *Lettre sur le magnétisme animal adressée à M. Bailly*. Paris: Pierre-J. Duplain, 1784.

Gaubil, Antoine, trans. *Le Chou-king: Un des livres sacrés des Chinois, qui renferme les fondements de leur ancienne histoire, les principes de leur gouvernement & de leur morale*. Edited by Joseph de Guignes. Paris: N. M. Tilliard, 1770.

Hager, Joseph. *An Explanation of the Elementary Characters of the Chinese: With an Analysis of Their Ancient Symbols and Hieroglyphics*. London: Richard Phillips, 1801.

Hager, Joseph. *Description des médailles chinoises du Cabinet Impérial de France: précédée d'un essai de numismatique chinoise*. Paris: Imprimerie Impériale, 1805.

Hager, Joseph. *Monument de Yu, Ou la plus ancienne inscription de la Chine*. Paris: Treuttel et Wurtz, 1802.

Hegel, Georg Wilhelm Friedrich. *Elements of the Philosophy of Right*. Edited by Allen W. Wood. Translated by H. B. Nisbet. Cambridge: Cambridge University Press, 1991.

Hegel, Georg Wilhelm Friedrich. *Hegel: The Letters*. Translated by Clark Butler and Christiane Seiler. Bloomington: Indiana University Press, 1984.

Hegel, Georg Wilhelm Friedrich. *Lectures on the History of Philosophy*. Translated by E. S. Haldane. London: Kegan Paul, Trench, Trübner, 1892.

Hegel, Georg Wilhelm Friedrich. *Lectures on the Philosophy of Religion: Determinate Religion*. Edited by Peter C. Hodgson. Translated by R. F. Brown, Peter C. Hodgson, and J. M. Stewart, with H. S. Harris. Berkeley: University of California Press, 1987.

Hegel, Georg Wilhelm Friedrich. *The Philosophy of History*. Translated by J. Sibree. Rev. ed. New York: Colonial Press, 1900.

Helman, Isidore Stanislas Henri. *Abrégé historique des principaux traits de la vie de Confucius*. Paris: Helman and Ponce, 1788.

Hervier, Charles. *Lettre sur la découverte du magnétisme animal à Court de Gébelin*. Couturier: Beijing [Paris?], 1784.

Holmes, Samuel, and Louis Langlès. *Voyage en Chine et en Tatarie, a la suite de l'ambassade de Lord Macartney*. Paris: Delance et Lesueur, 1805.

Hua Guan 華冠. *Hongwu xingle tu* 弘旿行樂圖 [Painting of Hongwu making merry]. 1785.

Humboldt, Wilhelm von. *Lettre à M. Abel-Rémusat*. Paris: Dondey-Dupré, 1827.

Jiang Fan 江繁. *Siyiguan kao* 四譯館考 [Investigations of the Translation Bureau] (1696). Reprint ed. Beijing: *Beijing tushuguan guji zhenben congkan*, vol. 59, 2000.

Ji Yun 紀昀. *Yuewei caotang biji* 閱微草堂筆記 [Notes from the Thatched Cottage of Close Observation] (1800). Edited by Bei Yuan 北原 et al. Beijing: Zhongguo huaqiao chubanshe, 1994.

Journal des Sçavans. Paris: 1665–1792.

Journal encyclopédique. Vol. 4. Part 1. Paris: Bouillon, 1788.

Julien, Stanislas. *Le Livre de la voie et de la vertu: Composé dans le VIe siècle avant l'ère chrétienne*. Paris: Imprimerie Royale, 1842.

Kant, Immanuel. *Practical Philosophy*. Edited by Mary J. Gregor. Cambridge: Cambridge University Press, 1996.

Klaproth, Julius. *Inschrift des Yü*. Halle: Waisenhausbuchhandlung, 1811.

Klaproth, Julius. *Leichenstein auf dem Grabe der chinesischen Gelehrsamkeit des Herrn Joseph Hager*. Berlin: Waisenhausdruckerei, 1811.

Leibniz, Gottfried Wilhelm. Letters to Joachim Bouvet, 1697–1707. In "Leibniz-Bouvet Correspondence," translated and annotated by Alan Berkowitz and Daniel J. Cook. http://leibniz-bouvet.swarthmore.edu/.

Leibniz, Gottfried Wilhelm. *Writings on China*. Edited by Daniel J. Cook and Henry Rosemont. Chicago: Open Court, 1994.

Locke, John. *Second Treatise of Government* (1690). Edited by C. B. Macpherson. Indianapolis, IN: Hackett, 1980.

Lo-Looz, Robert de. *Recherches physiques et métaphysiques sur les influences célestes, sur le magnétisme universel, et sur le magnétisme animal: dont on trouve la pratique de temps immémorial chez les Chinois*. London: Couturier, 1788.

Lo-Looz, Robert de. *Les Militaires au-delà du Gange*. Paris: Bailly, 1770.

Longobardo, Niccolò. *Traité sur quelques points de la religion des Chinois*. Paris: Louis Guérin, 1701.

Macartney, George, and John Barrow. *Some Account of the Public Life, and a Selection from the Unpublished Writings, of the Earl of Macartney. . . . Journal of an Embassy from the King of Great Britain to the Emperor of China*. London: T. Cadell and W. Davies, 1807.

Mao Huijian 毛会建. Yu Bei 禹碑. Xi'an: 1666. (BnF MS Chinois 1170).

Mellet, Louis-Raphaël-Lucrèce de Fayolle, comte de. "Première lettre, de M. le comte de Mellet, maréchal de camp, Paris, ce 8 Octobre 1787." In Armand-Marie-Jacques Chastenet de Puységur, *Du magnétisme animal: considéré dans ses rapports avec diverses branches de la physique générale*, 383–387. Paris: Cellot, 1807.

Mellet, Louis-Raphaël-Lucrèce de Fayolle, comte de. "Recherches sur les Tarots, et sur la divination par les cartes des Tarots" (1781). In *Monde primitif, analysé et comparé avec le monde moderne, considéré dans divers objets*, edited by Antoine Court de Gébelin, 395–410. Paris: Court de Gébelin, 1781.

Mesmer, Franz Anton. *Mémoire sur la découverte du magnétisme animal*. Geneva: Didot le jeune, 1779.

Montmignon, Jean-Baptiste, ed. *Choix des lettres édifiantes: Écrites des missions étrangères; avec des additions, des notes critiques, et des observations pour la plus grande intelligence de ces lettres*. Paris: Maradan, 1808.

Montucci, Antonio. *Letters to the Editor of the Universal Magazine, on Chinese Literature: Including Strictures on Dr. Hager's Two Works, and the Reviewers' Opinions Concerning Them*. London: Knight and Compton, 1804.

Montucci, Antonio. *Remarques philologiques sur les voyages en Chine de M. de Guignes*. Berlin: Montucci, 1809.

Montucci, Antonio. *Réponse à une lettre imprimée et signée Julius v. Klaproth*. Berlin: 1810.

Montucci, Antonio. *The Title Page Reviewed: The Characteristic Merits of the Chinese Language*. London: Montucci, Spilsbury, and Snowhill, 1801.

Morrison, Robert, and Eliza Morrison. *Memoirs of the Life and Labours of Robert Morrison*. London: Longman, Orme, Brown, and Longmans, 1839.

Nougaret, Pierre-Jean-Baptiste. *Tableau mouvant de Paris, ou variétés amusantes, ouvrage enrichi de notes historique & critiques, & mis au jour*. Vol. 2. London: Thomas Hookham, Libraire, la Veuve Duchesne, 1787.

Parrenin, Dominique. "Lettre du Père Parennin à M. Dortous de Mairan . . . à Pékin, ce 11 août 1730." In *Lettres édifiantes et curieuses*, vol. 3, edited by Louis Aimé-Martin, 645–62. Paris: Société du Panthéon Littéraire, 1843.

Phillips, Sir Richard, and John Abraham Heraud, eds. *Monthly Magazine, Or, British Register*. Vol. 9. London: R. Phillips, 1800.

Prévost, Antoine François, ed. *Histoire générale des voyages*. Vol. 7. The Hague: Pierre de Hondt, 1749.

Puységur, Armand-Marie-Jacques de Chastenet de. *Du magnétisme animal: considéré dans ses rapports avec diverses branches de la physique générale*. Paris: Cellot, 1807

Puységur, Jacques-François-Maxime de Chastenet de, et al. *État actuel de l'art et de la science militaire à la Chine: Tiré des livres militaires des Chinois. Avec diverses observations sur l'étendue & les bornes des connoissances militaires chez les Européens*. Paris: Didot l'aîné, 1773.

Quesnay, François. *Œuvres Économiques et Philosophiques*. Edited by Auguste Oncken. Frankfurt: Joseph Baer and Jules Peelman, 1888.

Qinding qian sou yan shi 欽定千叟宴詩 [Imperially commissioned poems from the Feast of the Thousand Old Men] (1785). *Siku quanshu* 四庫全書, *jibu* 集部, *zongjilei* 總集類, *juan* 25.

Qing zhong qian qi Xiyang Tianzhujiao zai Hua huodong dang'an shiliao 清中前期西洋天主教在華活動檔案史料 [Archives concerning western Catholic missions in China from the early to mid- Qing period]. Edited by *Zhonggui di yi lishi dang'an guan* 中國第一歷史檔案館. Beijing: Zhonghua shuju, 2003.

Rabaut-Saint-Étienne, Jean-Paul. *Lettres à Monsieur Bailly sur l'histoire primitive de la Grèce*. Paris: de Bure l'aîné, 1787.

Rabaut-Saint-Étienne, Jean-Paul. *Lettres sur l'histoire primitive de la Grèce*. Paris: de Bure l'aîné, 1787.

Roussel, René Louis de, ed. *État militaire de France pour l'année 1772*. Paris: Guillyn, 1772.

Ruan Yuan 阮元, ed. *Chouren zhuan* 疇人傳 [Biographies of mathematicians and astronomers]. Yangzhou: Wenxuan lou, 1799.

Ruan Yuan 阮元. *Xiaocanglang bitan* 小滄浪筆談 [Brush talk of the Blue Wave Tributary]. Yangzhou: Wenxuan lou, 1842.

Ruan Yuan 阮元. *Yanjingshi ji* 揅經室集 [Collection of the Study of Classics Research] (1823). Reprint ed. Edited by Zhang Yuanji 張元濟 et al. Shanghai: Commercial Press, 1919.

Saint-Allais, Nicolas Viton de, et al. *Nobiliaire de France*. Vol. 11, part 1. Paris: Bachelin-Deflorenne, 1876.

Saint-Esprit, Jérôme Delandine de. *Vie de S. A. R. Charles Ferdinand d'Artois, duc de Berry*. Paris: L.-E. Herhan, 1820.

Saint-Martin, Louis-Claude de ("Le Philosophe Inconnu"). *Des erreurs et de la vérité, ou les hommes rappelés au principe universel de la science*. Edinburgh: 1775.

Shen Chu 沈初. *Lanyuntang shiwen ji* 蘭韻堂詩文集. [Collection of poems and texts from the Hall of Orchid Rhyme]. Qianlong period.

Silvestre de Sacy, Antoine-Isaac. "Deshautesrayes, Michel-Ange-André Le Roux." In *Biographie universelle*, 180–82. Paris: Imprimeur du Roi, 1814.

Sirey, Jean-Baptiste, and Le Moine Devilleneuve. *Recueil général des lois et des arrêts, en matière civile, criminelle, commercial et de Droit public*. Vol. 29. Paris: Lachevardière, 1829.

Société archéologique de Touraine. *Bulletin trimestriel de la Société Archéologique de Touraine*. Vol. 12. Tours: Péricat, 1900.

Tardy de Montravel, A. A. *Essai sur la théorie du somnambulisme magnétique*. London [Paris?]: 1786.

Thiéry, Luc-Vincent. *Guide des amateurs et des étrangers voyageurs à Paris*. Paris: Hardouin and Gattey, 1787.

Turgot, Anne Robert Jacques. *Œuvres de Turgot*. Paris: Guillaumin, 1844.

Ventavon, Jean-Matthieu. "Extrait d'une lettre de M. de Ventavon, missionnaire à Pékin, en date du 25 novembre 1784." In *Lettres édifiantes et curieuses*, vol. 4, edited by Louis Aimé-Martin, 306–10. Paris: Société du Panthéon Littéraire, 1843.

Voltaire (François-Marie Arouet). *Œuvres complètes, Correspondence and Related Documents*. Edited by Thomas Besterman. Geneva: Institut et Musée Voltaire, 1968–1977.

Voltaire (François-Marie Arouet). *Œuvres complètes de Voltaire*. Oxford: Voltaire Foundation, 1968. ARTFL Encyclopédie Projet, edited by Robert Morrissey.
Lettres philosophiques (1730).
Essai sur les mœurs et l'esprit des nations (1756).
Fragment sur l'histoire générale (1773).
La Philosophie de l'Histoire (1764).
Lettres chinoises, indiennes et tartares (1775).
Questions sur l'Encyclopédie (C–E) (1772).
Le Siècle de Louis XIV (1751).

Wang Chang 王昶. *Chunrongtang ji* 春融堂集 [Collection of the Hall of the Spring Thaw]. Shunan shushe: 1808.

Wang Chang 王昶, ed. *Huhaiwen zhuan* 湖海文傳 [Collected works of lakes and seas]. Jingxun tang: 1837.

Wang Chang 王昶. *Jinshi cuibian* 金石萃編 [Compendium of seals and stones]. Jingxun tang: 1805.

Xu Chunfu 徐春甫. *Gujin Yitong* 古今醫統 [System of ancient and modern medicine] (16th century). ("*Traité général de médecine, avec un historique*," MS Chinois 5087–5095, BNF).

Zangfu bu 臟腑部 [Organs section]. *Gujin tushu jicheng* 古今圖書集成 [Complete collection of ancient and modern figures and texts], *Minglun huibian* 明倫彙編 *Renshi dian* 人事典, *ce* 386, *juan* 21 (1726). Reprint ed. Shanghai: Zhonghua Publishing Company, 1934.

Zhao Yi 趙翼. *Yanpu zaji* 簷曝雜記 [Miscellaneous record of Sunning on the Eaves] (Late eighteenth century). Reprint ed. Beijing: Zhonghua shuju, 1982.

Published Sources Written after 1850

Actes du IIe Colloque International de Sinologie, Les rapports entre la Chine et l'Europe au temps des lumières. Paris: Les Belles Lettres, Cathasia, 1980.

Actes du IIIe Colloque International de Sinologie: Appréciation par l'Europe de la tradition chinoise à partir du XVIIe siècle. Paris: Les Belles Lettres, Cathasia, 1983.

Adas, Michael. *Machines as the Measure of Men: Science, Technology, and Ideologies of Western Dominance* (1989). Reprint ed. Ithaca, NY: Cornell University Press, 2015.

Afinogenov, Gregory. *Spies and Scholars: Chinese Secrets and Imperial Russia's Quest for World Power*. Cambridge, MA: Harvard University Press, 2020.

Allen, Amy. *The End of Progress: Decolonizing the Normative Foundations of Critical Theory*. New York: Columbia University Press, 2017.

Amiable, Louis. *La Loge des Neuf sœurs*. Edited by Charles Porset. Paris: Edimaf, 1989.

Andrews, Bridie. *The Making of Modern Chinese Medicine, 1850–1960*. Honolulu: University of Hawaii Press, 2015.

App, Urs. *Arthur Schopenhauer and China: A Sino-Platonic Love Affair*. Edited by Victor H. Mair. Philadelphia: Sino-Platonic Papers, 2010.

App, Urs. *The Birth of Orientalism: Encounters with Asia*. Philadelphia: University of Pennsylvania Press, 2010.

Armogathe, Jean-Robert. "Voltaire et la Chine: Une mise au point." In *La mission française de Pékin aux XVIIe et XVIIIe siècles*, 26–39. Paris: Belles Lettres, 1974.

Bailey, Sarah Pulliam. "Tarot Cards are Having a Moment with Help from the Pandemic." *Washington Post*, December 10, 2021.

Baker, Keith Michael. *Condorcet: From Natural Philosophy to Social Mathematics*. Chicago: University of Chicago Press, 1975.

Baker, Keith Michael. *Inventing the French Revolution: Essays on French Political Culture in the Eighteenth Century*. New York: Cambridge University Press, 1990.

Barbier, Frédéric, et al. *Dictionnaire des imprimeurs, libraires et gens du livre à Paris: A–C*. Geneva: Droz, 2007.

Barnes, Linda L. *Needles, Herbs, Gods, and Ghosts: China, Healing, and the West to 1848*. Cambridge, MA: Harvard University Press, 2009.

Barnett, Lydia. *After the Flood: Imagining the Global Environment in Early Modern Europe*. Baltimore, MD: Johns Hopkins University Press, 2019.

Barthélemy-Saint Hilaire, Jules. *M. Victor Cousin: Sa vie et sa correspondance*. Paris: Hachette, 1895.

Barzun, Jacques. "Diderot as Philosopher." *Diderot Studies* 22 (1986): 17–25.

Bates, David. "Super-Epistemology." In *The Super-Enlightenment: Daring to Know Too Much*, edited by Dan Edelstein, 53–74. Oxford: Voltaire Foundation, 2010.

Bates, David. "The Mystery of Truth: Louis-Claude de Saint-Martin's Enlightened Mysticism." *Journal of the History of Ideas* 61, no. 4 (2000): 635–55.

Becker, Carl L. *The Heavenly City of the Eighteenth-Century Philosophers* (1932). Reprint ed. New Haven, CT: Yale University Press, 2003.

Belhoste, Bruno. *Paris savant: Parcours et rencontres au temps des Lumières*. Paris: A. Colin, 2011.

Belhoste, Bruno, David Armando, and Stéphane Lamassé. "Harmonia Universalis." LabEx Hastec project, with IHMC, Centre Koyré, CARE, ISPF, C.R.I.S.E.S., and AHRHA. 2016. https://harmoniauniversalis.univ-paris1.fr/#/.

Belhoste, Bruno, and Nicole Edelman, eds. *Mesmer et mesmérismes: Le magnétisme animal en contexte*. Paris: Omniscience, 2015.

Bell, Daniel A. *The China Model: Political Meritocracy and the Limits of Democracy*. Princeton, NJ: Princeton University Press, 2015.

Bell, David. "The PowerPoint Philosophe: Waiting for Steven Pinker's Enlightenment." *Nation*, April 2, 2018.

Berlin, Isaiah. "The Counter-Enlightenment." In *Dictionary of the History of Ideas*, vol. 2, edited by Philip Wiener, 100–112. New York: Scribner, 1973.

Berlin, Isaiah. *The Roots of Romanticism*. 2nd ed. Edited by Henry Hardy. Princeton, NJ: Princeton University Press, 2013.

Bernard-Maître, Henri. "Catalogue des objets envoyés de Chine par les missionnaires de 1765 à 1786." *Bulletin de l'Université l'Aurore* 9 (1948): 33–34.

Bevilacqua, Alexander. *The Republic of Arabic Letters*: Cambridge, MA: Harvard University Press, 2018.

Blavatsky, Helena Petrovna, ed. *The Theosophist*. Vol. 1. Madras: Theosophical Society, 1879–1880.

Boas, Franz. *A Franz Boas Reader: The Shaping of American Anthropology, 1883–1911*. Edited by George W. Stocking Jr. Chicago: University of Chicago Press, 1974.

Bogdan, Henrik, and Martin P. Starr, eds. *Aleister Crowley and Western Esotericism*. Oxford: Oxford University Press, 2012.

Bol, Peter K. *Neo-Confucianism in History*. Cambridge, MA: Harvard University Press, 1997.

Bowler, Peter J., and Iwan Rhys Morus. *Making Modern Science: A Historical Survey*. Chicago: University of Chicago Press, 2005.

Breen, Benjamin. *The Age of Intoxication: Origins of the Global Drug Trade*. Philadelphia: University of Pennsylvania Press, 2019.

Brook, Timothy. "The Sinology of Joseph Needham." *Modern China* 22, no. 3 (1996): 340–48.

Brown, Tristan G. "A Mountain of Saints and Sages: Muslims in the Landscape of Popular Religion in Late Imperial China." *T'oung Pao* 105, no. 3–4 (2019): 437–91.

Brown, Tristan G. "The Veins of the Earth: Property, Environment, and Cosmology in Nanbu County, 1865–1942." PhD diss., Columbia University, 2017. https://clio.columbia.edu/catalog/12848655.

Burson, Jeffrey D. "Between Power and Enlightenment: The Cultural and Intellectual Context for the Jesuit Suppression in France." In *The Jesuit Suppression in Global Context*, edited by Jeffrey D. Burson and Jonathan Wright, 40–64. Cambridge: Cambridge University Press, 2015.

Burson, Jeffrey D. "Chinese Novices, Jesuit Missionaries and the Accidental Construction of Sinophobia in Enlightenment France." *French History* 27, no. 1 (2013).

Burson, Jeffrey D. "Unlikely Tales of Fo and Ignatius: Rethinking the Radical Enlightenment through French Appropriation of Chinese Buddhism." *French Historical Studies* 38, no. 3 (2015): 391–420.

Burson, Jeffrey D., and Jonathan Wright, eds. *The Jesuit Suppression in Global Context: Causes, Events, and Consequences*. Cambridge: Cambridge University Press, 2015.

Burson, Jeffrey D., and Jonathan Wright. "Towards a New History of the Eighteenth-Century Jesuit Suppression in Global Context." In *The Jesuit Suppression in Global Context*, edited by Jeffrey D. Burson and Jonathan Wright, 1–10. Cambridge: Cambridge University Press, 2015.

Bury, John Bagnell. *The Idea of Progress: An Inquiry into Its Origin and Growth* (1932). Reprint ed. New York: Dover Publications, 1955.

Bussières, Georges. *Henri Bertin et sa famille: La production nobiliaire du Ministre, ses ancêtres, son Intendance à Lyon, ses ministères*. Périgueux: Imprimerie de la Dordogne, 1906.

Butler, Alison. *Victorian Occultism and the Making of Modern Magic: Invoking Tradition*. Basingstoke: Palgrave McMillan, 2011.

Butterfield, Herbert. *The Whig Interpretation of History*. London: G. Bell and Sons, 1931.

Campbell, Mary B., Lorraine Daston, Arnold Ira Davidson, John Forrester, and Simon Goldhill. "Enlightenment Now: Concluding Reflections on Knowledge and Belief." *Common Knowledge* 13, no. 2–3 (2007): 429–50.

Cams, Mario. *Companions in Geography: East-West Mapping of Qing China*. Leiden: Brill, 2017.

Cañizares-Esguerra, Jorge. *How to Write the History of the New World: Histories, Epistemologies, and Identities in the Eighteenth-Century Atlantic World*. Stanford, CA: Stanford University Press, 2001.

Carey, Daniel, and Lynn Festa, eds. *The Postcolonial Enlightenment: Eighteenth-Century Colonialism and Postcolonial Theory*. Oxford: Oxford University Press, 2013.

Carey, Daniel, and Sven Trakulhun. "Universalism, Diversity, and the Postcolonial Enlightenment." In *The Postcolonial Enlightenment: Eighteenth-Century Colonialism and Postcolonial Theory*, edited by Daniel Carey and Lynn Festa, 254–89. Oxford: Oxford University Press, 2013.

Cartier, Michel. "Introduction." In *La Chine entre amour et haine: Actes du VIIIe Colloque de sinologie de Chantilly*, edited by Michel Cartier, 7–13. Paris: Desclée de Brouwer, 1998.

Cartier, Michel, ed. *La Chine entre amour et haine: Actes du VIIIe Colloque de sinologie de Chantilly*. Paris: Desclée de Brouwer, 1998.

Cartier, Michel. "Le despotisme chinois Montesquieu et Quesnay, lecteurs de Du Halde." In *La Chine entre amour et haine: Actes du VIIIe Colloque de sinologie de Chantilly*, edited by Michel Cartier, 15–32. Paris: Desclée de Brouwer, 1998.

Cavanaugh, Gerald J. "Turgot and the 'Encyclopédie.'" *Diderot Studies* 10 (1968): 23–33.

Chakrabarty, Dipesh. *Provincializing Europe: Postcolonial Thought and Historical Difference* Princeton, NJ: Princeton University Press, 2000.

Chaves, Jonathan. "Still Hidden by Spirits and Immortals: The Quest for the Elusive 'Stele of Yu the Great.'" *Asia Major* 3rd series 26, no. 1 (2013): 1–22.

Chen Duxiu 陳獨秀. *Duxiu wencun* 獨秀文存 (Collected works of Chen Duxiu) (1922). Reprint ed. Hefei: Anhui renmin chubanshe, 1987.

Chen Pei-Kai and Michael Lestz, with Jonathan Spence. *The Search for Modern China: A Documentary Collection*. New York: W. W. Norton, 1999.

Chen Zhongkai 陳忠凱 et al., eds. *Xi'an Beilin Bowuguan cang beike zongmu tiyao* 西安碑林博物館藏碑刻總目提要 (Annotated catalogue of the stele inscriptions in the collection of the Xi'an Stele Forest Museum). Beijing: Xi'an zhuang shuju, 2006.

Cheng, Anne. "Abel-Rémusat et Hegel: sinologie et philosophe dans l'Europe du XIXe siècle." At *Jean-Pierre Abel-Rémusat et ses successeurs*, symposium, June 11–13, 2014, Collège de France, Paris.

Cheng, Anne. "Philosophy and the French Invention of Sinology: Mapping Aca-

demic Disciplines in Nineteenth Century Europe." *China Report* 50, no. 1 (2014): 11–30.
Cheng, Anne. "'Y a-t-il une philosophie chinoise?': Est-ce une bonne question?" *Extrême-Orient, Extrême-Occident* 27, no. 27 (2005): 5–12.
Ching, Julia, and Willard G. Oxtoby, eds. *Discovering China: European Interpretations in the Enlightenment*. Rochester, NY: University of Rochester Press, 1992.
Ching, Julia, and Willard G. Oxtoby. "Introduction." In *Discovering China: European Interpretations in the Enlightenment*, edited by Julia Ching and Willard G. Oxtoby, xi–xxxi. Rochester, NY: University of Rochester Press, 1992.
Choi, Kee Il, Jr. "Father Amiot's Cup: A Qing Imperial Porcelain Sent to the Court of Louis XV." In *Writing Material Culture History*, edited by Anne Gerritsen and Giorgio Riello. New York: Bloomsbury Publishing, 2014.
Chu, Petra ten-Doesschate, and Ning Ding, eds. *Qing Encounters: Artistic Exchanges between China and the West*. Los Angeles: Getty Research Institute, 2015.
Clark, Frederic. *The First Pagan Historian: The Fortunes of a Fraud from Antiquity to the Enlightenment*. New York: Oxford University Press, 2020.
Clarke, John James. *The Tao of the West: Western Transformations of Taoist Thought*. London: Routledge, 2000.
Cohen, I. Bernard. "The Eighteenth-Century Origins of the Concept of Scientific Revolution." *Journal of the History of Ideas* 37, no. 2 (1976): 257–88.
Cohen, Paul A. *Between Tradition and Modernity: Wang T'ao and Reform in Late Ch'ing China*. Cambridge, MA: Harvard University Press, 1974.
Collège de France. "Intellectual History of China." Accessed May 10, 2022. https://www.college-de-france.fr/site/en-intellectual-history-china/index.htm.
Conrad, Sebastian. "Enlightenment in Global History: A Historiographical Critique." *American Historical Review* 117, no. 4 (2012): 999–1027.
Cook, Daniel J., and Henry Rosemont Jr. "The Pre-Established Harmony between Leibniz and Chinese Thought." *Journal of the History of Ideas* 42, no. 2 (1981): 253–67.
Cordier, Henri. *Histoire générale de la Chine et de ses relations avec les pays étrangers depuis les temps les plus anciens jusqu'à la chute de la dynastie mandchoue*. Paris: Paul Geuthner, Paris, 1920.
Cordier, Henri. "La Suppression de la Compagnie de Jésus et la mission de Péking." *T'oung Pao*, 2nd series, 17, no. 3 (1916): 271–347.
Cordier, Henri. "Les Chinois de Turgot." In *Mélanges d'histoire et de géographie orientales*, vol. 2, 31–39. Paris: Maisonneuve, 1920.
Cordier, Henri. "Les études chinoises sous la Révolution et l'Empire." *T'oung Pao*, 2nd series, 19, no. 2 (May 1918): 59–103.
Cordier, Henri. "Un orientaliste allemand: Jules Klaproth." *Comptes rendus des séances de l'Académie des inscriptions et belles-lettres* 61, no. 4 (1917): 297–308.
Coutel, Charles. "Utopie et Perfectibilité: Significations de l'Atlantide chez Condorcet." In *Condorcet: Homme des Lumières et de la Révolution*, edited by Anne-Marie Chouillet and Pierre Crépel, 99–107. Fontenat-aux-Roses: ENS Éditions, 1997.
Crabtree, Adam. *Animal Magnetism, Early Hypnotism, and Psychical Research, 1766–1925: An Annotated Bibliography*. Millwood, NY: Kraus International Publications, 1988.

Crook, Malcolm, William Doyle, and Alan Forrest, eds. *Enlightenment and Revolution: Essays in Honour of Norman Hampson*. Aldershot: Ashgate, 2004.

Crowley, Aleister. "The 'Worst Man in the World' Tells the Astounding Story of His Life." *Sunday Dispatch*, June 18, 1933.

Cruikshank, Julie. "Are Glaciers 'Good to Think With'? Recognising Indigenous Environmental Knowledge." *Anthropological Forum* 22, no. 3 (2012): 239–50.

Darnton, Robert. *Mesmerism and the End of the Enlightenment in France*. Cambridge, MA: Harvard University Press, 1968.

De Bary, William Theodore. "Neo-Confucian Cultivation and the Seventeenth-Century 'Enlightenment.'" In *The Unfolding of Neo-Confucianism*, edited by William Theodore de Bary, 141–216. New York: Columbia University Press, 2019.

De Bary, William Theodore, and Irene Bloom, eds. *Sources of Chinese Tradition, Volume I: From Earliest Times to 1600*. 2nd ed. New York: Columbia University Press, 1999.

De Bary, William Theodore, and Richard Lurfrano, eds. *Sources of Chinese Tradition, Volume II: From 1600 Through the Twentieth Century*. 2nd ed. New York: Columbia University Press, 2000.

Decker, Ronald, Thierry Depaulis, and Michael Dummett. *A Wicked Pack of Cards: The Origins of the Occult Tarot*. London: Duckworth, 1996.

Dehergne, Joseph. "Les historiens jésuites du Taoïsme." In *La Mission française de Pékin aux XVIIe et XVIIIe siècles: La Chine au temps des Lumières 2*, 59–67. Paris: Les Belles Lettres, Cathasia, 1976.

Dehergne, Joseph. "Une grande collection: Mémoires concernant les Chinois (1776–1814)." *Bulletin de l'École française d'Extrême-Orient* 72, no. 1 (1983): 267–98.

Demiéville, Paul. "Aperçu historique des études sinologiques en France." *Acta Asiatica* (September 1966): 56–100.

Détrie, Muriel. "L'évolution de l''Europe Chinoise' de la fin du XVIIIe siècle au début du XXe siècle." In *Idées de la Chine au XIXe siècle*, edited by Marie Dollé and Geneviève Espagne, 19–38. Paris: Les Indes Savants, 2014.

Détrie, Muriel. "L'image du Chinois dans la littérature occidentale au XIXe siècle." In *La Chine entre amour et haine: Actes du VIIIe Colloque de sinologie de Chantilly*, edited by Michel Cartier, 403–29. Paris: Desclée de Brouwer, 1998.

Dollé, Marie, and Geneviève Espagne, eds. *Idées de la Chine au XIXe siècle: Entre France et Allemagne*. Paris: Les Indes Savants, 2014.

Dollé, Marie, and Geneviève Espagne. "Introduction." In *Idées de la Chine au XIXe siècle: Entre France et Allemagne*, edited by Dollé and Espagne, 7–14. Paris: Les Indes Savants, 2014.

Douthat, Ross. "The Edges of Reason." *New York Times*, February 25, 2018.

Du Plessis, Eric. "L'influence de la Chine sur la pensée française au dix-huitième siècle: État présent des travaux." *Dalhousie French Studies* 43 (1998): 145–60.

Dyck, Anne-Lise. "La Chine hors de la philosophie: essai de généalogie à partir des traditions sinologique et philosophique françaises au XIXe siècle." *Extrême-Orient, Extrême-Occident* 27, no. 27 (2005): 13–47.

Edelstein, Dan. "Hyperborean Atlantis: Jean-Sylvain Bailly, Madame Blavatsky, and the Nazi Myth." *Studies in Eighteenth-Century Culture* 35 (2006): 267–91.

Edelstein, Dan. "Introduction to the Super-Enlightenment." In *The Super-*

Enlightenment: Daring to Know Too Much, edited by Dan Edelstein, 1–34. Oxford: Voltaire Foundation, 2010.

Edelstein, Dan. "Jean-Sylvain Bailly (1736–1793)." *The Super-Enlightenment: A Digital Archive*. Accessed July 18, 2022. https://exhibits.stanford.edu/super-e/feature/jean-sylvain-bailly-1736-1793.

Edelstein, Dan. *The Enlightenment: A Genealogy*. Chicago: University of Chicago Press, 2010.

Edelstein, Dan, ed. *The Super-Enlightenment: Daring to Know Too Much*. Oxford: Voltaire Foundation, 2010.

Elisseeff, Danielle. *Nicolas Fréret (1688–1749): Réflexions d'un humaniste du XVIIIe siècle sur la Chine*. Paris: Collège de France, Presses Universitaires de France, 1978.

Elliott, Mark C. "Abel-Rémusat, la langue mandchoue et la sinologie." At *Jean-Pierre Abel-Rémusat et ses successeurs*, symposium, June 11–13, 2014, Collège de France, Paris.

Elliott, Mark C. "The Limits of Tartary: Manchuria in Imperial and National Geographies." *Journal of Asian Studies* 69, no. 3 (2000): 603–46.

Elliott, Mark C. *The Manchu Way: The Eight Banners and Ethnic Identity in Late Imperial China*. Stanford, CA: Stanford University Press, 2001.

Elman, Benjamin A. *From Philosophy to Philology: Intellectual and Social Aspects of Change in Late Imperial China*. Cambridge, MA: Council on East Asian Studies, Harvard University, 1984.

Elman, Benjamin A. *On Their Own Terms: Science in China, 1550–1900*. Cambridge, MA: Harvard University Press, 2005.

Elshakry, Marwa. "When Science Became Western: Historiographical Reflections." *Isis* 101, no. 1 (2010): 98–109.

Elvin, Mark. "Vale Atque Ave." In Joseph Needham and Kenneth Girdwood Robinson, *Science and Civilisation in China: Volume VII, Part II*, xxiv–xliii.

Étiemble, René. "Les concepts de Li et de K'i dans la pensée européenne au XVIIIe siècle." *Mélanges Alexandre Koyré II* (1964): 144–59.

Étiemble, René. *L'Europe Chinoise II: De la Sinophilie à la Sinophobie*. Paris: Gallimard, 1989.

Fairbank, John King. *Trade and Diplomacy on the China Coast: The Opening of the Treaty Ports, 1842–1854*. Cambridge, MA: Harvard University Press, 1953.

Fan, Fa-ti. "The Global Turn in the History of Science." *East Asian Science, Technology and Society* 6 (2012): 249–58.

Fang Weigui. "Transferts de savoirs et représentations de la Chine en France et en Allemagne au XIXe siècle." In *Idées de la Chine au XIXe siècle*, edited by Marie Dollé and Geneviève Espagne, 39–58. Paris: Les Indes Savants, 2014.

Ferguson, Niall. *Empire: The Rise and Demise of the British World Order and the Lessons for Global Power*. New York: Basic Books, 2002.

Festa, Lynn, and Daniel Carey. "Introduction: Some Answers to the Question: 'What is Postcolonial Enlightenment?'" In *The Postcolonial Enlightenment: Eighteenth-Century Colonialism and Postcolonial Theory*, edited by Daniel Carey and Lynn Festa, 1–36. Oxford: Oxford University Press, 2013.

Findlen, Paula, ed. *Early Modern Things*. New York: Routledge, 2012.

Finlay, John. "Henri Bertin and the Commerce in Images between France and China in the Late Eighteenth Century." In *Qing Encounters: Artistic Exchanges between China and the West*, edited by Petra ten-Doesschate Chu and Ning Ding, 79–94. Los Angeles: Getty Research Institute, 2015.

Finlay, John. *Henri Bertin and the Representation of China in Eighteenth-Century France*. New York: Routledge, 2020.

Foucault, Michel. *The Foucault Reader*. Edited by Paul Rabinow. New York: Pantheon Books, 1984.

Frankel, Charles. *The Faith of Reason: The Idea of Progress in the French Enlightenment*. New York: Columbia University Press, 1948.

Freud, Sigmund. *Civilization and its Discontents*. Edited by James Strachey. New York: W. W. Norton, 1962.

Fröhlich, Thomas. "Introduction: Progress, History, and Time in Chinese Discourses after the 1890s." In Thomas Fröhlich and Axel Schneider, *Chinese Visions of Progress, 1895 to 1949*, 1–40. Leiden: Brill, 2020.

Fröhlich, Thomas, and Axel Schneider. *Chinese Visions of Progress, 1895 to 1949*. Leiden: Brill, 2020.

Fukuyama, Francis. *The End of History and the Last Man*. London: Penguin, 1992.

Garfield, Jay L., and Bryan W. Van Norden. "If Philosophy Won't Diversify, Let's Call It What It Really Is." *New York Times*, May 11, 2016. https://www.nytimes.com/2016/05/11/opinion/if-philosophy-wont-diversify-lets-call-it-what-it-really-is.html.

Gascoigne, John. *Encountering the Pacific in the Age of Enlightenment*. Cambridge: Cambridge University Press, 2014.

Gates, Bill. "My New Favorite Book of All Time." *GatesNotes: The Blog of Bill Gates*, January 26, 2018. https://www.gatesnotes.com/books/enlightenment-now.

Gaukroger, Stephen. *The Collapse of Mechanism and the Rise of Sensibility: Science and the Shaping of Modernity, 1680–1760*. New York: Oxford University Press, 2010.

Gay, Peter. *The Enlightenment: An Interpretation* (1966–1969). Reprint ed. New York: W. W. Norton, 1995.

Gay, Peter. *The Party of Humanity: Essays in the French Enlightenment*. New York: Alfred A. Knopf, 1964.

Gerritsen, Anne. *The City of Blue and White: Chinese Porcelain and the Early Modern World*. Cambridge: Cambridge University Press, 2020.

Gerritsen, Anne, and Stephen McDowall. "Material Culture and the Other: European Encounters with Chinese Porcelain, ca. 1650–1800." *Journal of World History* 23, no. 1 (2012): 87–113.

Gerritsen, Anne, and Giorgio Riello, eds. *Writing Material Culture History*. New York: Bloomsbury Publishing, 2014.

Gerth, Hans Heinrich, and Charles Wright Mills. "Introduction." In Max Weber, *From Max Weber: Essays in Sociology*, edited by Gerth and Mills, 3–74. New York: Oxford University Press, 1946.

Gillispie, Charles Coulston. *Science and Polity in France: The Revolutionary and Napoleonic Years*. Princeton, NJ: Princeton University Press, 2004.

Ginsberg, Morris. "Progress in the Modern Era." In *Dictionary of the History of Ideas*, vol. 3, edited by Philip P. Wiener. New York: Charles Scribner's Sons, 1973.

Giovannetti-Singh, Gianamar. "Writers of the Lost Ark: Reconstructing the Fight for Primacy in the Jesuit China Mission from the *Acta Pekinensia*, 1658–1707." *Modern Intellectual History*, forthcoming.

Girardot, Norman T. "James Legge and the Strange Saga of British Sinology and the Comparative Science of Religions in the Nineteenth Century." *Journal of the Royal Asiatic Society*, 3rd series, 12, no. 2 (2002): 155–65.

Girardot, Norman T. *The Victorian Translation of China: James Legge's Oriental Pilgrimage*. Berkeley: University of California Press, 2002.

Godwin, Joscelyn. *The Theosophical Enlightenment*. Albany: State University of New York Press, 1994.

Golvers, Noël. *Libraries of Western Learning for China: Circulation of Western Books between Europe and China in the Jesuit Mission (ca.1650–1750)*. Vol. 2, *Formation of Jesuit Libraries*. Louvain: Ferdinand Verbiest Institute K. U. Leuven, 2013.

Gordin, Michael. *On the Fringe: Where Science Meets Pseudoscience*. New York: Oxford University Press, 2021.

Grafton, Anthony. *Defenders of the Text: The Traditions of Scholarship in an Age of Science, 1450–1800*. Cambridge, MA: Harvard University Press, 1994.

Graham, Angus Charles. *Yin-Yang and the Nature of Correlative Thinking*. Singapore: Institute of East Asian Philosophies, 1986.

Grmek, Mirko Drazen. "Les reflets de la sphygmologie chinoise dans la médecine occidentale." *Biologie médicale*, numéro hors de série (1962).

Gunder Frank, Andre. *ReORIENT: Global Economy in the Asian Age*. Berkeley: University of California Press, 1998.

Guy, Basil. *The French Image of China before and after Voltaire*. Geneva: Institut et Musée Voltaire, 1963.

Guy, R. Kent. *The Emperor's Four Treasuries: Scholars and the State in the Late Ch'ien-Lung Era*. Cambridge, MA: Harvard University Asia Center, 1987.

Han Qi 韓琦. "Baijin de 'Yijing' yanjiu he Kangxi shidai de 'Xixue Zhongyuan' shuo." 白晉的「易經」研究和康熙時代的「西學中源」說 [Bouvet's *Yijing* studies and the Kangxi reign period 'Chinese origin of Western studies' theory]. *Hanxue yanjiu* 漢學研究 16, no. 1 (1998): 185–201.

Han Qi, "The Jesuits and their Study of Chinese Astronomy and Chronology in the Seventeenth and Eighteenth Centuries." In *Europe and China: Science and Arts in the 17th and 18th Centuries*, edited by Luís Saraiva, 71–79. Singapore: World Scientific Publishing, 2012.

Harder, Hermann. "La question du 'gouvernement' de la Chine au XVIIIe siècle." In *Actes du IIIe Colloque International de Sinologie: Appréciation par l'Europe de la tradition chinoise à partir du XVIIe siècle*, 80–91. Paris: Les Belles Lettres, Cathasia, 1983.

Harrison, Henrietta. "Chinese and British Diplomatic Gifts in the Macartney Embassy of 1793." *English Historical Review* 133, no. 560 (2018): 65–97.

Harrison, Henrietta. *The Perils of Interpreting: China and the Rise of the British Empire*. Princeton, NJ: Princeton University Press (forthcoming).

Harrison, Peter. *The Fall of Man and the Foundations of Science*. Cambridge: Cambridge University Press, 2007.

Hartman, Janine. "Ideograms and Hieroglyphs: The Egypto-Chinese Origins Controversy in the Enlightenment." *Dalhousie French Studies* 43 (1998): 101–18.

Harvey, David Allen. "The Lost Caucasian Civilization: Jean-Sylvain Bailly and the Roots of the Aryan Myth." *Modern Intellectual History* (2014): 279–306.

Hazard, Paul. *The Crisis of the European Mind, 1680–1715*. Translated by J. Lewis May. New York: New York Review Books, 2013.

Henderson, John B. "Ch'ing Scholars' Views of Western Astronomy." *Harvard Journal of Asiatic Studies* 46, no. 1 (1986): 121–48.
Hermans, Michel. "Biographie de Joseph-Marie Amiot." In *Les Danses rituelles chinoises d'après Joseph-Marie Amiot: Aux sources de l'ethnochorégraphie*, edited by Yves Lenoir and Nicolas Standaert, 11–77. Namur: Presses Universitaires de Namur, 2005.
Hervey de Saint-Denys, Marie-Jean-Léon Lecoq, marquis de. *Les rêves et les moyens de les diriger: Observations pratiques*. Paris: Amyot, 1867.
Hevia, James L. *Cherishing Men from Afar: Qing Guest Ritual and the Macartney Embassy of 1793*. Durham, NC: Duke University Press, 1995.
Hodgson, Peter C. "Editorial Introduction." In Georg Wilhelm Friedrich Hegel, *Lectures on the Philosophy of Religion: Determinate Religion*. Edited by Peter C. Hodgson. Translated by R. F. Brown, Peter C. Hodgson, and J. M. Stewart, with H. S. Harris, 1–90. Berkeley: University of California Press, 1987.
Hoefer, Jean Chrétien Ferdinand, ed. *Nouvelle biographie générale*. Paris: Firmin Didot, 1852–1856.
Horkheimer, Max, and Theodor Adorno. *Dialectic of Enlightenment: Philosophical Fragments* (1944). Stanford, CA: Stanford University Press, 2007.
Hostetler, Laura. "Qing Connections to the Early Modern World: Ethnography and Cartography in Eighteenth-Century China." *Modern Asian Studies* 34, no. 3 (2000): 623–62.
Hsia, Florence. "Athanasius Kircher's *China Illustrata* (1667)." In *Athanasius Kircher: The Last Man Who Knew Everything*, edited by Paula Findlen, 383–404. New York: Routledge, 2004.
Hsia, Florence. "Chinese Astronomy for the Early Modern Reader." *Early Science and Medicine* 13, no. 5 (2008): 417–50.
Hsia, Florence. *Sojourners in a Strange Land: Jesuits and Their Scientific Missions in Late Imperial China*. Chicago: University of Chicago Press, 2009.
Hsia, R. Po-chia. *A Jesuit in the Forbidden City: Matteo Ricci, 1552–1610*. Oxford: Oxford University Press, 2010.
Hsia, R. Po-chia. "Jesuit Survival and Restoration in China." In *Jesuit Survival and Restoration: A Global History, 1773–1900*. Edited by Robert A. Maryks and Jonathan Wright, 245–61. Leiden: Brill, 2014.
Hsia, R. Po-chia. "The End of the Jesuit Mission in China." In *The Jesuit Suppression in Global Context*, edited by Jeffrey D. Burson and Jonathan Wright, 100–16. Cambridge: Cambridge University Press, 2015.
Hu Jintao 胡锦涛. "Full text of Hu's report at the 18th Party Congress." Beijing, November 8, 2012. *China Daily*, November 18, 2012. https://www.chinadaily.com.cn/china/19thcpcnationalcongress/2012-11/18/content_29578562.htm.
Hu, Minghui. *China's Transition to Modernity: The New Classical Vision of Dai Zhen*. Seattle: University of Washington Press, 2015.
Hu, Minghui. "Provenance in Contest: Searching for the Origins of Jesuit Astronomy in Early Qing China, 1664–1705." *International History* 24, no. 1 (2002): 1–36.
Huard, Pierre, and Ming Wong. "Les enquêtes Françaises sur la science et la technologie chinoises au XVIIIe siècle." *Bulletin de l'École française d'Extrême-Orient* 53, no. 1 (1966): 137–226.
Huard, Pierre, and Ming Wong. "Mesmer en Chine: Trois lettres médicales du R.P.

Amiot, rédigées à Pékin, de 1783 à 1790." *Revue de Synthèse* series 3, 81, no. 17–18 (1960).

Hucker, Charles O. *A Dictionary of Official Titles in Imperial China*. Stanford, CA: Stanford University Press, 1985.

Hummel, Arthur W., ed. *Eminent Chinese of the Ch'ing Period (1644–1912)*. Washington, DC: United States Government Printing Office, 1943.

Huttmann, William. "A Notice of Several Chinese-European Dictionaries." *Asiatic Journal and Monthly Register for British India and its Dependencies* 12 (1821): 240–44.

Iliffe, Rob. *Priest of Nature: The Religious Worlds of Isaac Newton*. New York: Oxford University Press, 2017.

Israel, Jonathan I. "Admiration of China and Classical Chinese Thought in the Radical Enlightenment (1685–1740)." *Taiwan Journal of East Asian Studies* 4, no. 1 (2007): 1–25.

Israel, Jonathan I. *Enlightenment Contested: Philosophy, Modernity, and the Emancipation of Man, 1670–1752*. Oxford: Oxford University Press, 2006.

Jacobsen, Stefan Gaarsmand. "Chinese Influences or Images? Fluctuating Histories of How Enlightenment Europe Read China." *Journal of World History* 24, no. 3 (September 2013): 623–60.

Jacobsen, Stefan Gaarsmand. "Physiocracy and the Chinese Model." In *Thoughts on Economic Development in China*, edited by Ma Ying and Hans-Michael Trautwein, 12–34. London: Routledge, 2013.

Jami, Catherine. "Pékin au début de la dynastie Qing: Capital des savoirs impériaux et relais de l'Académie royale des sciences de Paris." *Revue d'Histoire Moderne et Contemporaine* 55, no. 2 (2008): 43–69.

Jami, Catherine. "Portrait of the Emperor as an Enlightened Monarch: The French Translation of Kangxi's Collection of the Investigation of Things in Leisure Time (1779)." Presentation, biennial conference of the European Society for the History of Science, September 14–17, 2018, London.

Jami, Catherine. *The Emperor's New Mathematics: Western Learning and Imperial Authority During the Kangxi Reign (1662–1722)*. Oxford: Oxford University Press, 2012.

Jean-Pierre Abel-Rémusat et ses successeurs. Deux cents ans de sinologie française en France et en Chine. Symposium, June 11–13, 2014, Collège de France, Paris. https://www.college-de-france.fr/site/pierre-etienne-will/symposium-2013-2014.htm.

Jensen, Lionel M. *Manufacturing Confucianism: Chinese Traditions and Universal Civilization*. Durham, NC: Duke University Press, 1998.

Johns Hopkins Medicine. "Acupuncture." Accessed May 13, 2022. https://www.hopkinsmedicine.org/health/wellness-and-prevention/acupuncture.

Josephson-Storm, Jason Ananda. *The Myth of Disenchantment: Magic, Modernity, and the Birth of the Human Sciences*. Chicago: University of Chicago Press, 2017.

Jung, Carl Gustav. "Foreword to the Second German Edition." In *The Secret of the Golden Flower: A Chinese Book of Life*. Edited by Richard Wilhelm. New York: Harcourt, Brace, & World, 1962.

Kim, Young Kun. "Hegel's Criticism of Chinese Philosophy." *Philosophy East and West* 28, no. 2 (1978): 173–80.

Kitson, Peter J. *Forging Romantic China: Sino-British Cultural Exchange 1760–1840*. Cambridge: Cambridge University Press, 2013.

Kleutghen, Kristina. *Imperial Illusions: Crossing Pictorial Boundaries in the Qing Palaces*. Seattle: University of Washington Press, 2015.

Kornicki, P. F. "Review: Julius Klaproth and His Works." *Monumenta Nipponica* 55, no. 4 (2000): 579–91.

Krahl, Joseph. *China Missions in Crisis: Bishop Laimbeckhoven and His Times, 1738–1787*. Rome: Gregorian University Press, 1964.

Kuhn, Thomas. *The Structure of Scientific Revolutions*. Chicago: University of Chicago Press, 1962.

Kui, Wong Kwok. "Hegel's Criticism of Laozi and its Implications." *Philosophy East and West* 61, no. 1 (2011): 56–79.

Kuiper, Koos. *The Early Dutch Sinologists (1854-1900): Training in Holland and China, Functions in the Netherlands Indies*. Leiden: Brill, 2017.

Laamann, Lars Peter. *Christian Heretics in Late Imperial China: Christian Inculturation and State Control, 1720–1850*. New York: Routledge, 2006.

Lach, Donald F. "Leibniz and China." *Journal of the History of Ideas* 6, no. 4 (1945): 436–55.

Lach, Donald F. "The Sinophilism of Christian Wolff (1679–1754)." In *Discovering China: European Interpretations in the Enlightenment*, edited by Julia Ching and Willard G. Oxtoby, 119–26. Rochester, NY: University of Rochester Press, 1992.

Lach, Donald F., and Edwin van Kley. *Asia in the Making of Europe*. 3 vols. Chicago: University of Chicago Press, 1965–1993.

Lackner, Michael. "Jesuit Figurism." In *China and Europe: Images and Influences in Sixteenth to Eighteenth Centuries*, edited by Thomas H. C. Lee, 129–50. Hong Kong: Chinese University Press, 1991.

Landry-Deron, Isabelle. "Bertin, Henri-Léonard." In *Dictionnaire Des Orientalistes de Langue Française*, edited by François Pouillon, 110–11. Paris: Karthala, 2008.

Landry-Deron, Isabelle. "De Guignes, Joseph." In *Dictionnaire Des Orientalistes de Langue Française*, edited by François Pouillon, 468. Paris: Karthala, 2008.

Landry-Deron, Isabelle. *La Preuve par la Chine: La "Description" de J.-B. Du Halde, jésuite, 1735*. Paris: Éditions de l'École des Hautes Études en Sciences Sociales, 2002.

Landry-Deron, Isabelle. "Le *Dictionnaire Chinois, Français et Latin* de 1813." *T'oung Pao* 101, no. 4–5 (2015): 407–40.

Landry-Deron, Isabelle. "Les outils de l'apprentissage du chinois en France en 1814 et les efforts d'Abel-Rémusat pour les améliorer." At *Jean-Pierre Abel-Rémusat et ses successeurs*, symposium, June 11–13, 2014, Collège de France, Paris.

Latour, Bruno. *Facing Gaia: Eight Lectures on the New Climatic Regime*. Translated by Catherine Porter. Cambridge, MA: Polity Press, 2017.

Latour, Bruno, and Steve Woolgar. *Laboratory Life: The Construction of Scientific Facts*. Beverly Hills: Sage Publications, 1979.

Laudan, Larry. "The Demise of the Demarcation Problem." In *Physics, Philosophy and Psychoanalysis*, eds. Robert S. Cohen and Larry Laudan, 111–27. Dordrecht: D. Reidel, 1983.

Lenoir, Yves, and Nicolas Standaert, eds. *Les Danses rituelles chinoises d'après Joseph-Marie Amiot: Aux sources de l'ethnochorégraphie*. Namur: Presses Universitaires de Namur, 2005.

Leung, Cécile. *Étienne Fourmont, 1683–1745: Oriental and Chinese Languages in Eighteenth-Century France*. Leuven: Leuven University Press, 2002.

Levitin, Dmitri. *Ancient Wisdom in the Age of the New Science*. Cambridge: Cambridge University Press, 2015.

Levy, Jim. "Joseph Amiot and Enlightenment Speculation on the Origin of Pythagorean Tuning in China." *Theoria* 4 (1989): 63–88.

Lewis, Gwynne. "Henri-Léonard Bertin and the Fate of the Bourbon Monarchy: The 'Chinese Connection.'" In *Enlightenment and Revolution: Essays in Honour of Norman Hampson*, edited by Malcolm Crook, William Doyle, and Alan Forrest, 69–90. Aldershot: Ashgate, 2004.

Lewis, Gwynne. *Madame de Pompadour's Protégé: Henri Bertin and the Collapse of Bourbon Absolutism c.1750–1792*. Gloucester: Emlyn Publishing, 2011.

Lhôte, Jean-Marie. *Le Tarot*. Paris: Berg, 1983.

Li Junzhi 李濬之. *Qing huajia shishi* 清畫家詩史 [A poetic history of Qing artists]. Li Shuzhi, 1906.

Li Qiang. "The Idea of Progress in Modern China: The Case of Yan Fu." In Thomas Fröhlich and Axel Schneider, *Chinese Visions of Progress, 1895 to 1949*, 103–31. Leiden: Brill, 2020.

Li Shi 李湜. *Zijin danqing—qinggong Huihua de chuangzuo yu shoucang* 紫禁丹青一清宮繪畫的創作與收藏 [Forbidden City painting—collecting and creating paintings in the Qing palace]. Beijing: Zhongguo guoji guangbo chubanshe, 2008.

Li Yu'an 李玉安 and Huang Zhengyu 黄正雨, eds. *Zhongguo cangshujia tongdian* 中國藏書家通典 [Encyclopedia of Chinese book collectors]. Hong Kong: China International Culture Press, 2005.

Lilla, Mark. "What is Counter-Enlightenment?" In *Isaiah Berlin's Counter-Enlightenment*, edited by Joseph Mali and Robert Wokler, 1–11. Philadelphia: American Philosophical Society, 2003.

Lilti, Antoine. "La civilisation est-elle européenne? Écrire l'histoire de l'Europe au XVIIIe siècle." In *Penser l'Europe au XVIIIe siècle: Commerce, civilisation, empire*, edited by Antoine Lilti and Céline Spector, 139–66. Oxford: Voltaire Foundation, 2014.

Lilti, Antoine. *The Invention of Celebrity*. Translated by Lynn Jeffress. Cambridge: Polity Press, 2017.

Littré, Émile. *Dictionnaire de la langue française*. 2nd ed. Paris: Hachette, 1872–1877.

Liu Dun 劉鈍. "Qingchu lisuan dashi Mei Wending" 初歷算大師梅文鼎 [Mei Wending, early Qing master of astronomical calculation]. *Ziran bianzhengfa tongxun* 1 (1986).

Lloyd, G. E. R., and Nathan Sivin. *The Way and the Word: Science and Medicine in Early China and Greece*. New Haven, CT: Yale University Press, 2003.

Long Yun 龙云. "Qian Deming yanjiu—18 shiji yi wei chu zai zhongfa wenhua jiaohuichu de chuanjiaoshi" 錢德明研究——18 世紀一位處在中法文化交匯處的傳教士 [Research on Joseph-Marie Amiot—a missionary at the confluence of Chinese and French culture in the eighteenth century]. PhD diss., Peking University, 2010.

Long Yun 龙云. *Qian Deming: 18 shiji Zhong-Fa jian de wenhua shizhe* 钱德明: 18世纪中法间的文化使者 (Qian Deming: A cultural intermediary between China and France in the eighteenth century). Beijing: Beijing daxue chubanshe, 2015.

Lowell, Julia. *The Opium War: Drugs, Dreams, and the Making of Modern China*. London: Picador, 2011.

Lowenthal, David. *The Past Is a Foreign Country—Revisited*. Cambridge: Cambridge University Press, 2015.

Lundbaek, Knud. "Notes on Abel Rémusat." In *Actes du VIIe Colloque International de Sinologie de Chantilly: Échanges culturels et religieux entre la Chine et l'Occident*, edited by Edward Malatesta, Yves Raguin, and Adrianus C. Dudink, 207–21. Paris: Ricci Institute, 1995.

Lundbaek, Knud. "Notes sur l'image du Néo-Confucianisme dans la littérature européenne du XVIIe siècle." In *Actes du IIIe Colloque International de Sinologie: Appréciation par l'Europe de la tradition chinoise à partir du XVIIe siècle*, 130–76. Paris: Les Belles Lettres, Cathasia, 1983.

Lundbaek, Knud. "The Image of Neo-Confucianism in Confucius Sinarum Philosophus." *Journal of the History of Ideas* 44, no. 1 (1983): 19–30.

Lynn, Michael R. "Divining the Enlightenment: Public Opinion and Popular Science in Old Regime France." *Isis* 92, no. 1 (2001): 34–54.

Malatesta, Edward, and Yves Raguin, eds. *Images de La Chine: Le Contexte Occidental de La Sinologie Naissante*. San Francisco: Ricci Institute for Chinese-Western Cultural History, 1995.

Mali, Joseph, and Robert Wokler, eds. *Isaiah Berlin's Counter-Enlightenment*. Philadelphia: American Philosophical Society, 2003.

Marchand, Suzanne L. *German Orientalism in the Age of Empire: Religion, Race, and Scholarship*. Cambridge: Cambridge University Press, 2010.

Marshall, P. J. "Introduction." In *The Oxford History of the British Empire: Volume II: The Eighteenth Century*, edited by P. J. Marshall and Alaine Low, 1–27. Oxford: Oxford University Press, 1998.

Marshall, P. J., and Alaine Low, eds. *The Oxford History of the British Empire: Volume II: The Eighteenth Century*. Oxford: Oxford University Press, 1998.

Maverick, Lewis A. "Chinese Influences Upon the Physiocrats." *Economic History* (1938): 54–67.

Mazlish, Bruce. *Civilization and Its Contents*. Stanford, CA: Stanford University Press, 2004.

McAnally, Henry. "Antonio Montucci." *Modern Language Quarterly* 7, no. 1 (1946): 65–81.

McMahon, Darrin. *Enemies of the Enlightenment: The French Counter-Enlightenment and the Making of Modernity*. Oxford: Oxford University Press, 2001.

Merchant, Carolyn. *The Anthropocene and the Humanities: From Climate Change to a New Age of Sustainability*. New Haven, CT: Yale University Press, 2020.

Mercier-Faivre, Anne-Marie. *Un supplément à L'Encyclopédie: Le "Monde primitif" d'Antoine Court de Gébelin, suivi d'une édition du "Génie allégorique et symbolique de l'Antiquité" extrait du "Monde primitif" (1773)*. Paris: Honoré Champion, 1999.

Messling, Markus. "Repräsentation und Macht: Selbstkritik der Philologie in Zeiten ihrer Ermächtigung." In *Sprachgrenzen—Sprachkontakte—kulturelle Vermittler: Kommunikation zwischen Europäern und Außereuropäern (16.-20. Jahrhundert)*, edited by Mark Häberlin and Alexander Keese, 247–60. Stuttgart: Franz Steiner Verlag, 2010.

Meynard, Thierry. *The Jesuit Reading of Confucius: The First Complete Translation of the Lunyu (1687) Published in the West*. Leiden: Brill, 2015.

Millar, Ashley Eva. "Revisiting the Sinophilia/Sinophobia Dichotomy in the

European Enlightenment through Adam Smith's 'Duties of Government.'" *Asian Journal of Social Science* 38 (2010): 716–37.

MIT Libraries. "Distinctive Collections." https://libraries.mit.edu/distinctive-collections/collections/visual-collections.

Mokyr, Joel. *A Culture of Growth: The Origins of the Modern Economy*. Princeton, NJ: Princeton University Press, 2016.

Mosca, Matthew. "Empire and the Circulation of Frontier Intelligence: Qing Conceptions of the Ottomans." *Harvard Journal of Asiatic Studies*, 70, no. 1 (2010): 147–207.

Mosca, Matthew. *From Frontier Policy to Foreign Policy: The Question of India and the Transformation of Geopolitics in Qing China*. Stanford, CA: Stanford University Press, 2013.

Mote, Frederick W. "The Case for the Integrity of Sinology." *Journal of Asian Studies* 23, no. 4 (1964): 531–34.

Mungello, David E. "An Introduction to the Chinese Rites Controversy." In *The Chinese Rites Controversy: Its History and Meaning*, edited by David E. Mungello, 3–14. Sankt Augustin: Institut Monumenta Serica and the Ricci Institute for Chinese-Western Cultural History, 1994.

Mungello, David E. *Curious Land: Jesuit Accommodation and the Origins of Sinology*. Honolulu: University of Hawaii Press, 1989.

Mungello, David E. *Leibniz and Confucianism: The Search for Accord*. Honolulu: University of Hawaii Press, 1977.

Mungello, David E. "The Reconciliation of Neo-Confucianism with Christianity in the Writings of Joseph de Prémare, S. J." *Philosophy East and West* 26, no. 4 (1976): 389–410.

Mülhan, Klaus. *Making China Modern: From the Great Qing to Xi Jinping*. Cambridge, MA: Belknap Press, 2019.

Nappi, Carla. *The Monkey and the Inkpot: Natural History and its Transformations in Early Modern China*. Cambridge, MA: Harvard University Press, 2009.

Naquin, Susan. *Shantung Rebellion: The Wang Lun Uprising of 1774*. New Haven, CT: Yale University Press, 1981.

Needham, Joseph. *Science and Civilisation in China: Volume I: Introductory Orientations*. Cambridge: Cambridge University Press, 1954.

Needham, Joseph. *The Grand Titration: Science and Society in East and West* (1969). Reprint ed. Oxford: Routledge, 2005.

Needham, Joseph, and Christoph Harbsmeier. *Science and Civilisation in China, Volume VII: The Social Background, Part I, Language and Logic*. Cambridge: Cambridge University Press, 1998.

Needham, Joseph, and Kenneth Girdwood Robinson. *Science and Civilisation in China: Volume VII: The Social Background, Part II, General Conclusions and Reflections*. Cambridge: Cambridge University Press, 2004.

Needham, Joseph, with Wang Ling. *Science and Civilisation in China: Volume II: History of Scientific Thought*. Cambridge: Cambridge University Press, 1956.

Needham, Joseph, with Wang Ling. *Science and Civilisation in China: Volume III: Mathematics and the Sciences of the Heavens and the Earth*. Cambridge: Cambridge University Press, 1959.

Needham, Joseph, Ho Ping-Yü and Lu Gwei-Djen. *Science and Civilisation in China:*

Volume V: Chemistry and Chemical Technology, Part III, Spagyrical Discovery and Invention. Cambridge: Cambridge University Press, 1976.

Newman, William R. *Promethean Ambitions: Alchemy and the Quest to Perfect Nature*. Chicago: University of Chicago Press, 2004.

Nii Yoko 新居洋子. *Iezusu kaishi to fuhen no teikoku: zaiKa senkyōshi niyoru bunmei no honyaku* イエズス会士と普遍の帝国：在華宣教師による文明の翻訳 [Jesuit missionaries and universal empire: translation of civilization by missionaries in China]. Nagoya: Nagoya daigaku shuppankai, 2017.

Nii Yoko 新居洋子. "Jūhasseiki ni okeru Chūgoku to Yōroppa no shisō kōryū: zaiKa Iezusu kaishi Amio no hōkoku wo chūshin ni 18" 世紀における中国とヨーロッパの思想交流：在華イエズス会士アミオの報告を中心に [Sino-European intellectual interactions in the eighteenth century: the case of Amiot, a Jesuit in China]. PhD diss., University of Tokyo, 2014.

Nisbet, Robert. "Idea of Progress: A Bibliographical Essay." *Liberty Fund*. 1979. https://oll4.libertyfund.org/page/idea-of-progress-a-bibliographical-essay-by-robert-nisbet.

Nisbet, Robert. "Turgot and the Contexts of Progress." *Proceedings of the American Philosophical Society* 119, no. 3 (1975): 214–22.

Norman, Larry F. *The Shock of the Ancient: Literature and History in Early Modern France*. Chicago: University of Chicago Press, 2011.

O'Brien, George Dennis. *Hegel on Reason and History*. Chicago: University of Chicago Press, 1975.

Obama, Barack. "Remarks by President Obama in Address to the United Nations General Assembly." New York, September 24, 2014. https://obamawhitehouse.archives.gov.

Obringer, Frédéric. "Jean-Pierre Abel-Rémusat, médecin et sinologue." At *Jean-Pierre Abel-Rémusat et ses successeurs*, symposium, June 11–13, 2014, Collège de France, Paris.

Osterhammel, Jürgen. *Unfabling the East: The Enlightenment's Encounter with Asia*. Princeton, NJ: Princeton University Press, 2018.

Outram, Dorinda. *The Enlightenment*. 4th ed. New York: Cambridge University Press, 2019.

Palmeri, Frank. *State of Nature, Stages of Society: Enlightenment Conjectural History and Modern Social Discourse*. New York: Columbia University Press, 2016.

Park, Katherine, and Lorraine Daston. "Introduction." In *The Cambridge History of Science, Volume III: Early Modern Science*, edited by Katherine Park and Lorraine Daston, 1–17. New York: Cambridge University Press, 2006.

Park, Peter K. J. *Africa, Asia, and the History of Philosophy: Racism in the Formation of the Philosophical Canon, 1780–1830*. Albany: State University of New York Press, 2013.

Park, Peter K. J. "Leibniz and Wolff on China." In *Germany and China: Transnational Encounters since the Eighteenth Century*, edited by Joanne Miyang Cho and David M. Crowe, 21–38. New York: Palgrave Macmillan, 2014.

Pédro, Georges. "Henri-Léonard Bertin et le développement de l'agriculture au siècle des Lumières." *Comptes Rendus Biologies* 335, no. 5 (2012): 325–33.

Perdue, Peter C. *China Marches West: The Qing Conquest of Central Eurasia*. Cambridge, MA: Belknap Press, 2010.

Perkins, Franklin. *Leibniz and China: A Commerce of Light*. Cambridge: Cambridge University Press, 2004.

Peyrefitte, Alain. *The Immobile Empire*. Translated by Jon Rothschild. New York: Knopf, 1992.

Pfister, Louis. *Notices biographiques et bibliographiques sur les jésuites de l'ancienne mission de Chine*. Shanghai: Imprimerie de la Mission Catholique, 1932.

Pinkard, Terry. *Hegel: A Biography*. Cambridge: Cambridge University Press, 2001.

Pinker, Steven. *Enlightenment Now: The Case for Reason, Science, Humanism, and Progress*. New York: Viking, 2018.

Pinot, Virgile. *La Chine et la formation de l'esprit philosophique en France, 1640–1740* (1932). Reprint ed. Geneva: Slatkine Reprints, 1971.

Pinot, Virgile. "Les physiocrates et la Chine au XVIIIe siècle." *Revue d'histoire moderne et contemporaine* 8, no. 3 (1906/1907): 200–14.

Pitts, Jennifer. "The Global in Enlightenment Historical Thought." In *A Companion to Global Historical Thought*, edited by Prasenjit Duara, Viren Murthy, and Andrew Sartori, 184–96. Chichester, UK: Wiley Blackwell, 2010.

Pocock, John Greville Agard. *Barbarism and Religion, Volume 1: The Enlightenments of Edward Gibbon, 1737–1764*. New York: Cambridge University Press, 1999.

Pocock, John Greville Agard. *Barbarism and Religion, Volume 2: Narratives of Civil Government*. New York: Cambridge University Press, 2001.

Pocock, John Greville Agard. *Barbarism and Religion, Volume 4: Barbarians, Savages and Empires*. New York: Cambridge University Press, 2005.

Pomeranz, Kenneth. *The Great Divergence: Europe, China, and the Making of the Modern World Economy*. Princeton, NJ: Princeton University Press, 2000.

Porset, Charles. *Franc-maçonnerie et religions dans l'Europe des Lumières*. Paris: H. Champion, 2006.

Porter, David. *Ideographia: The Chinese Cipher in Early Modern Europe*. Stanford, CA: Stanford University Press, 2001.

Porter, David. *The Chinese Taste in Eighteenth-Century England*. Cambridge: Cambridge University Press, 2010.

Porter, Jonathan. "The Scientific Community in Early Modern China." *Isis* 73, no. 4 (1982): 529–44.

Pouillon, François, ed. *Dictionnaire Des Orientalistes de Langue Française*. Paris: Karthala, 2008.

Pregadio, Fabrizio. "Religious Daoism." 2016. *Stanford Encyclopedia of Philosophy* (Fall 2020). Edited by Edward N. Zalta. https://plato.stanford.edu/archives/fall2020/entries/daoism-religion/.

Press, Steven. *Rogue Empires: Contract and Conmen in Europe's Scramble for Africa*. Cambridge, MA: Harvard University Press, 2017.

Principe, Lawrence M. *The Aspiring Adept: Robert Boyle and His Alchemical Quest*. Princeton, NJ: Princeton University Press, 1998.

Principe, Lawrence M. "The End of Alchemy? The Repudiation and Persistence of Chrysopoeia at the Académie Royale des Sciences in the Eighteenth Century." *Osiris* 29, no. 1 (2014): 96–116.

Qian Mu 錢穆. *Zhongguo jin sanbai nian xueshushi* 中國近三百年學術史 [History of Chinese scholarship of the last three centuries] (1937). Reprint ed. Beijing: Jiuzhou chubanshe, 2011.

Qian Mu 錢穆. *Zhongguo sixiangshi* 中國思想史 [An intellectual history of China] (1937). Reprint ed. Beijing: Jiuzhou chubanshe, 2012.

Raina, Dhruv. "Betwixt Jesuit and Enlightenment Historiography: Jean-Sylvain Bailly's History of Indian Astronomy." *Revue d'histoire des mathématiques* 9 (2003): 253–306.

Raj, Kapil. *Relocating Modern Science: Circulation and the Construction of Knowledge in South Asia and Europe, 1650–1900*. New York: Palgrave Macmillan, 2007.

Redding, Paul. "Hegel." 1997; revised 2015. *Stanford Encyclopedia of Philosophy*. Edited by Edward N. Zalta. https://plato.stanford.edu/entries/hegel.

Reill, Peter. "The Hermetic Imagination in the High and Late Enlightenment." In *The Super-Enlightenment: Daring to Know Too Much*, edited by Dan Edelstein, 37–51. Oxford: Voltaire Foundation, 2010.

Ribas, Albert. "Leibniz' 'Discourse on the Natural Theology of the Chinese' and the Leibniz-Clarke Controversy." *Philosophy East and West* 53, no. 1 (2003): 64–86.

Rinaldi, Bianca Maria. *Ideas of Chinese Gardens: Western Accounts, 1300–1860*. Philadelphia: University of Pennsylvania Press, 2016.

Riskin, Jessica. "Mr. Machine and the Imperial Me." In *The Super-Enlightenment: Daring to Know Too Much*, edited by Dan Edelstein, 75–94. Oxford: Voltaire Foundation, 2010.

Riskin, Jessica. "Pinker's Pollyannish Philosophy and Its Perfidious Politics." *Los Angeles Review of Books*, December 15, 2019.

Riskin, Jessica. *Science in the Age of Sensibility: The Sentimental Empiricists of the French Enlightenment*. Chicago: University of Chicago Press, 2002.

Robbins, Jim. "Native Knowledge: What Ecologists Are Learning from Indigenous People." Yale School of the Environment: *Yale Environment 360*, April 26, 2018. https://e360.yale.edu/features/native-knowledge-what-ecologists-are-learning-from-indigenous-people.

Roberts, John Anthony George. "L'image de La Chine Dans *l'Encyclopédie*." *Recherches Sur Diderot et Sur l'Encyclopédie* 22, no. 1 (1997): 87–108.

Rochemonteix, Camille de. *Joseph Amiot et les derniers survivants de la mission française à Pékin (1750–1795)*. Paris: Alphonse Picard et fils, 1915.

Roger, Jacques. *Buffon: A Life in Natural History*. Translated by Sarah Lucille Bonnefoi. Ithaca, NY: Cornell University Press, 1997.

Romano, Antonella. *Impressions de Chine: L'Europe et l'englobement du monde (XVIe–XVIIe siècle)*. Paris: Fayard, 2016.

Rose, Ernst. "China as a Symbol of Reaction in Germany, 1830–1880." *Comparative Literature* 3, no. 1 (1951): 57–76.

Roubekas, Nickolas Panayiotis. *An Ancient Theory of Religion: Euhemerism from Antiquity to the Present*. New York: Routledge, 2017.

Rowbotham, Arnold H. "The Jesuit Figurists and Eighteenth-Century Religious Thought." In *Discovering China: European Interpretations in the Enlightenment*, edited by Julia Ching and Willard G. Oxtoby, 39–53. Rochester, NY: University of Rochester Press, 1992.

Rowbotham, Arnold H. "Voltaire, Sinophile." *PMLA* 47, no. 4 (1932): 1050–65.

Rowe, William T. "Introduction: The Significance of the Qianlong-Jiaqing Transition in Qing History." *Late Imperial China* 32, no. 2 (2011): 74–88.

Rule, Paul. "Moses or China?" In *Images de La Chine: Le Contexte Occidental de La*

Sinologie Naissante, edited by Edward Malatesta and Yves Raguin, 303–32. San Francisco: Ricci Institute for Chinese-Western Cultural History, 1995.

Rule, Paul. "The Tarnishing of the Image: From Sinophilia to Sinophobia." In *La Chine entre amour et haine: Actes du VIIIe Colloque de sinologie de Chantilly*, edited by Michel Cartier, 89–109. Paris: Desclée de Brouwer, 1998.

Rusk, Bruce. *Goulou Feng Bei: Stele of Goulou Peak, Ink Squeeze of Mao Huijian's Stele of King Yu of 1666*. Melbourne: Quirin Press, 2016.

Said, Edward. *Culture and Imperialism*. New York: Vintage Books, 1993.

Said, Edward. *Orientalism*. New York: Pantheon Books, 1978.

Sarton, George. *Sarton on the History of Science*. Edited by Dorothy Stimson. Cambridge, MA: Harvard University Press, 1962.

Saussy, Haun. *Great Walls of Discourse and Other Adventures in Cultural China*. Cambridge, MA: Harvard University Asia Center, 2001.

Schaffer, Simon. "The Accomplishment of Facts at the End of the Enlightenment." Forthcoming.

Schaffer, Simon. "The Asiatic Enlightenments of British Astronomy." In *The Brokered World: Go-Betweens and Global Intelligence, 1770–1820*, edited by Simon Schaffer et al., 49–104. Sagamore Beach, MA: Science History Publications, 2009.

Schaffer, Simon. "The Astrological Roots of Mesmerism." *Studies in History and Philosophy of Biological and Biomedical Sciences* 42, no. 2 (2010): 158–68.

Schaffer, Simon. "Late Enlightenment Crises of Facts: Mesmerism and Meteorites." *Configurations* 26, no. 2: 119–48.

Schaffer, Simon. "Newton on the Beach: The Information Order of *Principia Mathematica*." *History of Science* 47 (2009): 243–76.

Schaffer, Simon, Lissa Roberts, Kapil Raj, and James Delbourgo, eds. *The Brokered World: Go-Betweens and Global Intelligence, 1770–1820*. Sagamore Beach, MA: Science History Publications, 2009.

Schäfer, Dagmar. *The Crafting of the 10,000 Things: Knowledge and Technology in Seventeenth-Century China*. Chicago: University of Chicago Press, 2011.

Schiebinger, Londa. *Plants and Empire: Colonial Bioprospecting in the Atlantic World*. Cambridge, MA: Harvard University Press, 2007.

Schlesinger, Jonathan. *A World Trimmed with Fur: Wild Things, Pristine Places, and the Natural Fringes of Qing Rule*. Stanford, CA: Stanford University Press, 2017.

Schwab, Raymond. *La Renaissance orientale*. Paris: Payot, 1950.

Schwarcz, Vera. *The Chinese Enlightenment: Intellectuals and the Legacy of the May Fourth Movement of 1919*. Berkeley: University of California Press, 1986.

Scott, James C. *Seeing Like a State: How Certain Schemes to Improve the Human Condition Have Failed*. New Haven, CT: Yale University Press, 1998.

Sela, Ori. *China's Philological Turn: Scholars, Textualism, and the Dao in the Eighteenth Century*. New York: Columbia University Press, 2018.

Sela, Ori. "Confucian Scientific Identity: Qian Daxin's (1728–1804) Ambivalence Toward Western Learning and its Adherents." *East Asian Science, Technology, and Society* 6 (2012): 147–66.

Shapin, Steven. *A Social History of Truth: Civility and Science in Seventeenth-Century England*. Chicago: University of Chicago Press, 1994.

Shapin, Steven, and Simon Schaffer. *Leviathan and the Air-Pump: Hobbes, Boyle and the Experimental Life*. Princeton, NJ: Princeton University Press, 1985.

Shank, John Bennett. *The Newton Wars and the Beginning of the French Enlightenment*. Chicago: University of Chicago Press, 2008.

Silvestre de Sacy, Jacques, and Michel Antoine. *Henri Bertin dans le sillage de la Chine: 1720–1792*. Paris: Éditions Cathasia, les Belles Lettres, 1970.

Sivin, Nathan. "Copernicus in China: or, Good Intentions Gone Astray" (1973). In *Science in Ancient China: Researches and Reflections*, 63–122. Aldershot: Variorum, 1995.

Sivin, Nathan. "Taoism and Science." In *Medicine, Philosophy and Religion in Ancient China: Researches and Reflections*, 1–73. Aldershot: Variorum, 1995.

Sivin, Nathan. "Why the Scientific Revolution Did Not Take Place in China—Or Didn't It?" *Chinese Science* 5 (1982): 45–66.

Skinner, G. William. "What the Study of China Can Do for Social Science." *Journal of Asian Studies* 23, no. 4 (1964): 517–22.

Smith, Blake Evan. "Myths of Stasis: South Asia, Global Commerce and Economic Orientalism in Late Eighteenth-Century France." PhD diss., Northwestern University, 2017. https://arch.library.northwestern.edu/concern/generic_works/m039k4989.

Smith, Pamela. *Entangled Itineraries: Materials, Practices, and Knowledges Across Eurasia*. Pittsburgh, PA: University of Pittsburgh Press, 2010.

Smith, Richard J., and Danny Wynn Ye Kwo, eds. *Cosmology, Ontology, and Human Efficacy: Essays in Chinese Thought*. Honolulu: University of Hawaii Press, 1993.

Smith, Richard J. *Fathoming the Cosmos and Ordering the World: The Yijing (I Ching, or Classic of Changes) and Its Evolution in China*. Charlottesville: University of Virginia Press, 2008.

Song, Shun-Ching. *Voltaire et la Chine*. Aix-en-Provence: Université de Provence, 1989.

Söderblom Saarela, Mårten. *The Early Modern Travels of Manchu: A Script and its Study in East Asia and Europe*. Philadelphia: University of Pennsylvania Press, 2022.

Spadafora, David. *The Idea of Progress in Eighteenth-Century Britain*. New Haven, CT: Yale University Press, 1990.

Spivak, Gayatri. *A Critique of Postcolonial Reason: Toward a History of the Vanishing Present*. Cambridge, MA: Harvard University Press, 1999.

Standaert, Nicolas, ed. *Handbook of Christianity in China*. Vol. 1. Leiden: Brill, 2001.

Standaert, Nicolas. "Jesuit Accounts of Chinese History and Chronology and Their Chinese Sources." *East Asian Science, Technology, and Medicine* 35 (2013): 11–88.

Standaert, Nicolas. *The Intercultural Weaving of Chinese Texts: Chinese and European Stories about Emperor Ku and His Concubines*. Leiden: Brill, 2016.

Statman, Alexander. "Fusang: The Enlightenment Story of the Chinese Discovery of America." *Isis* 107, no. 1 (2016): 1–25.

Statman, Alexander. "The First Global Turn: Chinese Contributions to Enlightenment World History." *Journal of World History* 30, no. 3 (2019): 363–92.

Stolzenberg, Daniel. *Egyptian Oedipus: Athanasius Kircher and the Secrets of Antiquity*. Chicago: University of Chicago Press, 2013.

Strickmann, Michael. *Chinese Magical Medicine*. Edited by Bernard Faure. Stanford, CA: Stanford University Press, 2002.

Svachula, Amanda. "Like a Virgo: How the Times Covers Astrology." *New York Times*, August 31, 2019.

Terrall, Mary. *The Man Who Flattened the Earth: Maupertuis and the Sciences in the Enlightenment*. Chicago: University of Chicago Press, 2002.

Thomas, Keith. *Religion and the Decline of Magic: Studies in Popular Beliefs in Sixteenth and Seventeenth Century England* (1971). Reprint ed. New York: Oxford University Press, 1997.

Thompson, D. Gillian. "French Jesuits 1756–1814." In *The Jesuit Suppression in Global Context*, edited by Jeffrey D. Burson and Jonathan Wright, 189–98. Cambridge: Cambridge University Press, 2015.

Thoraval, Joël. "De la magie à la 'raison': Hegel et la religion chinoise." In *La Chine entre amour et haine: Actes du VIIIe Colloque de sinologie de Chantilly*, edited by Michel Cartier, 111–41. Paris: Desclée de Brouwer, 1998.

Tocqueville, Alexis de. *The Old Regime and the Revolution*. Translated by John Bonner. New York: Harper and Brothers, 1856.

Toews, John E. "Berlin's Marx: Enlightenment, Counter-Enlightenment, and the Historical Construction of Cultural Identities." In *Isaiah Berlin's Counter-Enlightenment*, edited by Joseph Mali and Robert Wokler, 163–76. Philadelphia: American Philosophical Society, 2003.

"Trésor de la Langue Française informatisé." *Analyse et traitement informatique de la langue française*. Accessed May 16, 2022. http://atilf.atilf.fr/.

Trump, Donald J. "Inaugural Address." Washington, DC, January 20, 2017. https://www.govinfo.gov/features/presidential-inaugural-addresses.

Unschuld, Paul U. *Medicine in China: A History of Ideas*. Berkeley: University of California Press, 1985.

Van Kley, Edwin J. "Europe's 'Discovery' of China and the Writing of World History." *American Historical Review* 76, no. 2 (1971): 358–85.

Vande Walle, Willy, and Noël Golvers. *The History of the Relations between the Low Countries and China in the Qing Era (1644–1911)*. Leuven: Leuven University Press, 2003.

Varouxakis, Georgios. "The Godfather of 'Occidentality': Auguste Comte and the Idea of 'The West.'" *Modern Intellectual History* (October 11, 2017): 1–31. https://doi.org/10.1017/S1479244317000415.

Verhaeren, Hubert. *Catalogue de la bibliothèque du Pé-t'ang*. Beijing: Imprimerie des Lazaristes, 1949.

Viatte, Auguste. *Les sources occultes du Romantisme*. Paris: Honoré Champion, 1928.

Vickers, Brian. *Occult Scientific Mentalities in the Renaissance*. Cambridge: Cambridge University Press, 1986.

Vissière, Isabelle, and Jean-Louis Vissière. "Un carrefour culturel: la mission française de Pékin au XVIIIe siècle." In *Actes du IIIe Colloque International de Sinologie: Appréciation par l'Europe de la tradition chinoise à partir du XVIIe siècle*, 211–22. Paris: Les Belles Lettres, Cathasia, 1983.

Waley-Cohen, Joanna. "China and Western Technology in the Late Eighteenth Century." *American Historical Review* 98, no. 5 (1993): 1525–44.

Waley-Cohen, Joanna. "Commemorating War in Eighteenth-Century China." *Modern Asian Studies* 30, no. 4 (1996): 869–99.

Walf, Knut. "Fascination and Misunderstanding: The Ambivalent Western Reception of Daoism." *Monumenta Serica* 53 (2005): 273–86.

Walker, Daniel Pickering. *The Ancient Theology: Studies in Christian Platonism from the Fifteenth to the Eighteenth Century*. Ithaca, NY: Cornell University Press, 1972.

Waldren, William S. "Buddhism in Psychology and Psychotherapy." *Oxford Bibliographies*, last modified September 13, 2010. https://www.oxfordbibliographies.com/view/document/obo-9780195393521/obo-9780195393521-0130.xml.

Wallon, M. Henri. "Notice historique sur la vie et les travaux d'Aignan-Stanislas Julien." *Mémoires de l'Institut de France, Académie des inscriptions et belles-lettres* 31 (1884): 409–58.

Walravens, Helmut. "Julius Klaproth: His Life and Works with Special Emphasis on Japan." *Japonica Humboldtiana* 10 (2006): 177–91.

Walravens, Helmut. "Julius Klaproth, Stanislas Julien, et les débuts de la sinologie européenne." In *Idées de la Chine au XIXe siècle*, edited by Marie Dollé and Geneviève Espagne, 145–56. Paris: Les Indes Savants, 2014.

Wang, Robin. *Yinyang: The Way of Heaven and Earth in Chinese Thought and Culture*. New York: Cambridge University Press, 2012.

Watson, Walter. "Interpretations of China in the Enlightenment: Montesquieu and Voltaire." In *Actes du IIe Colloque International de Sinologie, Les rapports entre la Chine et l'Europe au temps des lumières*, 16–37. Paris: Les Belles Lettres, Cathasia, 1980.

Weber, Max. *From Max Weber: Essays in Sociology*. Edited by Hans Heinrich Gerth and Charles Wright Mills. New York: Oxford University Press, 1946.

Weber, Max. *The Vocation Lectures*. Edited by David Owen and Tracy B. Strong. Indianapolis, IN: Hackett, 2004.

Wei, Betty Peh-T'i. *Ruan Yuan, 1764–1849: The Life and Work of a Major Scholar-Official in Nineteenth-Century China before the Opium War*. Hong Kong: Hong Kong University Press, 2006.

Weller, Emil Ottokar. *Die falschen und fingierten Druckorte: Repertorium der seit Erfindung der Buchdruckerkunst unter falscher Firma erschienenen deutschen, lateinischen und französischen Schriften*. Leipzig: Wilhelm Engelmann, 1864.

Westman, Robert. *The Copernican Question: Prognostication, Skepticism, and the Celestial Order*. Berkeley: University of California Press, 2011.

Will, Pierre-Étienne. "Abel-Rémusat, l'Orientaliste." At *Jean-Pierre Abel-Rémusat et ses successeurs*, symposium, June 11–13, 2014, Collège de France, Paris.

Winterer, Caroline. "Buck Up, Everyone! We Are Riding along the Enlightenment's Long Path of Progress." *Washington Post*, February 23, 2018.

Witek, John W. *Controversial Ideas in China and Europe: A Biography of Jean-François Foucquet*. Rome: Institutum Historicum S. I., 1982.

Witek, John W. "Jean-François Foucquet and the Chinese Books in the French Royal Library: A Preliminary Survey." In *Actes du IIe Colloque International de Sinologie, Les rapports entre la Chine et l'Europe au temps des lumières*, 145–83. Paris: Les Belles Lettres, Cathasia, 1980.

Witek, John W. "Manchu Christians and the Sunu Family." In *Handbook of Christianity in China*, vol. 1, edited by Nicolas Standaert, 444–48. Leiden: Brill, 2001.

Withers, Charles W. J., and David N. Livingstone, eds. *Geography and Enlightenment*. Chicago: University of Chicago Press, 2007.

Wolff, Larry. "Discovering Cultural Perspective: The Intellectual History of Anthropological Thought in the Age of Enlightenment." In *The Anthropology of the*

Enlightenment, eds. Wolff and Cipolloni, 3–32. Stanford, CA: Stanford University Press, 2007.

Wolff, Larry, and Marco Cipolloni, eds. *The Anthropology of the Enlightenment*. Stanford, CA: Stanford University Press, 2007.

Wolloch, Nathaniel. "Joseph de Guignes and Enlightenment Notions of Material Progress." *Intellectual History Review*, 21, no. 4 (2011): 435–48.

Wright, Arthur. "The Study of Chinese Civilization." *Journal of the History of Ideas* 21, no. 2 (1960): 233–55.

Wu Boya 吳伯婭. "*Cong xin chuban de qingdai dang'an kan Tianzhujiao chuan hua shi*" 從新出版的清代檔案看天主教傳華史 [The history of Catholicism in China, based on newly published Qing archival documents]. *Qingshi luncong* 清史論叢. Beijing: Chinese Academy of Social Sciences, 2005.

Wu Huiyi. "Alien Voices under the Bean Arbour: How an Eighteenth-Century French Jesuit Translated the Doupeng xianhua 豆棚閒話 as the 'Dialogue of a Modern Atheist Chinese Philosopher.'" *T'oung Pao* 103, no. 1–3 (2017): 155–205.

Wu Huiyi. *Traduire la Chine au XVIIIe siècle: Les jésuites traducteurs de textes chinois et le renouvellement des connaissances européennes sur la Chine (1687–ca. 1740)*. Paris: Honoré Champion, 2017.

Wu Huiyi. "'The Observations We Did in the Indies and in China': The Shaping of French Jesuits' Knowledge of China by Other Non-Western Regions." *East Asian Science, Technology and Medicine* 46 (2018): 47–88.

Xi Jinping 习近平. "Full Text of Xi Jinping's Report at 19th CPC National Congress." Beijing, October 18, 2017. *China Daily*, updated November 4, 2017. https://www.chinadaily.com.cn/china/19thcpcnationalcongress/2017-11/04/content_34115212.htm.

Xu Minglong 許明龍. *Huang Jialüe yu Faguo zaoqi hanxue* 黃嘉略與法國早期漢學 [Arcadio Huang and the beginning of French Sinology]. Beijing: Shangwu yinshuguan, 2014.

Yang Danxia 楊丹霞. "Hongwu de shuhua jiaoliu yu chuangzuo" 弘旿的書畫交流與創作 [The exchange and creation of Hongwu's painting and calligraphy]. *Forbidden City* 紫禁城. Beijing: Palace Museum, 2005.

Yan Fu 严复. *Yan Fu ji* 严复集 (Selected works of Yan Fu). Edited by Wang Shi 王栻. Beijing: Zhonghua shuju, 1986.

Yang Danxia 楊丹霞. "Hongwu ji qi 'jiang shan gong cui tu'" 弘旿及其江山 [Hongwu and his "Painting of Rivers and Mountains Surrounded by Emerald Green."] *Shoucangjia* 收藏. Beijing: Capital Museum, 2003.

Yang Zhongxi 楊鐘羲. *Xueqiao shihua* 雪橋詩話 [Snow Bridge poetry talks]. Qiushuzhai congshu: early twentieth century.

Yates, Frances Amelia. *Giordano Bruno and the Hermetic Tradition*. Chicago: University of Chicago Press, 1964.

Yuan Chen. "Legitimation Discourse and the Theory of the Five Elements in Imperial China." *Journal of Song-Yuan Studies* 44, no. 1 (2014): 325–64.

Zarrow, Peter. "An Anatomy of the Utopian Impulse in Modern Chinese Political Thought, 1890–1940." In Thomas Fröhlich and Axel Schneider, *Chinese Visions of Progress, 1895 to 1949*, 165–205. Leiden: Brill, 2020.

Zhang Chunjie. "From Sinophilia to Sinophobia: China, History, and Recognition." *Colloquia Germanica* 41, no. 2 (2008): 97–110.

Zhang, Qiong. *Making the New World Their Own: Chinese Encounters with Jesuit Science in the Age of Discovery*. Leiden: Brill, 2015.

Zhang Zhidong 張之洞 et al., eds. *Guangxu Shuntian Fu zhi* 光緒順天府志 [Guangxu period gazetteer of the Capital prefecture]. 1886.

Zurndorfer, Harriet T. "Comment la science et la technologie se vendaient à la Chine au XVIIIe siècle." *Études chinoises* 7, no. 2 (1988): 59–90.

Zurndorfer, Harriet T. "Orientalism, Sinology, and Public Policy: Baron Antoine Isaac Silvestre de Sacy and the Foundation of Chinese Studies in Post-Revolutionary France." In *Images de La Chine: Le Contexte Occidental de La Sinologie Naissante*, edited by Edward Malatesta and Yves Raguin, 175–92. San Francisco: Ricci Institute for Chinese-Western Cultural History, 1995.

Archival Sources

ARCHIVES DES MISSIONS ÉTRANGÈRES DE PARIS, PARIS, FRANCE

MS 438. "Chine, Lettres, 1780–1787."

ARCHIVES JÉSUITES DE LA PROVINCE DE FRANCE [AJPF], VANVES, FRANCE

Brotier 134.
Brotier 135.
Henri Cordier.
Joseph Bernard Maître 50–52.
Joseph Bernard Maître 69.
Joseph Bernard Maître 84.
Vivier 1.

ARCHIVES NATIONALES [AN], PIERREFITTE-SUR-SEINE, FRANCE

F^17, 1047. "Desvoyes."
F^17, 1188. "Commission Temporaire des Arts."
291 AP/1. "Papiers de Marius-Jean-Baptiste-Nicolas d'AINE."

BIBLIOTHÈQUE DE LA SOCIÉTÉ DE L'HISTOIRE DU PROTESTANTISME FRANÇAIS [BSHPF], PARIS, FRANCE

MS 316. "Papiers Paul Rabaut, Lettres 1777–1783."
MS 317. "Papiers Paul Rabaut, Lettres 1783–1790."
MS 318. "Papiers Paul Rabaut, Lettres à Court de Gébelin."
MS 361. "Papiers Court de Gébelin."
MS 367. "Lettres de Rabaut-Saint-Étienne à Court de Gébelin."
MS 622. "Papiers Court de Gébelin, Recueil de pièces et extraits divers."

BIBLIOTHÈQUE INTERUNIVERSITAIRE DE LA SORBONNE [BIU], PARIS, FRANCE

MS VC 214. "Correspondance générale de Victor Cousin."
MS VC 232. "Cinq lettres de Georg Wilhelm Friedrich Hegel à Victor Cousin."

BIBLIOTHÈQUE INTERUNIVERSITAIRE DE SANTÉ [BIS], PARIS, FRANCE

MS 2270. "Mémoire pour servir à l'histoire du somnambulisme."
MS 2446. "Copies de lettres à sujet médical rédigés à Pékin adressées à M. Desvoyes pseudonyme de l'abbé Louis Augustin Bertin" [mislabeled].

BIBLIOTHÈQUE NATIONALE DE FRANCE [BNF], PARIS, FRANCE

Bréquigny 1. "Mélanges sur la Chine et les Chinois."
Bréquigny 2. "Mélanges sur la Chine et les Chinois."
Bréquigny 3. "Mélanges sur la Chine et les Chinois."
Bréquigny 5. "Mélanges sur la Chine et les Chinois."
Bréquigny 7. "Mélanges sur la Chine et les Chinois."
Bréquigny 158. "Correspondance de Bréquigny I."
Bréquigny 160. "Correspondance de Bréquigny III."
Bréquigny 163. "Correspondance de Bréquigny VI."
Chinois 1170. "Copie du monument que le grand yu éleva sur la Montagne heng-chan . . . , 1777."
Chinois 5087–5095. "Traité général de médecine, avec un historique."
DELTA 11947. *Catalogue des Livres de Feu M. Court de Gébelin, de la Société Économique de Berne, de l'Académie Royale de la Rochelle, Président honoraire du Musée de Paris, etc.* Chez Musier: maison de M. Didot l'aîné, imprimeur, 1786.
Français 12305. "Correspondance Macquer."
Manchu 285. *Hymne mandchou chanté à l'occasion de la conquête du Jin-Chuan*, trans. Amiot.
NAF 279. "De Guignes, papiers divers."
NAF 2491. "Deshauterayes I, Papiers Intimes / l'Apocalypse / Dissertations."
NAF 2492. "Deshauterayes II, Histoire générale de la Chine."
NAF 2493. "Deshauterayes III: Histoire générale de la Chine II, Travaux Divers."
NAF 8872. "Correspondance d'Anquetil-Duperron."

BIBLIOTHÈQUE NATIONALE DE FRANCE, SITE ARSENAL, PARIS, FRANCE

MS 5790. "Extrait pour le Roy de lettres de Pékin, 1786–1787."

COLLÈGE DE FRANCE [CF], PARIS, FRANCE

14 CDF art 65a. "Sinologie, Chaire de Langue et littérature chinoises et tartare-mandchou."
15 CDF 142. "Joseph de Guignes."

15 CDF 187. "André Leroux des Hauterayes."
15 CDF 256. "Abel Rémusat."

CONGRÉGATION DE LA MISSION [CM], PARIS, FRANCE

164.II B, 1–10.
166.I B, 1–3.
171.II B.

FIRST HISTORICAL ARCHIVES OF CHINA [FHA] 中國第一歷史檔案館, BEIJING, CHINA

Junjichu shangyu dang 軍機處上諭檔 [Edicts of the Grand Council].
Zongguan neiwufu 總管內務府 [Records of the Imperial Households Department].

INSTITUT DE FRANCE [IF], PARIS, FRANCE

MS 1515. "80 lettres du P. Amiot à Bertin."
MS 1516. "80 lettres du P. Amiot à Bertin."
MS 1517. "80 lettres du P. Amiot à Bertin."
MS 1518. "Lettres de divers correspondants."
MS 1519. "Lettres de plusieurs missionnaires."
MS 1520. "Lettres des PP. Kô et Yang et de M. Brisson."
MS 1521. "24 Lettres de M. Bertin aux Missionnaires en Chine, 1764–1772."
MS 1522. "36 Lettres de Mr. Bertin aux Missionnaires en Chine, 1773–1778."
MS 1523. "28 Lettres de Mr. Bertin aux Missionnaires en Chine, 1779–1782."
MS 1524. "24 Lettres de Mr. Bertin aux Missionnaires en Chine, 1783–1788."
MS 1525. "60 lettres relatives aux missionnaires en Chine."
MS 1526. "30 mémoires par divers savants adressés aux missionnaires."
MS 5401. Fonds Henri Cordier, "Collection des Autographes."
MS 5409. Fonds Henri Cordier, "Notes sur la correspondance de Henri Bertin et sur les missions catholiques."

INSTITUT NATIONAL D'HISTOIRE DE L'ART, PARIS, FRANCE

MS 131. "Bertin et la Mission de Chine, 1762–1792."

NATIONAL LIBRARY OF CHINA, RARE BOOKS [NLC] 國家圖書館普通古籍, BEIJING, CHINA

MS 25072. Hongwu 弘旿. *Yaohua Daoren shichao* 瑤華道人詩抄 [Poetry collection of the Daoist of Illustrious Jade] [Edition B].
MS 25573. Yongzhong 永忠. *Yanfen shi gao* 延芬室稿.
MS 93051. Hongwu 弘旿. *Yaohua Daoren shichao* 瑤華道人詩抄 [Poetry collection of the Daoist of Illustrious Jade] [Edition A].

Index

Page numbers in italics refer to figures.